In the Shadow
of the Palms

In the Shadow of the Palms

More-Than-Human

Becomings in West Papua

SOPHIE CHAO

DUKE UNIVERSITY PRESS
Durham and London   2022

All rights reserved
Printed and bound by CPI Group (UK) Ltd, Croydon, CR0 4YY
Designed by Courtney Leigh Richardson
Typeset in Minion Pro and Avenir by Westchester Publishing Services

Library of Congress Cataloging-in-Publication Data
Names: Chao, Sophie, author.
Title: In the shadow of the palms : more-than-human becomings in
West Papua / Sophie Chao.
Description: Durham : Duke University Press, 2022. | Includes biblio-
graphical references and index.
Identifiers: LCCN 2021031629
ISBN 9781478015611 (hardcover)
ISBN 9781478018247 (paperback)
ISBN 9781478022855 (ebook)
Subjects: LCSH: Palm oil industry—Social aspects—Indonesia—Papua
Barat. | Palm oil industry—Environmental aspects—Indonesia—Papua
Barat. | Palm oil industry—Indonesia—Papua Barat. | Plantation
workers—Indonesia—Papua Barat—Social conditions. | Sustainable
development—Indonesia—Papua Barat. | Rural development—Indo-
nesia—Papua Barat. | Deforestation—Indonesia—Papua Barat. | BISAC:
SOCIAL SCIENCE / Anthropology / Cultural & Social | SCIENCE /
Environmental Science (see also Chemistry / Environmental)
Classification: LCC HD9490.5.P343 I5 2022 |
DDC 331.7/63385109951—dc23/eng/20220103
LC record available at https://lccn.loc.gov/2021031629

Cover art: Sago palms, color altered. Courtesy of the author.

Duke University Press gratefully acknowledges the University of
Sydney, which provided funds toward the publication of this book.
Funds were also provided by the Australian Research Council.

Publication of this book is supported by Duke University Press's
Scholars of Color First Book Fund.

Frontispiece. In the shadow of the palms. Sago fronds reflected in starch and water.
Photo by Sophie Chao.

To Jacob, for his humbling courage in all walks of life.

CONTENTS

Oil palm killed the sago
Oil palm killed our kin
Oil palm choked our rivers
Oil palm bled our land

Valuable like agarwood, sago is not
Expensive like red meranti, sago is not
Elegant like the frangipani, sago is not
Majestic like the banyan, sago is not
But life it brings and growth to share
Food it gives and water it cleanses
Shade it offers, rest it promises

So, jail me, shoot me, burn me, kill me
But bring my shattered bones to the sago grove
To rest among the suckers, to drink from cleaner rivers

Sago, sago, you first came into being
In a place called Timasoe
There, our children grew strong and bold
Our wives had shiny skin and abundant sweat
Our men were tall and fit
Timasoe, Timasoe, Timasoe
You are west of the cassowary mound near Doeval
East of the last bend of the Milavo tributary
North of the juniper bushes
Where my ancestor Khiano gave birth to Yom
A sacred place, a peaceful place

Where wild deer and pigs and birds came
For water and shade and protection from the rain

Oil palm killed the sago
Oil palm killed our kin
Oil palm choked our rivers
Oil palm bled our land

Timasoe, Timasoe, Timasoe
Dare I visit you now?
With sorrow and shame, I tread your soil
My bones weak from riding trucks
My skin grey from eating rice
My hands bloodstained from the dollar bills

Timasoe, you are now a bare and barren place
Lodged between the Trans-Irian Highway and plantation blocks
Between roads and dust, you stand
Hostage to oil palm, the settler palm, you weep

For no sago here will grow
No rivers here will flow
No gentle winds shall blow
No songs tomorrow know
Our bones your earth shall stow

—The song of Marcus Gebze, elder of Mirav village, West Papua

Nausea. Anger. Grief. Driving through oil palm plantations with my Marind companions in rural West Papua brought home to me the boundless devastation and disciplined monotony of industrial monocrops as no high-resolution drone footage or glossy environmental magazine ever could. Endless rows of oil palms surrounded us, silently condemning our clandestine vehicle. A cortege of trucks rumbled into the horizon, dragging loads of felled woods amid shrouds of stubborn red dust. The palm oil processing plant, looming on higher ground, spewed smoke and steam throughout the day and night. Illegal land-clearing fires consumed the forest, blanketing the landscape in a choking haze.

Hunched beside the road, young plantation laborers watched us drive by with dull gazes. Paraquat, a deadly herbicide, trickled down from rusty canisters strapped to the women's backs, the blue-green venom seeping into their exposed skin. Banned in many countries because of its toxic effects, no antidote exists for this lethal chemical. I thought of babies never to come. The faces of my friends, huddled in the bed of the truck, were caked in dust and watched the landscape unfurl, weeping. Infants retched from the stench of mill effluents as we jolted down dirt roads without stopping so as to avoid attracting the attention of military men employed by the companies to guard their plantations. Bunches of oil palm fruit lay strewn along roadsides, piles of moldering blood-red and coal-black, shot through with razor-sharp thorns. Bulldozers and chainsaws ripped through isolated patches of the remaining vegetation. Silhouetted against the bleary sun, pesticide-spraying helicopters zigzagged back and forth above us, spreading a milky veil of hazy toxins.

Crouched in the back of one of the trucks in late July 2015, Paulus Mahuze, a Marind clan head from Khalaoyam village in the West Papuan regency of Merauke, explained to me how oil palm had arrived in his homeland.[1] On August 11, 2010, a delegation of government representatives from Jakarta, led by then minister of agriculture Ir. H. Suswono, had officiated an inauguration

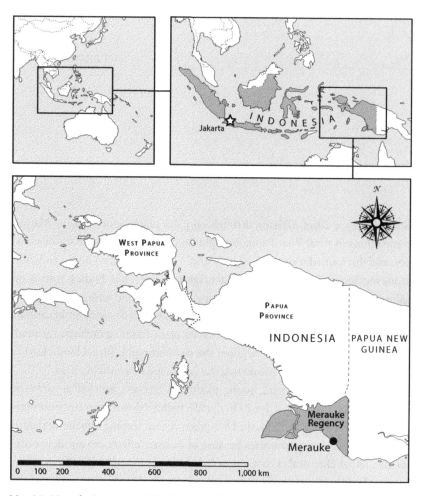

Map I.1. Merauke Regency in West Papua, Indonesia. Map by Geoffrey Wallace.

ceremony in the nearby village of Sirapu. They were launching the Merauke Integrated Food and Energy Estate (MIFEE), a USD $5 billion agribusiness scheme intended to promote the country's self-sufficiency in basic foodstuffs and make Indonesia a net-food-exporting nation. Papuans from across the region were invited to the event, including Marind community members from villages along the upper reaches of the Bian River, where I undertook my fieldwork. Paulus described the ceremony:

It was a hot day. There was dust (abu) everywhere, raised by the government convoys and military trucks.[2] The dust stung our eyes and made the children cry. The government brought oil palm (sawit) company bosses with them from pusat ("the center," or Jakarta). They gave us instant noodles, pens, bottles of water. They also gave us cigarettes—the expensive kind. They talked a lot about MIFEE. MIFEE this, MIFEE that . . . but we didn't understand what MIFEE was. We did not know what oil palm was because oil palm does not live in our forests. Then, the government officials and the oil palm bosses left. They never returned to the village. They promised us money and jobs. They said MIFEE would provide us with food. I thought that they would plant yams, vegetables, and fruit trees. Instead, they planted oil palm. They planted oil palm everywhere they could. They turned the whole forest into oil palm. They cut down all the sago to plant oil palm. This is what happened. Since then, everything is abu-abu ("gray" or "uncertain").

By May 2011, the Indonesian government had allocated some two million hectares of land in Merauke to thirty-six domestic and international corporations for the development of oil palm, timber, and sugarcane plantations. Vast swaths of forest had been felled or burned. Major watercourses had been diverted to irrigate the newly established monocrops. Today, Paulus's home village of Khalaoyam, along with several others along the Upper Bian River, are encircled by oil palm plantations that cover several hundred thousand hectares of former forest and extend north into the neighboring regency of Boven Digul. As we enter the third decade of the third millennium, dozens more companies are applying for operational permits. Agribusiness continues to expand relentlessly across the region.

I first visited the Upper Bian in 2011, while working as a project officer for the UK-based human-rights organization Forest Peoples Programme. At the time, I was undertaking field investigations with nongovernmental organizations (NGOs) and church institutions to document the social and environmental impacts of oil palm developments in Merauke. These investigations revealed that agribusiness projects were being designed and implemented without the free, prior, and informed consent of Indigenous Marind (see Awas MIFEE 2012; Ginting and Pye 2013; Ito, Rachman, and Savitri 2014). Military-corporate collusion was rampant. Consultations, when undertaken, presented projects as a fait accompli and offered limited information to communities on the potential risks to their food security, land rights, and economic livelihoods. Oil palm projects were routinely framed in corporate and government rhetoric as key to national interests, regional economic growth, and the "development" (pembangunan) of West Papuans into modern, civilized subjects. Yet employment opportunities for local Marind proved limited, as companies preferred to bring in their own labor force or hire migrants. Other grievances shared by Marind villagers included unfulfilled corporate social responsibility schemes, critical levels of water pollution, endemic biodiversity loss, and deforestation through illegal burning.

Oil palm developments in Merauke thus exemplified vividly what anthropologists have called the "dispossessory dynamics" of agribusiness expansion—a process premised on and perpetuating structural violence in the form of land alienation, growing poverty, intergenerational displacement, and precarious rural livelihoods (Li 2017a; Tsing 2005; West 2016). The plantations also represented a classic case of "land-grabbing," or the large-scale acquisition of land in the Global South for agricultural development, intensified by the food, fuel, and finance crisis of 2008 (Borras and Franco 2011; Edelman, Oya, and Borras 2015; D. Hall 2011). In this regard, the dispossessory dynamics of agribusiness in Merauke were not radically dissimilar to what I had witnessed in other parts of the Indonesian archipelago where oil palm is industrially cultivated, and most notably in Sumatra and Kalimantan. However, the particular *ways* in which this dispossession was being experienced on the ground differed.

Very early on, I was struck by how Upper Bian Marind conceptualized the arrival of oil palm. The stories I heard in the field were not about global markets, corporate interests, or food security. Nor did they primarily revolve around the issue of rights—land, human, or Indigenous. Instead, cryptic statements abounded in villagers' reflections on their present condition, which were invariably preceded by the temporal marker "since oil palm arrived." Oil palm, people told me, was a modern totem that had made time stop. The forest had become a world of straight lines, haunted by a rapacious and foreign plant-being. Cassowaries and crocodiles were turning into plastic and weeping like humans as their native habitats disappeared. At night, oil palm depleted the flesh and fluids of dreamers in their sleep. Meanwhile, the skin of animals and plants was drying out as oil palm sapped wetness from the earth and devoured the forest.

These narratives challenged my activist habitus. They also stimulated my curiosity. Eventually, they brought me to leave the world of human rights advocacy and undertake long-term ethnographic fieldwork among Upper Bian Marind. These early experiences thus marked the beginning of a long personal and intellectual journey of encounter with difference—a difference whose many facets I will explore in the chapters that follow. Oil palm expansion, I came to realize, could not be framed as either a social or an ecological problem. Nor could it be addressed purely through the discourse of human rights or environmental justice. This expansion was radically reconfiguring Marinds' sense of place, time, and personhood—their bodies, their stories, even their dreams. It affected men, women, and children both present and to come who, together with their forest kin, appeared to be undergoing a more-than-human existential crisis—one that left no single sphere or species of life untouched. Many NGOs, including the one I worked for, targeted the Indonesian government, international corporations, and financial investors in their anti–oil palm campaigns. And yet the communities whose rights we advocated for seemed more interested in oil palm itself—where it comes from, what it wants, how it differs from native species, and why it is so destructive.

Against this backdrop, the book before you explores how Indigenous communities in an out-of-the-way place engage with the disruptive effects of an other-than-human actor.[3] Specifically, I ask: How do Marind experience, conceptualize, and contest the social and environmental transformations provoked by deforestation and oil palm expansion? How do these transformations reconfigure the relations of Marind to each other, to other species, and to their environment? And how do plant-human dynamics in the Papuan plantation nexus inform our understanding of more-than-human entanglements in an age of planetary unraveling?

Appreciating how oil palm transforms the interspecies relations, geographies, and temporalities of the Upper Bian requires that, like Marind, we take seriously the attributes of plants as particular kinds of agents. The villagers with whom I worked do not conceive of oil palm solely as a sessile object of human exploitation or a passive instrument of capitalist gain. Rather, widespread speculation over oil palm's affects and effects arises from the fact that the plant itself is seen (and feared) as a willful entity—one that is voracious, destructive, and alien. In the proliferating being of oil palm, the forces of neoliberal capitalism and settler-colonization resist conceptual abstraction and find a material grip. Violence reveals itself as a multispecies act.

ALONGSIDE MELTING GLACIERS, MARINE oil spills, and inundated islands, large-scale plantations are emblematic of an era characterized by the unprecedented magnitude of human activity on the planet.[4] Within the agribusiness industry, the palm oil sector is particularly notorious for its destructive environmental impacts. Palm oil represents one of just four commodities responsible for the majority of tropical deforestation and the second largest industry sector driving global warming (Global Forest Coalition 2017). Oil palm plantations dramatically reduce biodiversity and damage the habitats of endangered species. They undermine ecosystem services such as nutrient cycling, water purification, and soil stability. The adverse consequences of oil palm expansion on the livelihoods and land rights of Indigenous peoples and other local communities have also been extensively documented (see, inter alia, Andrianto, Komarudin, and Pacheco 2019; Colchester and Chao 2011, 2013; Gabriel et al. 2017; Li 2017b). Yet despite growing controversy over their social and environmental impacts, oil palm plantations continue to spread across the tropical belt, driven by economic development imperatives, renewable energy policies, and a growing world population. Integral to the global agroindustrial food system, palm oil remains the cheapest and most versatile vegetable oil on the market, present in over half of all packaged goods globally (World Wildlife Fund 2020).[5]

Scholars from a range of disciplines have condemned industrial plantations for subjecting cash crops to totalizing human control and for jeopardizing biodiverse forest ecologies.[6] Comparatively speaking, however, agribusiness has received less ethnographic attention than other environmentally destructive industries, such as mining and logging. Existing studies have focused primarily on the anthropogenic forces driving plantation expansion and the experiences of peasant groups involved (more or less willingly) in the plantation sector as laborers or smallholders.[7] The ways in which Indigenous communities in Merauke conceptualize and engage with monocrops provide an important counterpoint to these accounts. Marind are directly affected by the ecological destruction wrought by agribusiness, but most remain excluded from the sites and circuits of palm oil production. Few are, or wish to be, employed by local corporations. Indeed, many Marind are averse on moral grounds to agriculture, horticulture, and other forms of plant or animal domestication.

Perhaps most important of all, Upper Bian Marind do not primarily attribute the destructive impacts of oil palm expansion to human actors, technologies, and market forces—even as they are well aware of them. Instead, they attribute these effects to the volition and actions of oil palm itself. The blame that Marind place on oil palm is pivotal to this story. It is what makes it differ from other works on plantations and plant-human relations. It is what disrupts the human-centered focus of political economy approaches to the agroindustrial sector. It is also what brings into the picture other powerful entities that, like oil palm, are deemed by Marind to be introduced and invasive—the state, settlers, soldiers, and corporations.

And yet blaming oil palm is only part of this story. As much as they resent the plant for its radically destructive effects, Upper Bian Marind also pity oil palm for its subjection to totalizing human control. Others express curiosity about oil palm's origins, needs, and desires. Ambivalent affects and heterogeneous perspectives coalesce around this alien plant of unknown ways and wants. Taking seriously the conflicting meanings of oil palm prompts us to ask which lives and deaths matter within capitalist natures, to whom, and why.[8] It invites attention to justices alternately enabled or preempted by agroindustrial landscapes—environmental and social, restorative and intergenerational, human and more-than-human (Chao, Bolender, and Kirksey 2022).[9] It reveals the potential and limits of the "species" as a mode of analysis, relation, and practice. And it points to violence itself as a multispecies act—one in which humans are not always the perpetrators and nonhumans not always the victims.

In this book, both ethnographic description and conceptual abstraction help to reveal the granular textures of Marinds' changing lifeworld. I avoid imposing a carapace of theory atop the moving flesh of ethnography. Instead, I thread thick description and distilled abstraction in the manner of the barks and filaments that my Marind sisters artfully fashion into woven sago bags. Some of the concepts I deploy in this story are Marinds', and others are mine but draw from those of Marind. Some concepts are inspired by the work of Indigenous and critical race scholars and others stem from what might be considered the traditional Western canon of theory.[10] Moving back and forth between theorizing ethnography and ethnographizing theory, I respond to Black feminist scholar Zakkiyah Imam Jackson's (2013, 681) call to collapse the hierarchical distinction between Western theory and non-Western cosmology—a distinction that itself replicates and perpetuates the historical oppression of Indigenous and other marginalized peoples by (settler-)colonial regimes. In switching my analysis of Marind thought between Western eyes and Indigenous eyes, I work against the colonization of ethnography by theory when theory is taken to be "produced" by (and often for) the Global North, based on ethnographic realities that "happen" in the Global South.[11] Instead, I look for theory in "small places" (Agard-Jones 2013, 183) produced by peoples who persist in the face of imposed invisibility and who have something important to say about what it means to live under entrenched regimes of color and capital (see also Banivanua-Mar 2016; Hviding and Bayliss-Smith 2018; K. Teaiwa 2014; West 2016).

Attending to theory in small places reveals the agentive and imaginative capacities of people in the face of structural inequalities that are relative to, but never totally determined by, macrolevel forces. It draws attention to the critical vantage points held by communities at the margins of the world capitalist economy and the complex idioms through which they articulate ongoing processes of accumulation through dispossession.[12] In the context of the global ecological crisis, starting from the local allows us to appreciate the specificity of loss and potentiality in the very places where they materialize and come to matter. Rather than a study of the ontology *of* Marind, this is an account of Marind *as* ontologists of their own changing worlds—one that takes as its primary objective the acknowledgment of Indigenous creativity and the decolonization of anthropological thought and practice.[13]

This book adopts human-vegetal relations as a central lens for exploring the changing lifeworld of Upper Bian dwellers. In doing so, it contributes to a vast body of anthropological literature that has found in plants a fruitful entry point to understanding human cultural forms and social organization.[14] Alongside their material uses and ecological functions, plants in Indigenous and other horticultural communities are often endowed with a soul or other form of agentive consciousness. In Melanesia and Amazonia, for instance, plants may be personified as kin (notably as surrogate children) or classified as male or female and associated with particular personalities or traits—gentle, aggressive, ugly, or beautiful.[15] Some cultures correlate the substance and structure of particular plants to those of humans. The sexualization of plants is often linked to notions of fertility and procreation, giving rise to gender-inflected modes of plant cultivation, exchange, and consumption.[16] Stages of vegetal growth may be associated with the human life cycle and intergenerational reproduction or serve as the basis for broader divisions of seasonal or calendrical time.[17] In some societies, plants coaxed into maturation through ritual, magic, and song enable those who nurture and consume them to access sacred sources of knowledge or acquire other-than-human forms and faculties.[18]

Anthropological studies of plant-human relations have tended to focus on native vegetal lifeforms with a well-defined status within local cosmologies— for instance, taro, yams, and sago in Melanesia and cassava, tobacco, and ayahuasca in Amazonia. This book, on the other hand, focuses primarily on a plant—oil palm—that was only recently introduced into West Papua and that many Marind consider alien and invasive. I examine the ontology of oil palm by cross-pollinating classic environmental anthropological literature with insights derived from the *plant turn*, a budding interdisciplinary current that foregrounds the role of plants as communicative, sentient, and worldmaking actors.[19] The plant turn moves beyond the treatment of plants in purely representational and functional terms. It invites us to think and theorize *with*—rather than just about—vegetal lifeforms as agents in their own right. It also considers the historical, affective, and mimetic entanglements of humans and plants, in a practice that Theresa Miller (2019) calls "sensory ethnobotany."[20] In an age of rampant ecological crisis, scholars of the plant turn exhort us to "make allies" of vegetal beings to sustain our mutual dependencies and generate "new scenes of, and new ways to see" plant-people relations (Myers 2017a, 299–300).[21]

The storied relations of plants and people recounted in this book speak powerfully to the ethical urgency of reimagining interspecies entanglements in an age of planetary undoing. At the same time, the sites and subjects at the heart of this story—Indigenous lifeworlds and monocrop plantations—offer a critical counterpoint to the predominantly Western- and technocentric focus of the plant turn and related strands of thinking in the broader fields of the environmental humanities and posthumanism. Departing from the prevalent focus on scientific and conservationist perspectives within these currents, I ground my analysis in the theories, experiences, and knowledges of an Indigenous community whose social relations have always encompassed other-than-human beings but are now challenged by the occupying presence of a lethal vegetal lifeform.[22] In so doing, I seek to expand approaches for reimagining what is possible in more-than-human worlds that remain largely situated in the unmarked White space of Euro-American (settler-)colonialism.[23]

But this book also invites a more fundamental critique of posthumanist currents. On the one hand, Marind practice a posthuman ethic in positioning themselves as one kind of self among a plethora of agentive forest lifeforms. No "Great Divide" separates or elevates humans from "nature" in the Upper Bian. Rather, Marind come into existence through their corporeal, affective, and material connections to kindred plants and animals, within a broader ethos of relationality in which all lives and lifeforms are interdependent.[24] Yet Marind are also grappling with an other-than-human being—oil palm—that is invasive and destructive. Many of them actively resist the technocapitalist assemblages attempting to turn them into posthuman "cyborgs."[25] These assemblages include the plantation economy and its production-driven logic; the dreams of "modernity" promoted by the government and incarnated in oil palm; the racialized treatment of Papuans as primitive peoples in need of development; the commodified foodstuffs replacing Indigenous, sago-based foodways and ecologies; and the conversion of animate forests into homogeneous monocrops.[26] Together, these imposed transformations perpetuate the dispossession of Marind of their bodily and territorial sovereignty. Together, they alienate Marind from the multispecies relations that make Marind human in the first place.[27]

In this light, the posthumanist effort to decenter the human and practice multispecies love becomes problematic. It brings us into alliance with a plant whose entanglements with Marind and their forest lifeworld is neither desired nor conducive to multispecies thriving. These entanglements stem from a capitalist formation—the plantation—that is itself imbricated with imperial forms of violence, enslavement, and expropriation, including racialized hierarchies of humanness and attendant necrobiopolitical regimes.[28] Far from solely an economic production system, the plantation, in the words of Haitian anthropologist Michel-Rolph Trouillot, is also a race-making institution (2002, 200). As landscapes of empire, plantations remind us that environmental problems are indissociable from histories of colonialism, capitalism, and racism, which have rendered some beings less or differently human than others.

In the story that follows, the racial logics of capitalism and colonialism manifest in the paralyzing effects of state-corporate geographies, the asymmetric relations of Indigenous Marind to non-Papuan settlers, and the paternalistic rhetoric of progress surrounding agribusiness developments. These dynamics reveal how oil palm's relatively recent arrival exacerbates the ongoing subjection of Indigenous communities to racializing assemblages that render them subhuman and disposable before the law.[29] West Papuans today, independence activist Filep Karma (2014) notes, continue to be treated like half-animals.[30] Their imprisonment, killing, and torture are not only tolerated by the state but also at times celebrated (Hernawan 2015). Their right to self-determination continues to be denied and their lands and resources continue to be appropriated without consent (Chao 2019a). In arguable contrast to the postcolonial world, where the imperial logic of discrimination and displacement perdures as ruin and debris (see Stoler 2013, 2016), the racialized logic of settler-colonialism in West Papua is very much alive and well.[31]

Giving center stage to plants in a world where colonizing plants and people are destructive and racialized multispecies communities are their victims serves to challenge universalist assumptions of human exceptionalism—a widely critiqued concept in the posthumanist tradition. It demands that we approach posthumanism itself as a *plural* category of being—one alternately embraced or eschewed by communities positioned as subhuman under colonial and technoscientific regimes. It demonstrates the importance of attending to Indigenous epistemologies in appreciating which lifeforms are deemed loving or unloving, and consequently lovable or unlovable.[32] Never straightforward binaries, these categories reveal themselves to be species-inflected—but not always species-determined—modalities of being within the dispersed ontologies of the Upper Bian.

The poetics and politics of more-than-human entanglements in Merauke invite us in turn to rethink the notion of violence as solely or primarily an anthropogenic act. As my host father, Marcus Gebze, sings in the Prologue,

Marind inhabit a world held hostage by an invasive "settler palm" that kills the sago, murders their kin, chokes their rivers, and bleeds their land. This world demands that we take seriously the possibility of plants, not as amoral, but as *immoral*, subjects. It brings into question the prevalent characterization of plant-human dynamics as reciprocal and beneficial and of plants as nonappropriative, giving beings.[33] It also offers a sobering counterpoint to the celebration of more-than-human encounters as inherently conducive to multispecies intimacy and thriving.[34] Instead, the words and worlds of Upper Bian Marind draw attention to the potentially exclusionary and diminishing effects of more-than-human entanglements.[35] In doing so, these words and worlds provocatively reframe the assumed human monopoly on violence as potentially yet *another* instance of human exceptionalism. When a particular group of humans and their other-than-human kin are subjected to the damaging effects of a proliferating plant, we find ourselves forced to redefine violence itself as a multispecies act.

In elaborating this argument, I explore how oil palm—a literal *neophyte* (from the Greek words for *new* and *plant*)—becomes a potent object of wonder for Marind, which alternately indexes or challenges the ontological ruptures wrought by agribusiness expansion (cf. Scott 2016, 476). Such ruptures manifest in the dynamics of Marinds' everyday village life, in their material engagements with the environment, in their interactions with state and corporate entities, and in their dreams, which magnify the dystopic effects of oil palm on the landscape and its lifeforms. Across these and other settings, multiple diverse scales, subjects, and species coalesce or collide in generative friction (Tsing 2005).[36] The frictions I examine arise from Marinds' fraught encounters with colonialism and modernity, along with their associated actors, technologies, and claims to knowledge and power, including knowledge as power. They entail the substitution of sentient forests with technocapitalist plantations. They encompass the antagonistic relation between introduced cash crops and native species, whose respective proliferation and obliteration speak unsettling truths to Marind about their own fates and futures. Together, frictions in the plantation as *contact zone* reveal an ontological dissonance between the Marind lifeworld and the forces of agroindustrial capitalism, as incarnated in the sago palm and the oil palm, respectively.[37]

Exploring the dispersed ontologies of the oil palm and the sago palm brings me to examine the clashing visions, projects, and desires of state and corporate actors, on the one hand, and Indigenous actors, on the other—what emerges in these spaces of encounter, what is excluded from them, and what might be hoped of them.[38] Drawing from almost a decade of involvement in the land rights campaigns of Upper Bian Marind, I assess the obstacles faced by activists as they struggle to curb the proliferation of oil palm and protect their sago-based ecologies, foodways, and relations. I also demonstrate how Marind engage creatively with the ambiguity of oil palm to generate new possibilities of being for themselves and their forest kin.

In focusing on the radical ruptures engendered by oil palm in Merauke, the story that follows exemplifies what Sherry Ortner (2016) calls "dark anthropology"—an anthropology that attends to social experiences of oppression and injustice in the rise of global neoliberal capitalism. To this end, I explore the Indigenous modes of analysis and praxis through which Marind conceptualize and critique the ontological ruptures wrought by agribusiness expansion.[39] I situate these ruptures within broader processes of cosmological decline that manifest in the transformed bodies and relations of humans and other-than-humans.[40] I examine how plants themselves come to act as potent symbols for larger sociohistorical forces that shape the contested spaces of the plantation, forest, and village. I also attend to the moral and sensory dimensions of plant-human entanglements as they manifest in the tangible violence of the waking world and in the anxiogenic dreams that haunt Marind in the sleeping world. By interweaving political ecology with phenomenology, I seek to bring to light what Paige West calls the "affects of dispossession" (2020, 122), or the sensory and affective ways in which systemic loss, violence, and destruction are experienced by people in their situated relations to each other and to the more-than-human dwellers of unevenly shared and increasingly vulnerable environments.

At the same time, this story engages with dark anthropology's counterpart, or what Joel Robbins (2013) calls "anthropologies of the good." To this end, I explore the meaning of the good life among Marind in light of their conceptions of morality, relatedness, and interspecies care.[41] I investigate how beings in the forest participate in the (trans)formation of moral selves and relations through bodily exchanges of wetness and skin. I examine how the good coalesces in the affective textures of Marinds' relations with sago—a plant that my companions invariably describe in contrast to oil palm.[42] Following Unangax scholar Eve Tuck (2009), I also analyze how Upper Bian communities resist and refuse the darkness of the present and the precarity of futures both imposed and arrested through their daily interactions with human and other-than-human beings, their involvement in land rights campaigns and participatory mapping, and their emergent sense of solidarity as collective victims of the violence of oil palm.[43] These everyday imaginative acts in turn invite us to reflect critically and capaciously on the (im)possibility of hope in a present of impasse—a present when, as many Marind affirm, the arrival of oil palm has made time itself grind to a halt.[44]

The good and the bad form one of several counterpoints that animate the story. As entities that accrue meaning through their relationship to each other, counterpoints reveal how Marind creatively flesh out the categorial differences that matter as they forge a "Papuan Way" in the wake of ecological destruction.[45] The counterpoints I explore include the materiality of the landscape and its cartographic representation, the duality of body and mind, and the mirrored ontologies of human and bird shape-shifters. They encompass the opposed moralities of sago palm and oil palm, the gastropolitical divides embodied in rice and sago, and the fraught dynamics of oneiric possession and diurnal suffering. Other counterpoints include the interplay of interspecies violence and care, the opposed perspectives of plastic drones and forest birds, and the seemingly incompatible patterns of monocrop capitalistic production and multispecies social reproduction. The text before your eyes, which draws together multiple voices, utterances, and discourses that I gathered through my own intersubjective interactions with Marind in the field, is itself nothing less than contrapuntal.[46]

More than anything, however, the story I tell attends to the generative spaces that lie *between* the counterpoints of good and dark, or what Paulus Mahuze—the head of Khalaoyam village—described as the realm of abu-abu.[47] Many Marind in the Upper Bian referred to 2015—the year I started my fieldwork—as a time when the world became abu-abu, meaning "gray" or "uncertain." That year, the sky turned hazy and dark from the thick smoke raised by large-scale forest burning—a cheap and fast, if illegal, way of clearing land to make way for agribusiness concessions. As the ashes of incinerated vegetation dispersed across land and sky, 1.5 billion tons of greenhouse gas emissions were released from over 120,000 fires across the archipelago. The gray year was also one of severe drought caused by El Niño and aggravated by the diversion of major waterways to irrigate the newly established plantations. When the rains finally fell they were brief. By then, the waters of the Bian had turned gray from the daily discharge of toxic palm oil mill effluents.

Much like gray is neither black nor white but rather a mix of both and ashes are the barely tangible residue of irretrievably incinerated forms, the oil palm, the MIFEE project, and the future itself, were shrouded in menacing opacity during the year of ashes. Compensation payments and employment opportunities that had been promised to local communities had yet to materialize. Instead, cheap housing popped up across the landscape to house a sudden influx of Javanese laborers. New concession markers were erected in unexpected locations and without prior notice to local inhabitants. Despite sustained efforts, local and international advocacy initiatives were failing to slow agribusiness expansion.

At the same time, rumors that MIFEE might be relocated to other parts of West Papua rippled sporadically throughout the villages. Several oil palm companies were said to have gone bankrupt. Others had vanished after reportedly making a fortune by illegally logging the precious woods in their concessions and trading them on the international market.[48] While many Marind remained staunchly opposed to oil palm developments, others sought employment within the plantations or worked as intermediaries for the corporations. Opaque like the tenacious haze blanketing the parched landscape of the Upper Bian, oil palm itself lay at the heart of a material and ontic crisis of visibility.[49] Intense concerns and curiosity were exacerbated by uncertainties surrounding the plant's own abu-abu dispositions and desires.

Abu-abu, as I examine in this story, encompasses ambiguous affects and atmospheres, things and beings, and spatialities and temporalities.[50] It is a condition of awkward existence distributed across unevenly situated human and other-than-human communities of life whose futures are threatened by intensifying agroindustrial landscape transformations. Amid such transformations, inhabiting abu-abu means living with opacity as a generalized and constitutive state of being. But abu-abu can also generate new becomings amid ruptured more-than-human meshworks. In certain instances, embracing abu-abu can even become a form of covert resistance—one that refuses the exclusions and erasures produced by fixed classificatory schemes intent on governing matter and meaning through reductionist logics of separation and stratification.[51]

In the Upper Bian, abu-abu manifests in the uncertain fate of the forest, the ambiguous efficacy of Indigenous maps, and the strange lives of village-bound cassowaries. Abu-abu shrouds the conflicting desires of Marind as they make do in a world of plastic foods, concrete totems, and deadly dreams. It haunts the clashing temporalities of the world before and after oil palm, the violence of imposed futures, and the shape-shifting beings that lurk within the murk. Abu-abu will follow us throughout this account, alternately foregrounding or subverting the contrapuntal dynamics of the Marind lifeworld.

Acts of resistance and refusal in the world of abu-abu are often mundane and rarely heroic. These acts, to borrow Elizabeth Povinelli's words, are "ordinary, chronic, and cruddy rather than catastrophic, crisis-laden, and sublime" (2011, 13). For some Marind, resistance takes the form of defiant, self-directed violence—the ripping of one's hair and the drawing of one's blood. For others, resistance means making maps that won't sit still, eating sago rather than rice, rebuffing the passage of time, or finding solace in the world of dreams. Perhaps most important, resistance among Marind entails an epistemic refusal to reduce oil palm to any singular or bounded ontology—good, bad, or other.

As destructive as it might be, oil palm *also* exists to Marind as an exploited victim, an object of curiosity, a pathway to an expanded world, and possibly even a kind of kin. In this shadowy world, where new beings subvert the realities and relations of Marind and their forest companions, the line between the good and the bad remains very much in the making. But before we enter the murky realm of abu-abu, let me flesh out the ethnographic setting where our story unfolds.

MERAUKE REGENCY IS LOCATED at the southern tip of Indonesian Papua, a region that borders Papua New Guinea to the east, the Boven Digoel and Mappi regencies to the north, and the Arafura Sea to the south and west. In ecological terms, the area is composed of low-lying and generally flat peat land, grassland, and dense swamp forests. In the inland back plains, serpentine rivers heave to the cadence of monsoonal rains, giving rise seasonally to Papua's most extensive wetlands. A range of resident and migratory birds, including waterfowl and waders, inhabit this zone of the TransFly EcoRegion. Larger animals, including cassowaries, tree kangaroos, possums, and crocodiles, populate its forests and rivers.

Plant life in Merauke is equally diverse, with monsoon forests containing an exceptionally high number of endemic plants unique to the region. A mixture of Phragmites, tall sedge grasses, and low-swamp grasses flourish in the permanent marshes, while semipermanent to seasonal Melaleuca swamp forests occupy terrains on higher ground.[52] Riverbanks and mangroves are home to dense groves of sago, a pinnate-leaved palm of the tropics known in Western taxonomy as *Metroxylon sagu* Rottbøll and as <u>dakh</u> and sagu in Marind and Indonesian, respectively.[53] Today, these biodiverse ecosystems are increasingly being replaced with monocrops of oil palm, a plant known scientifically as *Elaeis guineensis* Jacquin and as kelapa sawit or simply sawit in Indonesian.[54]

The villages of Khalaoyam, Bayau, and Mirav, where I undertook my fieldwork, are three of eight settlements lying along the upper reaches of the Bian River in Merauke's inland subdistricts of Ulilin and Muting. They sit within the customary territories of Marind, a vast triangle that stretches two hundred kilometers eastward along the coast from the Muli Strait in the west to twenty-five kilometers east of Merauke City and two hundred kilometers inland beyond the banks of the Fly River in Papua New Guinea. The villages are home to approximately six hundred households who self-identify in the Upper Bian dialect as Marind Bian (Marind of the Bian River) or Marind-deg (Marind of the forest).[55]

Each Marind clan (bawan) is related to a group of species whom they call grandparents (amai) or siblings (namek). Clans and their amai share descent from dema, or ancestral creator spirits, who drew them out of fissures in the earth at the beginning of time.[56] Many Marind names take the form of the plant or animal amai followed by "-ze," meaning "child of." For instance, the Mahuze clan are "children of the dog" and the Balagaize clan are "children of the crocodile." The interactions of amai and Marind are anchored in principles of exchange and care. Amai grow to support their human kin by providing them with food and other resources. In return, humans must exercise respect and perform rituals as they encounter amai in the forest, recall their stories, and hunt, gather, and consume them. These reciprocal acts of nurture enable humans and amai to partake in a shared community of life within the ecocosmology of the forest.

The communities of Khalaoyam, Bayau, and Mirav derive their subsistence primarily from hunting, fishing, and gathering. Sago flour, the staple starch food, is supplemented with forest tubers and roots (mainly taro and yam), fish, and forest game such as Rusa deer, lorises, possums, cassowaries, fowl, kangaroos, crocodiles, and wild pigs. Fruit including rambutans, papayas, bananas, golden apples, traditional mangoes, figs, watery rose apples, langsat, kedondong, jackfruit, and coconuts are also obtained from the forest, alongside leaves, roots, barks, resins, and saps that are used to make medicinal ointments and concoctions.[57] With large-scale deforestation underway, however, access to these foods has become difficult. Today, imported foodstuffs, such as government-subsidized rice, cooking oil, sugar, coffee, tea, instant noodles, and cookies, are increasingly consumed in the villages of the Upper Bian. These goods are also offered by oil palm companies as part of their land compensation or social welfare packages and now constitute an important component in Marinds' diet.

Aside from Marind, a minority of other Papuan ethnic groups live in the villages along the Upper Bian, such as Jair, Auyu, Muyu, and Wambon. Kinship connections across these settlements, as well as with villages in the northern regency of Boven Digoel and across the border in Papua New Guinea, are close. Community members travel regularly up and down the river and through the forest to visit relatives and attend customary rituals and meetings. These movements, however, are increasingly hindered by the establishment of privatized and strictly guarded monocrops along the national border. Most Marind in the region are Catholic, with a minority of Protestants concentrated in the upstream villages of Wam and Pasior. The Upper Bian is also home to a small population of primarily Muslim transmigrants (orang trans) and spontaneous migrants (pendatang), originating from Java, Makassar, Nusa Tenggara, and

Maluku. This non-Papuan population has increased significantly over the last decade, as settlers relocate to Merauke to work as laborers and harvesters in the newly established oil palm plantations.

Accounts produced by anthropologists, explorers, and administrators during the Dutch colonial years frequently portrayed Marind as violent and invasive warmongers.[58] According to archival materials from the British Public Record Office dating from around 1891 to 1903, Marind—whom the British and Dutch administration called *Tugeri*—were renowned throughout the region for their frequent headhunting raids on the neighboring Wasi and Buji tribes (MacGregor 1893a, 1893b).[59] Joining forces to go out on war parties, Marind reportedly managed to venture far east into what Europeans designated as British territory, west to Frederik Hendrik Island (now Yos Sudarso Island), and north across the Digul River. Headhunting and the adoption of children from raided communities purportedly enabled Marind to expand their territorial control and increase their population. At the same time, trade, intermarriage, cultural exchange, and ritual cooperation with other ethnic groups remained widespread.

Repeated Marind incursions eventually led the British administration to request that the Dutch establish a physical presence in the region. In February 1901, the governor of the Dutch East Indies demarcated Merauke as an onderafdeling (subdistrict) under the afdeling (district) of Southern New Guinea, and an official outpost was founded in Merauke City in February 1902. Three years later, the Missionaries of the Sacred Heart established themselves in the coastal village of Wendu and its surrounds. The mission gradually spread up north, reaching Okaba in 1910 and the hinterlands of the Upper Bian two decades later. The first inland mission was established in present-day Mirav in 1930.

However, the advance of foreign missionaries, colonial administrators, scientific expeditions, and traders in the Upper Bian was hindered by a landscape of semipermanent swamps and thick forests, the prevalence of various mosquito-borne diseases, and the purported reputation of Marind as lawless headhunters. The large body of colonial and ethnographic literature concerning the coastal—rather than inland—Marind reflects the limited influence of external actors in the hinterland. This includes Dutch ethnologist Jan van Baal's detailed monograph, *Dema: Description and Analysis of Marind-Anim Culture (South New Guinea)*, which, as van Baal acknowledges and as my own fieldwork confirmed, is primarily a description of coastal Marind groups (1966, 12–13).[60]

By the 1930s, many Marind ritual practices had been abolished by the Dutch administration—for instance, ceremonies that marked the transition of children across age-groups and headhunting expeditions that once sustained Marind populations through the adoption of children from raided tribes (Boelaars 1981; Corbey 2010; Ernst 1979). Nevertheless, informal activities in the forest remained key indicators of children's growth into <u>anim</u>, or "humans"—the first capture of game among boys, for instance, and the first weaving of sago bags among girls. Despite the sedenterization of Marind into "model villages" (model kampong) and the establishment of "civilizing schools" (beschavings scholen), village and school absenteeism was prevalent and Marind continued to regularly travel to the forest with their kin and children (Derksen 2016, 129).[61] The <u>dema</u> were recast by missionaries as primitive fetishes to be abandoned in the age of Christianity. But the <u>dema</u> lived on in the forest and continued to exert their influence on the landscape and its diverse dwellers. From the early twentieth century onward, the coastal Marind adopted introduced horticultural techniques such as rice paddy cultivation and entertained a lively (albeit at times animus-filled) trade in copra and iron with Chinese, Javanese, and Makassarese merchants (Swadling 1996, 178; Verschueren 1970, 57).[62] In contrast, and with the notable exception of the plume-trade boom of 1908–1924, traditional modes of subsistence in the interior continued largely unaffected throughout much of Dutch rule (Garnaut and Manning 1974, 15–17; van der Veur 1972, 277).[63] Horticultural projects initiated in this period were small-scale and located near the coast and the city rather than the hinterland.

Even today, the Upper Bian remains relatively less urbanized compared to coastal Merauke. Telecommunication services are available only a few hours a night in Mirav and there is no telephone signal in Khalaoyam or Bayau. Roads and other infrastructure in the region are minimal. Settlements consist of rows of rickety wooden houses with one or two small kiosks that provide limited basic supplies. Villages receive some income from government-support programs such as GERBANGKU and RESPEK and from the sale of nontimber forest products in Merauke City. Such income, however, is scarce and sporadic. Government funds only occasionally reach the villages and community members' access to urban markets is impeded by distance and transportation costs. Compensation payments for lands surrendered to oil palm companies represent another one-off source of income for some villagers. Averaging just under 5 USD per hectare, these payments are disproportionately less than the value of lands that were ceded (Al Jazeera 2020; Forest Peoples Programme, PUSAKA, and Sawit Watch 2013). Education rates in the Upper Bian are low, with less than half the population completing high school, 1 percent attending university, and 13 percent receiving no formal education (Basik 2017, 46). In the province with the lowest Human Development Index of the nation, infant mortality rates are high, life expectancies are thirty-five years for men and thirty-eight years for women, and HIV infection rates are the second highest in Indonesia.

The modern history of West Papua, which I explore in greater detail in ensuing chapters, is notoriously violent and volatile.[64] The Dutch authorities transferred administration of the region to Indonesia on May 1, 1963. This was followed by the controversial Act of Free Choice in July–August 1969, which resulted in what many Papuans see as the forceful incorporation of West Papua into the Republic of Indonesia. In response to ongoing demands for independence, a Papuan Special Autonomy Law was passed in 2001 but then radically weakened under the rule of Megawati Sukarnoputri, when political and economic power were firmly redirected into the hands of the central government. Hopes for peaceful resolution of what Jason MacLeod, Rosa Moiwend, and Jasmine Pilbrow (2016, 8) call the longest-running and most violent political conflict in the South Pacific, grew in the buildup to the election of Joko Widodo ("Jokowi") in 2014. Soon thereafter, however, concerns were raised when the president appointed several contentious military commanders to West Papua and established a new transmigration program, prompting a renewed influx of settlers into the region (Munro 2015a; Wangge 2014).[65]

Little has changed on the ground for most West Papuans since Jokowi's election. Top-down extractive activities have exacerbated community impoverishment and ecological degradation. Government corruption, military-corporate collusion, and the criminalization of activists restrict Papuans' capacity to seek recognition of their rights to lands and livelihoods. Cultural and religious assimilation policies are compounded by a growing population disparity between Papuans and non-Papuans across the province (Elmslie 2017). This disparity is particularly marked in Merauke, where Papuans now represent less than 40 percent of the population (Ananta, Utami, and Handayani 2016, 472). The violence of the colonizing state perdures in the form of incarceration, harassment, torture, sexual violence, and brutal military responses to Indigenous social justice movements. Since 1969, military clampdowns have occurred every year on the first of December, when Papuans commemorate their stolen independence by raising their national flag, the Morning Star.[66]

The entrance of Jokowi into office has also seen an acceleration in the implementation of the Merauke Integrated Food and Energy Estate, the mega-scheme driving oil palm expansion in Merauke. Central to Indonesia's Masterplan for the Acceleration and Expansion of Economic Development 2011–2025, MIFEE is expected to encompass three regencies and connect Merauke to six other economic production centers in the Papua–Maluku Economic Corridor. Although originally designed as a paddy cultivation scheme, the actual composition of MIFEE is dominated by oil palm, timber, and pulp and paper operations. Today, oil palm plantations, planned or projected, extend across some 1.7 million hectares in Merauke regency and occupy over 20 percent of the Upper Bian area (Franky and Morgan 2015). Ranging from 20,000 to 100,000 hectares each and operated by thirty-six national and international corporations, plantations creep right up to the edge of the villages, encroaching on sago groves, hunting zones, sacred graveyards, and ceremonial sites.

In historical terms, MIFEE constitutes the latest development in a long process of top-down resource exploitation in West Papua. This exploitation has included large-scale pulp and paper production, timber plantations, endangered wildlife trafficking, and nickel, oil, coal, gas, copper, and gold mining (see Down to Earth 2011). Moreover, MIFEE sits within a long history of oil palm cultivation in Indonesia, dating back to the early 1900s, when the first monocrop estates were established in Deli, North Sumatra. While oil palm plantations expanded rapidly under Dutch colonial rule, palm oil yields suffered episodic plunges during the Japanese occupation in World War II, the struggle for Indonesian independence up to 1945, and following the nationalization of Dutch companies in the 1950s. With the establishment of the New Order under Suharto, full-scale government support for agribusiness development led to a tenfold expansion in oil palm monocrops within two decades, boosted by capital injections from the World Bank and the Asian Development Bank. State-owned and smallholder-managed agribusinesses were gradually subsumed within larger estates, operated by a handful of private conglomerates. Political decentralization and global market forces have done little to undermine the sustained flow of profit to this powerful politicobusiness oligarchy, whose rise to power was facilitated by the close ties of its magnates to Suharto's totalitarian regime.[67]

In 2006, Indonesia surpassed Malaysia as the top palm oil–producing country, and today it supplies some 61 percent (thirty-six million tons) of the world's palm oil (Indonesia Investments 2017). According to the Indonesian Central Bureau of Statistics, oil palm plantations in the country covered 12.3 million hectares as of 2017, representing a 1.1 million hectare increase from the preceding year (Badan Pusat Statistik 2017, 9). With arable land increasingly scarce in Sumatra and Java, the monocrop frontier is now moving east into West Papua, a region deemed attractive for its vast areas of unexploited lands and cheap labor force.[68] This expansion is further boosted by the government's annual crude palm oil production target of sixty million tons by 2045. Achieving this target will require developing an additional 8.2 million hectares of land, an area equivalent to the entire island of New Guinea (Saleh et al. 2018).

Across the national border, the oil palm sector is also expanding rapidly in Papua New Guinea (see Cramb and Curry 2012; McDonnell, Allen, and Filer 2017). Today, palm oil constitutes Papua New Guinea's most valuable agricultural export and oil palm plantations represent the largest source of nongovernment employment (Allen, Bourke, and McGregor 2009). In 2017, oil palm concessions accounted for 2.2 million of the 5 million hectares alienated through Special Agricultural and Business Leases, a legal process designed to enable customary landowners to exploit their land for business purposes and to participate in the cash economy (Gabriel et al. 2017).

As in Indonesia, oil palm plantations in Papua New Guinea usually take the form of private estates owned by mega-conglomerates of predominantly Malaysian origin, which are also active in other sectors such as sugar and beef production and logging (Filer 2013, 316; Gabriel et al. 2017, 219). Patronage politics has facilitated the allocation of land to these companies without the free, prior, and informed consent of local landowners, fueling horizontal disputes and community fragmentation on the ground (Filer 2011; Nelson et al. 2013). Increasingly reliant on palm oil for their income, many rural villagers struggle to maintain a livelihood balance between subsistence horticulture, small-scale business, and export crops (T. Anderson 2015; Koczberski and Curry 2005). Competition over land, resources, and benefits also provokes tension between incoming migrant workers and native inhabitants (Koczberski and Curry 2004).

Local communities in Papua New Guinea and Indonesia have resorted to an array of institutional mechanisms to seek redress for the violation of their land rights. These include cases submitted to national courts, transnational advocacy campaigns, and complaint mechanisms activated under the voluntary certification standard of the Roundtable on Sustainable Palm Oil (Pye and Bhattacharya 2013; Filer 2017; Gabriel et al. 2017). Similar land rights advocacy efforts have been underway in Merauke since 2011, when oil palm was introduced to the region under the MIFEE mega-project.[69] These efforts, however, have been mired in a dearth of accurate information about the corporations active in the area and by the often contradictory policies regulating land acquisition and development at the national and provincial levels. Repeated attempts to activate UN mechanisms and palm oil certification schemes have been hampered by bureaucratic red tape, ineffective redress mechanisms, and communities' limited knowledge of their rights under national and international laws. Poor infrastructures, high travel costs, land privatization, and a prevalent military presence make access to the area difficult and dangerous for NGOs and researchers. The politically volatile context of West Papua and the threat (whether real or perceived) posed by independence movements to the Indonesian State further impede the efforts of Marind to secure their rights to lands and livelihoods. Government surveillance has intensified in response to their campaigns, including in the form of interrogations, extrajudicial incarcerations, and harassment from the police and military.

BEFORE I OFFER AN outline of the story to come, allow me to dwell briefly on how this book came into being. Like the shape-shifting humans, animals, and plants that enliven it, this is a "becoming" book—both in terms of the places and peoples I describe and in terms of my own changing relationship with Marind over the last decade. The themes, subjects, and setting of this research were specific, selective, and situated—both by me and by my interlocutors in the field. Neither comprehensive nor timeless, the study I present is thus a necessarily partial and subjective reconstruction of the Upper Bian lifeworld.[70]

My fieldwork was facilitated by the Merauke Secretariat for Peace and Justice, the humanitarian branch of the Missionaries of the Sacred Heart and a key collaborator in my previous human rights advocacy work in the region. During my eighteen months in the Upper Bian, I divided my time equally among the settlements of Khalaoyam, Mirav, and Bayau, following the movements of local inhabitants and the practicalities of weather and transport.[71] The greatest portion of my fieldwork, however, was spent, not in the villages, but rather in the forest, in the company of Marind groups traveling to meet friends and kin, to forage, and to process sago. These expeditions were crucial to understanding Marinds' place-making practices and their relations to the forest and its diverse lifeforms. It was in the forest, for instance, that I was enskilled by my companions in the arts of pounding sago, sharing skin and wetness with the grove, walking, and listening to the voices of birds and rivers. Cultivating these bodily ways of knowing was central to my transformation from foreign friend to near-kin—a transformation that culminated when, finally, I learned to dream in the forest like Marind.

But this world was also a difficult and dangerous one to enter and navigate. Inter- and intracommunity tensions ran high in the Upper Bian at the time of my research. The slow violence of ecological degradation was compounded by the immediate violence of the everyday. While forests were being systematically decimated to make way for oil palm, over a dozen community members had been incarcerated for opposing agribusiness. Twenty-two local land rights activists had died under mysterious circumstances after receiving anonymous death threats. Many faced ongoing intimidation and harassment from the police and military. My own fieldwork was cut short after two nuns at the Franciscan nunnery in Mirav, where I would seek shelter whenever military surveillance intensified, were beaten and raped by company-hired thugs. At the same time, a growing number of local landowners were ceding lands (their own and others') to companies in exchange for cash. Elite co-optation, bribery, and inequitably distributed compensation were breeding conflict between Marind standing "for" (pro) or "against" (kontra) oil palm and the many more individuals who sat somewhere in between. Disputes within clans, villages, and households had taken the lives of five community members.

In a place where the haunting force of the state, military, and corporations manifests as both lawfare and lawlessness, I had to be enskilled by my companions in the arts of strategic concealment and cultivated (in)visibility. As a person of French and Chinese descent, my Eurasian physique proved both an advantage and a challenge. On the one hand, my Asian traits reduced my visibility in a region where the presence and activities of foreigners are strictly monitored. On the other hand, some Marind initially regarded me with suspicion as a possible government spy or Javanese migrant. Others voiced concerns that I was working as an undercover consultant for oil palm companies because of the associations they made between my Chinese origins and the world of "business." My role as a foreign researcher had to be disguised under other identities, both prearranged and improvised. Depending on the setting, I was alternately a nun finishing seminary in Jayapura, a voluntary English teacher from Korea, a cousin thrice removed of the local Dayak priest, or a first-time tourist and avid birdwatcher. My tools of data collection, too, had to be camouflaged under various guises. Notebooks written in Chinese and French, encrypted hard drives, and quadcopter drones made their way to and from the villages at the bottom of jute bags filled with salted fish and sago flour, which were then set aside for me to collect from trustworthy traders. Meanwhile, second-hand mobile phones recorded police patrols' conversations and decimated forests from inside carefully punctured cigarette boxes—some brands, not all—held out of passenger windows or balanced between my knees during strategically timed toilet breaks.

The longer I spent in the Upper Bian, the better I became at noticing and dealing with situations of potential danger to myself and my hosts. I learned to recognize undercover militia from their crooked right index finger—"it never recovers from pulling the trigger"—and identify spies from the smell of scented aftershaves available only in the city. I learned to time my movements against the rounds of plantation security patrols. I learned to wait for days for cars that never arrived because their drivers had been called in for police interrogations or that arrived unexpectedly packed with armed passengers. I gradually became accustomed to the 3 a.m. wakeup call of police truncheons banging violently on village doors. I discovered where the women and children would retreat when drunk plantation guards staggered through the village at dusk, shooting blanks, vomiting bile, and jeering in slurs at the Papuan "monkeys" and "dogs." I learned when to be quiet or feign ignorance, how to be part of the field and when to let go of it.

The hauntings of the field perdured long after I left West Papua—in threatening communiqués from the Indonesian National Intelligence Service, in trolls offering to facilitate my return to the region, and in oil palm itself. Indeed, the more I came to know this plant in the field, the more I realized its omnipresence beyond it—illustrating investment advertisements on the pages of Air Asia inflight magazines, spread out below me when I flew into Kuala Lumpur for my monthly visa renewal, printed on out-of-circulation 1,000 rupiah coins, growing in the botanical gardens of Bogor, Sydney, Paris, and Singapore—and of course, present in the foods and toiletries that I consumed daily.

This book, too, is a kind of haunting. I write it knowing that half the people cited have died of more or less natural causes and that many remain incarcerated for their activism. Others, meanwhile, have since joined the palm oil companies they once so vehemently opposed, eking out a precarious existence as seasonal fruit harvesters and pesticide sprayers. I write knowing that I can no longer share skin and wetness with my sisters in the grove or return to West Papua in the foreseeable future, at a time when state violence against West Papuans, while certainly not new, has become newly prominent.[72] I write wondering what Marind today remember of me and what they will make of me if ever I return. I write in the company of haunting thoughts and traces: the mysterious fate of an orphaned cassowary chick; the bones of children who died of malnutrition; the miscarriage I suffered shortly after my friends were brutally assaulted in the nunnery; and the oil palms, which, to this day, still visit me in my sleep.

This story, then, is written from a place of grief and loss. But it is also a story written out of defiance and responsibility—the responsibility not just to tell the story, but to *tell the story well* (cf. Tuhiwai Smith 2012, 343–58). Telling the story well required that I do justice to the heterogeneous ways in which Upper Bian Marind conceptualize the radical transformations taking place across their lands and forests. It involved dwelling in the pervasive grayness of an abu-abu world both imposed and contested. It meant foregrounding the complexity of Marinds' own theories of change and changing theories about worlds present, past, and to come. It entailed grappling with the limits of the textual medium in conveying the affective and phenomenological textures of landscapes at once aural, sonic, and oneiric. Telling the story well also invited attention to the difficulties in tacking back and forth between narratives of damage and defiance, of crisis and continuance, of suffering and survivance—the challenge, not of suppressing bitter stories in favor of hopeful ones, but of telling *better*, bitter stories.[73]

In attempting to tell this story well, I have sought to flesh out differences of all kinds as they play out among human, animal, and vegetal protagonists on the Papuan resource frontier and to attend to the *difference that difference(s) makes*—for an Indigenous community in an out-of-the-way place, for an anthropology beyond the human, and for all of us implicated in oil palm's lifeway as everyday consumers of palm oil and as situated dwellers of a wounded planet. A work of politically engaged anthropology, this book focuses primarily on socioecological topics. But it also speaks more broadly to changing possibilities of life in an age of earthly unraveling and to the kind of work that anthropologists can do to illuminate these possibilities.

At the same time, telling the story well has demanded a politics of refusal on my part—one that accounts for, and respects, meaningful silencings and erasures. The need to sustain relations of trust with my hosts, compounded by the precariousness of my presence in a region where foreigners' movements are strictly controlled, limited my insights into the perspectives of other relevant but potentially hostile actors. These included state and corporate representatives, the military, and non-Papuan settlers. Internal factions among Marind themselves demanded that I choose sides during my fieldwork. For instance, I had to avoid interacting with "pro"–oil palm villagers who had surrendered lands and sought employment in the palm oil sector. These individuals were widely criticized by those standing "against" oil palm, and whose views predominate in this account. In any case, "pro"–oil palm community members were difficult to find. Fearing retribution from fellow villagers, many had relocated with their families to plantation lodgings or to Merauke City.

As cultural theorist Eva Giraud notes, in an activist context, all political and ethical positions come with omissions that are necessary and necessitate acknowledgment (2019, 4). Ethnographic writing and analysis, too, involve making conscious decisions about what stories *not* to tell, in line with the refusals of our interlocutors in the field (Simpson 2016, 328, 331). Absent from this account, then, are the stories of Marind men laboring as plantation pesticide sprayers and fruit harvesters. Absent, too, are the stories of Marind women ostracized by their kin for selling their bodies in the city; stories of local politicians and agribusiness tycoons; stories of malnourished infants whose lives were too brief to be either remembered or retold; stories of eighteen-year-old Javanese soldiers thrust to the far-most end of the archipelago after pulling the short straw in the placement lottery; and stories of wombs and breasts burned by pesticides and shame. Largely absent also from this story are the perspectives of those who facilitated my research in Merauke—the local church and various NGOs. While I touch on them in passing, I have chosen, based on my deep indebtedness to those who made this research possible, not to delve into the conflicting politics at play between these actors and Marind communities.

In a similar vein, I have chosen to respect in my analysis the tendency toward cultural generalization, which constituted a dominant feature in the discourses of my interlocutors. This tendency speaks to the paramount importance Marind place on communal knowledge production and collective consensus when it comes to self-representation. Remaining faithful to how Marind themselves wished to be portrayed was all the more important in the context of their everyday subjection to top-down, exclusionary decision making; untransparent land appropriation; and paternalistic development rhetoric. At the same time, I have sought throughout my account to convey Marinds' own conundrums over what counts as cultural knowledge, for whom and to what effect, and how contestations over these matters were shaped by the personal backgrounds and life stories of the individuals concerned.

Finally, my attempt to follow the everyday life of Upper Bian villagers has excluded certain topics from the purview of this work. These include heterosexual and homosexual fertility rites, warfare, and headhunting, which represent central cultural themes among the language families of coastal south New Guinea.[74] My gender and age limited my access to these topics, which tend to be discussed only among male elders. These former practices also continue to reinforce primitivist stereotypes of West Papuans in Indonesia and are a source of shame for many of my companions, who were consequently reluctant to discuss them.[75] The second domain I do not explore in detail is religion. Marind have certainly been affected in various ways by decades of Christianity, yet the topic of religion was invariably eclipsed by my interlocutors' deep-seated concern and curiosity about oil palm—the plant that was relentlessly taking over their territories and destroying their treasured forests and groves. Moreover, religion was not a major part of everyday life in the Upper Bian. People would attend church services only occasionally, and sago expeditions in the forest always took precedence over religious events in the village. While discourses about religion did occur in the presence of representatives of the church, they quickly gave way to stories about ancestral spirits and forest kin when these representatives left the scene. In excluding religion from my analysis, I do not seek to downplay its impact on the Marind lifeworld. Rather, I seek to foreground the histories, presences, and beliefs that mattered to my interlocutors in the context of what *they* perceived to be the most important event of their time—the arrival of oil palm.

Many of the themes I explore in this book are embedded within the global phenomenon of climate change. Statements from Marind themselves speak powerfully to the uncanny ruptures characteristic of the present planetary crisis—rivers flowing upstream, worlds becoming plastic, or time coming to a stop. Indeed, my companions are acutely aware that there is something global about the local realities they inhabit—the transnational career of palm oil as capital, for instance, or the international demand for food and fuel that drives the expansion of this cash crop.[76]

Marind have their own idioms for describing these partial, interscalar connections: cosmological desiccation, unrestrained violence, colonizer plants, to name just a few. Marind also situate the ecological transformations of the present within a series of historical antecedents that have cumulatively eroded their relations with the more-than-human world. In grounding my analysis in Indigenous theories of continuity and change, I thus seek to give precedence to Marinds' *own* understandings of historicity without imposing climate change as a temporal framing that, as Anishinabe scholar Kyle Powys Whyte (2018a, 236) reminds us, Indigenous peoples did not create nor consent to and against which Indigenous peoples do not necessarily situate their existences and relations.[77]

The foregrounding of Indigenous modes of analysis and representation shaped the process through which this book was produced from the very outset. Practicing what Charles Hale (2006) calls "activist-research," I involved communities from the initial conception of the research topics to the drafting of ethics applications; the planning of fieldwork locations, timing, and activities; the selection of outlets where the data would be published; and the form and content of the book before you—what stories it would tell, in what order, and why.[78] I have respected my companions' wishes as far as possible with the exception of pseudonyms, which, while used throughout, remain contentious for many of those whom I have cited. One of these is Marcus, whose song opens this book and whose stance on pseudonyms muddies the ethical and political merits of established writing conventions. "The government and corporations have taken our land and forests," Marcus noted. "They have taken our food and future. We have lost everything. Yet still, you would take away our names?"

Practicing activist-research also brought me to support Marinds' land rights campaigns by facilitating human rights workshops in the field, training communities in participatory mapping, and producing a documentary on customary lands and livelihoods in the Upper Bian.[79] These activities, which I touch on in several chapters, highlighted the struggles Marind face in (re) claiming their rights and aspirations effectively in the presence of state and corporate audiences. They also speak to my own politics as an engaged anthropologist and to "engagement" itself as a means through which I sought to remain accountable for the many risks Marind took in accepting me into their world.[80]

The story that follows is structured around four contrapuntal couplets, each shot through with the grayness of abu-abu—place and maps, humans-turned-cassowary and cassowaries-turned-human, sago palm and oil palm, and hopelessness and hope-in-dreams.

Each chapter opens with the pleasures of story and description. I return to these descriptions throughout the chapters, interweaving them with scholarly concepts that have helped me grapple with the complexities of more-than-human worlds. This practice is intentional. Repetition and return help me avoid relegating Marinds' words and deeds to mere anecdote or illustrative vignette. Instead, I aim to keep these experiences alive and present—in the image of the refrains of Marind songs, the constant pounding of sago in the grove, and the layered sounds of birds and humans in the forest. By bringing scholarly insights into conversation with Marind voices, this evocation also represents a kind of nurturing of community. By saying the names, the positions, and the stakes and then repeating them so we never forget, I seek to nurture an ethos of intellectual inclusiveness and generosity. This ethos brings me to engage with theories across a broad disciplinary and durational scope. Beyond the realm of scholarly writing, it provides a much-needed shot of life to a discipline that is often dead through individuation.

The first chapter explores the making of the landscape in the Upper Bian. As they travel through their environment, Marind retrace the paths of their predecessors and create relations with each other and with organisms encountered along the way. These intersecting routes give rise to Marinds' sense of rootedness within the forest, seen as an animate realm cocreated with diverse, other-than-human lifeforms. Today, this dynamic landscape exists within a network of state and corporate nodes of control—roads, military garrisons, and plantations—which I describe as pressure points. I introduce these sites by describing a journey I undertook with a village elder, Darius, which culminated with Darius violently lacerating his own body in reaction to an armed security guard who refused us entry onto an oil palm plantation. The fraught significance of roads, military garrisons, and plantations, I demonstrate, arises from the tension between what they promise and what they destroy. I analyze how these pharmakonic pressure points, along with their inhabitants, exert an ambivalent force on both Marind and forest beings by enabling certain kinds of movement while simultaneously interrupting the flow of organisms that enlivens the forest.

The ambiguous effects of topographic pressure points resurface in a different guise in the context of mapping, a contested representational practice explored in the second chapter. Marind criticize government maps and their unnaturally straight lines because they epitomize the totalizing control of the state over the landscape and its inhabitants. Some also disapprove of drone-mapping technology because, like the state, drones impose a top-down but lifeless perspective on space. In contrast, Marind community members produce living maps that are shaped by the sounds and movements of forest organisms and their emplaced relationship to humans past and present. I illustrate this process by describing a mapping expedition that was guided by the song and flight of a bird and its storied relations to Marind mappers. Producing maps that refuse to sit still constitutes a form of resistance on the part of Marind to the state's hegemonic gaze. However, this dynamism also undermines the legitimacy of community maps in the context of advocacy. Among Marind themselves, cartographic conundrums abound over whose perspective, participation, and perception matter in the production of accurate and effective spatial representations. These three elements are in turn linked to Marind conceptions of personhood as a malleable and more-than-human attribute, as I examine in chapter 3.

I begin chapter 3 by analyzing *skin* and *wetness* as physical expressions of human and other-than-human beings' social and moral standing. Glossy skins and wet bodies communicate Marinds' capacity to become <u>anim</u>, or "human," through reciprocal exchanges of fluids with species and elements of the forest—from plants and animals to rivers and soils. In contrast, the poor or deteriorating condition of skin and wetness indicate an imbalance in social relations, which is now exacerbated by the expansion of monocrop plantations and their noxious chemical atmospheres. At the same time, the porosity of bodies produced through interspecies exchanges of skin and wetness puts humans at risk of perspectival capture by forest organisms. This hazard is heightened during skin-changing, when individuals adopt animal bodies and perspectives but then find themselves unable to retrieve their human flesh and fluids. Becoming <u>anim</u> thus reveals itself a precarious and potentially reversible process—one that depends on, but can also be undermined by, fluid encounters with other-than-human beings.

If oil palm challenges the possibility of sustaining interspecies kinships in the forest, it also gives rise to new and ambiguous interspecies relationships in the village. With agribusiness projects expanding relentlessly, a growing number of animals now approach Marind settlements in search of shelter and subsistence. As I explore in chapter 4, Marind are conflicted over how to interact with these creatures. Many pity domesticates because they have lost their "wildness" and behave like non-Papuan settlers, whom Marind deem alien because of their "modern" way of life and foreign origins. This transformation, in turn, evokes for Marind their own experiences of political oppression and ethnic domination as coerced citizens of the Indonesian State. Yet many domesticates appear to enjoy living in settlements and refuse to return to the wild. Similarly, a growing number of Marind are drawn by the promissory lure of modernity. Some appear resigned to their subjection to Indonesian rule. Those who decry their political domination realize that they themselves replicate the oppressive role of the state over Papuans by subjecting animals to human control. Domesticates thus provoke anxiety for Marind because they offer an all-too-faithful reflection of the ambiguous condition of their human keepers.

Chapter 5 presents a welcome hiatus from the oppressive violence that characterizes the world "after oil palm." Here I invite the reader back from the village to the forest to explore the intimate relations of Marind with the sago palm. I begin by following Marind in pigi kenal sagu, "going to know sago." This practice encompasses a range of activities through which community members affirm their social ties to each other and attune themselves to the lifeworld of sago and its symbiotes. By immersing themselves in the sounds, sights, and smells of the grove, Marind discover the storied lifeways of sago palms and how they intersect with those of humans and other organisms across time and space. The grove is also a gender-inflected realm where women celebrate their role as mothers based on affinities between their lifegiving form and fluids and those of the sago palm. Together, the physical, sensory, and affective dimensions of being-in-the-grove are what endow sago pith with a distinctive social taste. Eating and knowing sago are also politically imbued acts through which Marind affirm their identity as "sago people," in counterpoint to non-Papuan "rice people" and to the colonial-capitalist regimes that foreign beings and foods incarnate.

The storied lifeways of sago palms contrast markedly with the dispositions and effects that Marind attribute to oil palm, as I examine in chapter 6. While sago sustains the forest lifeworld, oil palm refuses relations with Marind and the diverse organisms whose lifeways it destroys. Alien and invasive, the plant pursues a solitary existence and devours land and water to further its proliferation. In the image of its own self-interested disposition, oil palm also breeds fragmentation within Marind communities over matters of compensation and land rights. Sago and oil palm thus emerge as two radically opposed extremes within an affectively and politically charged moral-vegetal spectrum. However, this seemingly stark counterpoint is complicated by the fact that Marind also pity oil palm for its subjection to totalizing forms of human control. Furthermore, many villagers express deep-seated curiosity about oil palm's unknown origins, needs, and lifeway. The ontic opacity of the plant thus intensifies its speculative affordances as a heterogeneous object of wonder.

Chapter 7 turns to the attritive effect of oil palm on time. In particular, I examine a prevalent statement among Upper Bian Marind—that since oil palm arrived, time has come to a stop. After outlining the episodic disruptions that characterize Marind historicity, I examine how the time-stopping effects of oil palm arise from the plant's modality of growth, its association with the future-oriented temporality of capitalist modernity, and its enlistment in the nation-building visions of the Indonesian State. By imposing its monotemporal growth on the formerly polytemporal forest ecosystems, oil palm obliterates the spatially experienced past of human and other-than-human organisms. This, in turn, forestalls the possibility of a meaningful present and thwarts the shared future of the forest's dwindling communities of life. Yet the halting of time can also be conceived as a form of resistance on the part of Marind to the promissory futures inflicted on them by the state. By rejecting hope—an inherently future-oriented disposition—Marind symbolically repudiate the temporal configurations on which externally imposed technocapitalist and nationalist visions of the future are premised.

The final chapter explores "being eaten by oil palm," a dysphoric mode of dreaming that has become increasingly common since the establishment of agroindustrial plantations. Marind consumed by oil palm frequently undergo experiences of harrowing torture in their sleep. Most dramatically, dreamers witness and experience their own deaths repeatedly from the perspectives of diverse forest beings whose existence, like their own, is jeopardized by agribusiness. Dreams of being eaten by oil palm thus constitute amplified projections of everyday anxieties triggered by the deleterious effects of oil palm on places, persons, and time. At the same time, these nocturnal experiences, along with their collective narration, enable the formation of new solidarities among people bound by their subjection to the violence of oil palm. Practices of communal dream interpretation in particular reveal the interpsychic significance of dreaming as a social activity that creates oneiric alliances between oil palm's victims, both human and other. Being eaten by oil palm thus becomes a powerful imaginative means through which Marind unearth hope amid the dystopic transformations haunting their waking and sleeping worlds.

Dreams of being eaten by oil palm, experienced both by Marind and myself, form interludes between the couplets of this story. I convey these accounts in the image of the disjointed experience of dreaming itself, an interstitial realm lying somewhere between the conscious and the unconscious, between the real and the imagined—a realm, to return to Paulus's words, of abu-abu, or grayness. These disturbing, haphazard dreams disrupt the narrative flow of my account. They trouble any semblance of holistic coherence or conclusiveness to the analysis presented therein. In doing so, dream interludes counter what Michael Taussig (2015, 5–7) calls "agribusiness writing"—a writing stripped of its capacity to provoke surprise and confusion, which entrenches the illusion of mastery over reality. The raw and harrowing accounts of these dreams are enhanced by the eschewal of literary embellishment. Their meanings remain in the making, inviting the reader's own contrapuntal interpretation.

1. Pressure Points

In late November 2016, I traveled to Merauke City to pick up Yustina, a young girl from Mirav village whose family had offered me their hospitality during my first visit to the Upper Bian some five years earlier. Yustina had recently been hospitalized after suffering several epileptic fits at the primary school where I volunteered as an English teacher. A few weeks prior to my journey, a group of Marind villagers, including Yustina's parents, Petrarchus and Cecilia, had held a peaceful land rights demonstration in a nearby oil palm concession. One of the protesters disappeared shortly afterward. A rumor spread claiming that the young man was being held captive in the plantation headquarters or in a nearby military garrison. In the buildup toward the commemoration of West Papua's stolen independence on the first of December, police surveillance was high across the regency.[1] Feeling reluctant to travel alone, I was relieved when a close friend, Darius Mahuze, offered to accompany me on my journey. Darius was a leading figure in the local land rights struggle. He was also familiar with the network of roads around Merauke, having traveled it frequently to visit his kin in neighboring settlements and meet corporate representatives in the city.

Figure 1.1. Scars on the land. Tread marks and dust-coated sago palms near Khalaoyam. Photo by Sophie Chao.

We left the village at dawn in a car that Darius had arranged to use. After passing the first of the sixteen military checkpoints lining the road from Mirav village to Merauke City, we stopped. Darius and the driver changed the car's license plate "to clear our tracks," Darius explained. We changed the plate two more times after the next set of military compounds. As we drove out of sight of the fourth, I saw a motorbike and rider on standby on the side of the road. "Get your stuff, Miss Sophie," Darius said; "now we change vehicles. It's less visible that way." Bidding our driver goodbye, we continued our journey, changing motorbikes three more times along the way. Darius drove while I sat nervously behind him, my oversized backpack bobbing up and down violently with every bump. I gazed at the hundreds of Indonesian flags displayed along the road at intervals of ten meters. Gaudily festooned provincial election campaign vehicles roared past without end, enveloping us in a choking shroud of dust and fumes.

Around noon, we stopped at the crossroad transmigrant town of Sora for coffee and lunch in a small, decrepit kiosk. The Javanese owner, a sturdy, middle-aged woman, watched us from a nearby table as we chose fish and vegetables from the self-service row of dishes. She started asking questions: "Where are you from? Where are you going? Where is *she* from?" Darius responded evasively. "Up north. Just traveling around. She's a teacher." Suddenly, a couple of soldiers entered the kiosk. I jumped when they slammed their FNC assault rifles down on the table next to ours. The soldiers hovered.

They did not sit down or order any food. I felt the lingering eyes of the armed men behind me. I was glad to be sitting facing the back wall, as Darius had advised me to do when we first walked into the kiosk. When the soldiers went outside to answer a phone call, Darius quickly heaped a spoonful of chili sauce on my rice and told me to eat with my right hand, Indonesian-style. "You need to look like a settler (pendatang)," he whispered. "You can pull it off—you've got Asian blood. Just let me do the talking."

Sure enough, the soldiers started asking questions. Darius introduced me as a cousin twice removed of Father Andreas, who was in charge of the parishes of the Upper Bian and well known across the region for his close ties to Romanus Mbaraka, then regent of Merauke. I was on the way to the city to meet the archbishop and discuss plans to initiate small-scale agriculture funds for transmigrant women in a nearby settlement. I had been living abroad for several years, hence my embarrassingly poor language skills and difficulty digesting chili sauce. The soldiers laughed. The atmosphere relaxed. We talked about faraway homes, the foods we missed, and the heat and drought afflicting Merauke.

Darius and I continued our journey. We stopped once along the way to stretch our legs but moved on quickly after a group of young Javanese men, dressed in plantation workers' uniforms, appeared out of the bush and asked if we had come down from Mirav. Despite the frequent vehicle changes, the information had somehow managed to follow us. These men, Darius told me, were spies (intel) paid by oil palm companies and the military to monitor community members' movements. By this time, the sun had started to set. Darius suggested we take a shortcut through a nearby plantation to reach the city. I knew this was risky, but my friend insisted. After all, the concession was established on his clan's customary land. After riding several hours through a landscape of decimated vegetation and burning forest, we arrived at a plantation checkpoint. The strident shout of a security guard stopped us in our tracks: "Halt! Where are you going? Report first!" "Idiot," muttered Darius. The guard strode over with his rifle in hand and barked, "Who are you? You don't have the right to pass here. This is oil palm road. Find another road. You don't have any business here. You've probably come to steal oil palm fruit. We know you Marind people. Where's your ID card?"

Darius removed his helmet slowly and turned the engine off. I could hear chainsaws buzzing in the distance. My companion grabbed his hair and, in one swift movement, ripped out a thick tuft of his frizzy locks. He handed them to the guard and said, "This is my identity. This is my land. I am Darius, head of the Mahuze clan, descendant of <u>dema</u> (ancestral spirit) Tete Kenepe, brother of the sago and dog. I own everything here for my ancestors inhabit this land.

You are the thieves. Sago is my kin, my _amai_. Sago is my grandparent. Who are your grandparents here? Oil palm? Oil palm is no kin. Oil palm is no one. Sago is everything."

The guard was visibly taken aback. He looked down at his dusty boots, holding on hesitantly to the clump of hair. Darius did not take his eyes off him. Eventually, the guard disappeared into the checkpoint and returned bearing a large, dog-eared book. Opening the book, he grumbled, "Whatever, just sign the register. Write your name, address, ID number, where you're going, where you've come from." Darius pulled out his machete. Without hesitation, he sliced his right hand from the tip of his thumb to the middle of his forearm and smeared the spurting blood along the page from one margin to the other, in one slow, controlled streak. "Here is my identity," Darius said. "My blood. The blood of my land you stand on. The land you stole from my _amai_, the sago and dog. Are you satisfied now?" The book lay open in the guard's hands as he retreated quickly to the check post. Carmine human ink bled through the pages of the register one by one, smudging words, numbers, signatures, and dates.

IN THIS CHAPTER, I explore the making of the Marind landscape. Marind often describe the forests, savannahs, and swamps of the Upper Bian as being "alive" (hidup) with the movements of humans (_anim_) and their animal and plant kin (_amai_). In precolonial times, human migration, movement, and resettlement were driven by hunting and foraging patterns, uxorilocal marriage, religiopolitical congregations, ritual feasts, headhunting, and warfare. Today, these events are commemorated through stories and songs and, more important, through the journeys of community members to the sites where they occurred. Just as important as the movements of humans are those of animals and plants with whom Marind share *relatedness*, a term deployed by Janet Carsten (2000) to describe relations achieved through cumulative social interactions that transcend intraspecies or biological affinities.

The relatedness of Marind clans to animals and plants determines the location and extent of their respective customary land rights. All clans may travel, hunt, and gather outside their territories so long as they seek permission from the landowning clan and carry out the required rituals for its _amai_. Often, land boundaries follow the trajectories of _amai_ through the landscape. The territories within them encompass the places where _amai_ and their human kin were created by _dema_ and the sites of their meaningful encounters in the past and the present (cf. Hviding and Bayliss-Smith 2018; Mondragón 2009). Some boundaries, for instance, follow the proliferation of plant _amai_ such as sago groves, coconut palm clusters, and bamboo forests. Others are enlivened by the actions of animals—for instance, the mound where the dog amai napped along the Bian, the river bend where the cassowary amai laid its first egg, or the soil depression where the heron amai nursed its broken wing before flying west to Kondo, home of the deceased. Plants and animals trace new paths of life across the landscape as they grow, migrate, propagate, and seek out subsistence. As Salom, a mother of six from Bayau explained during one of our many walks through the forest, "Amai (plant and animal siblings) are like anim (humans). They are always on the move." As they travel the forest, human and other-than-human wayfarers retrace past trajectories and create new ones, in a process of mutual and ongoing (re)generation. In turn, land boundaries shift as these more-than-human communities of life expand and contract across space. The flowing paths of amai thus intersect with those of humans in shared *wayfaring*, a term used by Tim Ingold to describe the movements of human and other-than-human beings that *produce* the environment through their interagentive dwelling (2011, 145–64). "Land boundaries," Salom continued, "are like bamboo stalks. They are not rigid. They move in the wind. They are alive."

In the image of the dynamic topography of the forest described by Salom, many of my companions emphasized that physical and collective movement is necessary to understand the landscape and one's own identity in relation to it.[2] For instance, people trace connections with each other and with plants and animals by walking the forest to the sites where these relations were achieved. Those who successfully identify shared kinship from traveling across the landscape together say that they "travel in the same canoe." This expression harks to the importance of canoes in myths recounting the peregrinations of ancestral spirits across the landscape, which in turn gave rise to clan formations and interclan relations. In the past, canoes were given the names of the clans who had traveled in them and constituted central symbols of common descent. The expression "to travel in the same canoe" also commemorates the watery body of the river itself as a fluid conduit for movement and relation making. It points to the importance of place as an identity-generating journey rather than a geographically determined location (cf. Bonnemaison 1994; Hau'ofa 2008, 81; Lipset 2014).[3]

Today, roads, military posts, and oil palm plantations, such as those that Darius and I encountered during our journey, form a topography of fear and control superimposed on the moral topography of the forest and its lively beings.[4] I describe these sites—and the institutions and actors associated with them—as pressure points in light of their alternately enabling and restricting effects on the movements of Marind and their forest kin. Depending on the circumstances in which they are encountered or activated, pressure points may alternately produce well-being or suffering in the bodies in which they are located.[5] Pressure points are thus pharmakonic—from the Ancient Greek *pharmakon* meaning "drug"—in that they constitute sites of uncertain and potentially dangerous effects. Devoid of a stable essence or identity, they generate difference precisely because they condense apposite meanings, emotions, and possibilities.[6] Pressure points and their constitutive geographies thus constitute our first entry point into the murky grayness of the world that Marind call abu-abu—a theme we will return to in later chapters.

Pressure points in the Upper Bian exert an ambiguous force on the forest landscape and its human and other-than-human dwellers in concomitantly bodily, affective, racialized, topographic, and infrastructural ways. On the one hand, pressure points are pregnant with different kinds of promise—employment and development from the plantation, security from the military, and access to new places and products via the road. At the same time, pressure points pin down the fluid body of the forest by restricting or subverting the flow of organisms that make the landscape "alive." Military garrisons, as organs of the nation-state, impose order *beyond* their boundaries through their pervasive surveillance of local inhabitants. Corporate plantations, as socially and spatially isolated enclaves, impose order *within* their boundaries—on the cash crops they control and the local communities and forest species they exclude. Both sites, along with the roads that connect them, exert their influence in distinctive yet complementary ways.

At the same time, the power of pressure points arises as much from their tangibly violent effects as from their contingent, and often illegible, affects as spaces of confusion, or abu-abu. These pressure points accrue further significance because they operate in a region where infrastructure is relatively recent and minimal, and therefore all the more powerful where it *does* exist. The ambiguous affects generated by particular topographic pressure points are further enhanced by their relation to *other* pressure points, which together embody the material, institutional, and political force of the state-military-corporate troika over Marind and the living landscape. Their suspended and often invisible power extends across physical and geographical bodies like nodes of a complex nervous system that is at once omnipresent and opaque.[7]

ROADS FOR MANY MARIND are spatial forms replete with contradictory yet conjoined affective significance. On the one hand, roads are conduits for foreign persons and foreign things that display characteristics assigned to "road belong cargo," a Pidgin English expression referring to the arrival of vast supplies of new goods from faraway places (see Lawrence 1964). Government officials and corporate representatives arrive by the road to promote agribusiness projects that will purportedly enhance community development and material wealth. The road also brings NGO activists from Jayapura (the capital of the province of Papua) and Jakarta. Full of candid revolutionary intentions and precious "modern knowledge," these educated urbanites raise hope among Marind for the recognition of their rights and the protection of their forest. I, too, as my companions often reminded me, arrived from a foreign land by way of the road, bearing books, maps, drones, and a desire, as a Basik-Basik elder and ritual healer, Pius, put it, to "know and share the story of Marind." Alongside inquisitive anthropologists, new kinds of manufactured goods and processed foods from distant and unknown places flow into the settlements by way of the road—mobile phones, instant noodles, plastic toys, and more.

Like the network of physical pressure points in the human anatomy, roads also connect villages to places and persons associated by Upper Bian Marind with modernity and progress. Individuals who traveled by road to Merauke City, for instance, often returned with exciting stories of the encounters and events that took place during their journey. Among them were Yustina's parents, Petrarchus and Cecilia, who reveled in describing to their relatives how they had successfully learned the modern habits of city dwellers, such as crossing at red lights, taking public transport, and using chopsticks. For Petrarchus, Cecilia, and others, roads bear a positive connotation. They enable Marind to access new forms of wealth and knowledge, to seek formal employment in urban areas, and to overcome the precarity of rural livelihoods. Taking the road thus becomes an opportunity to discover ways of life associated with an unknown and, for some, a *better* somewhere—the city, Jakarta, or even overseas.[8]

At the same time, roads are threatening pressure points that change people by taking them to alien places. Traveling the road disconnects travelers from their homes, forest, and kin, and therefore from their spatially embedded and relational sense of self. Those who stay behind worry over how long their relatives will be away, what they will eat when far from home, and whom they will spend time with. Marind women, in particular, often expressed fears that their traveling kin would "forget how to make sago" and "walk the forest"—that they would no longer "recognize the sounds of birds" or the "voices of their own children." When individuals return to the village, community members feed them vast quantities of sago, encourage them to walk the forest, and increase physical contact with them in the form of skin-rubbing and sweat-sharing through collective labor. These endeavors, Darius explained, ensure that individuals become resocialized into the community and that "they can become <u>anim</u> (human) again."[9]

The ambiguous effects of roads—a widely documented phenomenon across Melanesia—acquire heightened significance in the context of West Papua, where roads are often associated with danger and fear.[10] As my account of traveling with Darius conveys, those who frequent roads face unremitting physical and psychological pressures. For instance, travelers are routinely subjected to security inspections and reporting. They face demands for bribes from the police, military, and oil palm security staff as a rite of passage. Land Cruisers rumble down the road, bearing on the corner of the windshield a telling sticker with an eagle and an anchor—the logo of Kopassus, the dreaded Indonesian National Army Special Forces.[11] The passengers in these vehicles are strangely invisible behind the tinted windows. No one knows who they are or what they want. Meanwhile, endless convoys of military trucks crawl forward on the road, like obedient files of hunter-green scarabs, chugging smoke and oozing dark clots of leaking oil. Roads purport to enhance peoples' mobility, but the surveillance of the state, army, and spies along them is omnipresent. The oppressive effects of roads were powerfully captured by a Bayau elder, Ambrosius, to whom Darius and I recounted our journey on returning to the village. Ambrosius had two children studying at a high school in Merauke City and several relatives living in Jayapura, but it had been many years since he had visited them. Ambrosius described himself as a strong and fearless man, like his father and forefathers. There was no patch of forest or swamp where he had not hunted or walked. But on no account would Ambrosius leave the village for the city. When I asked him why, he replied: "There is no freedom in the road."

Roads also accrue meaning for Marind as destructive pressure points through the *absences* that they create. Indeed, when I asked my companions what the road brings, they often told me instead what the road had obliterated. For instance, community members would explain that because of the road, a certain type of tree, mammal, or freshwater spring "no longer is (su tiada)." Some described the eerie silence of long-disappeared birds, the razing of sago groves, and the absence of shelter, shade, and sustenance along the winding, empty spaces of the road. Roads also participate in transforming interweaving flows of more-than-human life into unidirectional flows of life-turned-capital. Day and night, Ambrosius recounted, overloaded trucks of illegally felled timber travel along the road to the city. The identities, stories, and relations of these centenary trees that have been turned into dismembered commodities are replaced by red numbers painted along their trunks that codify their value, type, destination, and use. Meanwhile, fires blaze violently by the side of the road or glow ember-like inland, incinerating the "skin" (or terrain) of the forest to make way for oil palm. As acres of vegetation are felled and extensive wetlands drained, a multiplying network of roads replaces the reciprocal meshwork of life of Marind and their other-than-human kin. For many villagers, the road thus becomes a space of remembrance of things gone and never to return, which conjures up mixed feelings of sadness and frustration. As David, Ambrosius's twenty-year-old nephew and a member of the local land rights movement, told me, "When I travel the road, I feel angry. It reminds me of loss. It reminds me of death. It is a kind of mourning."

Threatening presences and meaningful absences on the road coalesce in the figure of the bulldozer, whose passage signals the imminent destruction of the forest. Bulldozers are a source of consternation and fear for many Marind.[12] One of them is Evarius, an elder from Mirav whose hunting and foraging trips in the forest had on several occasions been curtailed by the arrival of bulldozers. Evarius described bulldozers to me as "a strange kind of faceless beast." They appeared to move like animate beings and were handled by humans, but they were also lifeless machines. Like oil palm, Evarius explained, bulldozers eat the forest to make way for monocrop plantations, but they are seemingly never satiated. The merciless mandibles of these voracious robotic arthropods leave in their trail a landscape of flattened sago groves, obliterated life, choking dust, and deep tread marks that Evarius and many other Marind called "scars on the land." Meanwhile, deer, wild pigs, cassowaries, and birds of paradise flee in the wake of the bulldozer. Silence and stasis replace the lively sounds and species of the vanishing forest.

OIL PALM PLANTATIONS ESTABLISHED in the last decade are a budding pressure point in the Upper Bian. Like the opportunities and mobilities afforded by the road, plantations come with the promise of material wealth, community welfare, and local employment for some Marind. For instance, companies design corporate social responsibility schemes to improve local education and village infrastructure and provide financial compensation packages for the lands that are ceded. Some offer occasional jobs to Marind as fruit harvesters, truck drivers, and security guards. Many community members who support the palm oil sector in these capacities see themselves as contributing to regional economic growth and national food security. For these individuals, oil palm plantations are enabling pressure points. They allow Marind to accede and participate in national and global flows of capital, markets, and wealth, while themselves furthering the development of Papua and the international standing of Indonesia as the world's top palm oil–producing country.[13]

At the same time, plantations threaten the liveliness of the landscape and its constitutive lifeforms by transforming forests into heavily guarded concessions.[14] Juliana, a young Marind woman and primary schoolteacher in Bayau, described how, prior to oil palm's arrival, she and her kin would spend around two thirds of the year in the forest, traveling to harvest sago and bivouacking in sago groves for two to five months at a time. "We were occupied by Indonesia," Juliana explained, "but whenever the military harassed us or the government bothered us, we escaped to the forest. We lost our freedom as Papuans. But we kept our freedom in the forest." Since 2010, however, Marinds' strategic mobility has been undermined by the relentless expansion of monocrops around their villages. Entrance to the concessions is strictly forbidden to nonpersonnel, including to landowners like Darius, on whose customary territories the plantations have been established. Whereas in the past, Marind traveled easily into Papua New Guinea to visit their relatives, the national border has now become a militarized oil palm frontier. Just as important, land clearing for oil palm means there is increasingly less forest for Marind to go *to*. Remnant patches are often located far from the villages, with the result that trips to the grove now only take place on average two to three times a year. These trips, Juliana noted, usually only last a couple of weeks to a month because communities are required to be in the villages to attend monthly oil palm promotional campaigns. Many trips are cut short by the arrival of bulldozers, chainsaw-wielding laborers, and plantation security patrols or the ominous buzz of pesticide-spraying helicopters overhead.

The physical constraints posed by plantations on Marind are heightened by the contrasting and seemingly relentless expansion of oil palm itself—both in the form of plantations proliferating across Merauke and in the form of a vegetable oil that travels across the globe to supply myriad foreign and unknown consumers. Paskalis, an elder of Mirav, commented on this as he perused a promotional pamphlet distributed by a local agribusiness company, which included a map of palm oil's global trade networks. Running his index finger from one continent to the other and back, Paskalis muttered, "Oil palm grows everywhere across the tropics. It is sold all over the world. But Marind can no longer travel the forest. We cannot even travel to the city. Palm oil, oil palm—they travel everywhere. But we are allowed nowhere."

As Paskalis's astute reflection suggests, plantations enable the movements of some forms of matter but not others. The global mobility of palm oil—a cosmopolitan cash crop and commodity—in particular acts as a stark counterpoint to Marinds' own immobilizing severance from the forest, their exclusion from privatized monocrops, and their limited capacity to make a living in the village, plantation, city, or beyond. Even then, young Marind men and women who *had* studied or worked outside West Papua were quick to note that physical mobility, where achieved, does not in itself preclude social immobility or discrimination.[15] One of them was Kristiana, a twenty-year-old girl who had spent two years studying nursing in the Javanese city of Bandung. Kristiana noted, "Even when we travel far away to get an education and find modern jobs, we are still treated like animals because of our black skin and curly hair. We are told that we smell. We are made fun of because of our accents. When we speak our language or celebrate our culture, we are accused of being independentists. So, even when we are far away, we stay among ourselves, because we are not wanted or accepted. In Indonesia and the world, everyone needs palm oil. But Papuans? No one needs Papuans."

The abu-abu force of plantation infrastructures is also temporally inflected for many Marind. Some of my companions, for instance, noted that the destruction of the forest to make way for roads and plantations was certainly harmful to them in the short term, but that it would ultimately be beneficial for future generations in terms of employment opportunities, social mobility, and access to the "modern" world. Such views were particularly prevalent among Marind women, including Claudina, a middle-aged woman from Bayau village who was pregnant with her fifth child at the time of my fieldwork. Rubbing her belly with eucalyptus oil at the end of a long day of sago processing, Claudina reflected, "If oil palm continues to expand, the children in our wombs will never know the forest. That breaks my heart, but it means they will never miss the forest like I do. They will only ever know oil palm. Oil palm will be their past and present—and probably also their future. Maybe they will suffer less that way." Claudina's elder sister, Isabela, who was tending the fire beside us, appeared less convinced. Despite numerous threats from the police and corporations, she and her husband remained at the forefront of the local land rights movement. "We cannot allow forest to be destroyed," Isabela insisted, "To do so would be to destroy our children's futures—*your* children's futures. What will you leave them? What will you feed them? How will *you* be remembered by them?"

But it is not only Marind, present and to come, who are affected by plantation proliferation. Monocrop concessions dominated by a single plant put pressure on native flora and fauna by creating impervious barriers to species migration across land and water. Few species can thrive in oil palm ecologies, which are characterized by low canopies, sparse undergrowth, unstable microclimates, high temperatures, and a toxic mélange of chemical fertilizers, herbicides, and pesticides. Conserved forest fragments within plantations and along their boundaries are susceptible to edge effects, whereby forest quality deteriorates over time. Gradually, species communities shift toward a simpler composition, dominated by a few common types. Robbed of their water, nutrients, and symbiotes, plants and animals who once thrived in the forest now wilt and starve. Wild pigs and cassowaries who venture onto the plantation to feed on oil palm fruit are hunted down by company workers for consumption or trade. Bamboo clusters and sago groves collapse as the soil is depleted of its minerals and nutrients. In the plantation, the calls of birds and beasts are replaced by a deathly silence, which is particularly eerie in the glaring heat of the midday sun. Sounds of life are replaced by sounds of death—roaring bulldozers, gnawing chainsaws, the crackle of illegal burning, and the rumble of overloaded trucks carrying oil palm fruit and timber.

The disfiguration of the landscape caused by oil palm produces anachronistic juxtapositions of buildings and beasts and of plants and people. Crude palm oil factories replace Marind <u>kanda</u>—traditional drums used in rituals and ceremonies—as towering effigies of modernity and capitalism. Animals find little to feed on in the barren plantations, yet free T-shirts from oil palm companies depict cassowaries feasting on palm oil fruit in incongruous symbiosis. Ancestral toponymies are irreverently supplanted by corporate logos displayed on bulky aluminum signboards. The glorifying names of many of these companies—for instance, Core Life Indonesia, Eternally Grateful Blessing, and Prosperous Glory—contrast vividly with the rampant destruction that they provoke. A deceptive lining of trees along plantation boundaries, which some Marind villagers call "fake forest" (hutan tipu), suggests that some vegetation is intact, but a five-minute walk inland reveals a decimated landscape of ashes and rubble. Meanwhile, communities are forbidden from entering their sago groves because they have been demarcated as conservation zones by agribusiness corporations.[16] Ironic twists and productive confusions abound in the emergent oil palm frontier (Tsing 2005, 33).

One particularly poignant encounter revealed the shared fate of humans and their plant and animal siblings in the face of oil palm plantations as deadly pressure points. In late March 2016, a group of Marind villagers and I traveled back to Khalaoyam by car after three consecutive days of meetings with corporate and government representatives in Merauke City. My companions were in low spirits. The company had refused to return the five thousand hectares of land recently taken from their clans without their consent. Half the forest on that land had already been decimated and the other half would be cleared by the end of the year. On the question of compensation, the company had remained frustratingly elusive.

As we reached the heart of the 100,000-hectare PT BIA plantation, we encountered a lone cassowary. The bird stood remarkably tall on its strong muscular legs. Its sleek, iridescent plumage glimmered in the afternoon light. Yet the bird appeared grotesque, looking distraught as it stood awkwardly in the middle of the road, terrified by approaching vehicles. We stopped the car and the cassowary cocked its head. It looked left to the licking flames of the burning forest, lit by plantation operators to make way for oil palm. Then, the bird turned to gaze at the vegetal barricade of thorny oil palms lining the plantation. Yohanes, who had acted as the group's key spokesperson during the negotiations in the city, whispered sadly: "See, its destiny is just like ours. It has lost its home and has no kin to flee to. It is in a strange place and no longer recognizes the forest." Miranda, Yohanes's younger sister, tightened her grip on my hand. She breathed heavily, her eyes full of anger and pain. "Everything has become strange since oil palm arrived," Miranda whispered. "Animals get lost in the plantations. Oil palm eats the land. People lose themselves on the road. Who knows—someday, the Bian River may start to flow upstream." Her voice trailed off.

ALONGSIDE ROADS AND PLANTATIONS, military garrisons constitute yet another node of control in the network of pressure points overlaid on the living anatomy of the forest. These establishments are omnipresent in the national border region of Merauke, with no less than six different military bodies based in and around Khalaoyam, Bayau, and Mirav.[17] On the one hand, the military are associated with the enhancement of security and the protection of Indonesia's national boundary with Papua New Guinea. Like acupuncture points activated to counter the negative effects of foreign objects and forces on the human anatomy, the military shields the body of the state and those of its citizens from external threats. The small handful of Marind who had enlisted for military service described with pride their role in supporting the unity and safety of the Indonesian nation. Many among them noted that military service had improved their quality of life. Some, for instance, showed me the shiny uniforms and boots they had received as part of their enlistment. Others described the "modern" food available to them in the army mess—rice, processed meat, and canned fish—as more nutritious and healthier than forest foods. Yet others emphasized that the income they received from the military was not only regular but also generous by local standards.

At the same time, military zones are a source of fear and frustration for many Marind. For instance, soldiers regularly subject community members to unwarranted harassment in the form of interrogations, temporary incarceration, ID confiscation, and surveillance of their movements in and across the villages. During military trainings, gunshots rip endlessly through the air, causing birds and other animals to flee in terror, with the result that hunting expeditions fail and communities go hungry. Many young women associate military garrisons, alongside plantations, with the threat of sexual harassment on the part of "dangerous" and "violent" men—soldiers and laborers. For others, the military symbolizes the continued oppression of West Papua since the region's forceful incorporation into Indonesia. One of these is Ambra, a teenage girl from Khalaoyam, whose father was being detained in a prison in Jayapura for his suspected involvement with the Free Papua Movement (OPM). As she watched a tank rumble down the road, coating nearby sago palms with a heavy sheet of dust, Ambra commented bitterly, "The national army is powerful, that's for sure. But this army is in the wrong country."

The negative associations of the military in geopolitical and historical terms are heightened in the context of oil palm expansion. As is common across West Papua, military personnel in Merauke often work in collusion with agribusiness corporations as land-clearing contractors, community liaison officers, security guards, and company shareholders. To many of my companions, the army thus constitutes the most visible arm of the state-corporate-military complex driving agroindustrial expansion in the region. Indeed, key attributes of military sites themselves are replicated in the state-like formation of the monocrop plantation.[18] For instance, terms used in the context of military-spatial organization are also deployed by oil palm companies to describe the layout of plantations as districts (rayon), divisions (divisi), echelons (eselon), building complexes (kompleks), and blocks (blok). Company staff are housed in barracks (barak) and the military are hired by agribusiness companies as oil palm patrols (patrol sawit). Like plantations, garrisons extend across vast swaths of land and are off-limits to nonpersonnel. Their entrances and exits are strictly guarded and trespassing is heavily sanctioned. Army and plantation headquarters are always located on elevated land so concessions and garrisons can be surveyed easily, from a distance and in every direction. In this regard, army and plantation headquarters function in the manner of the panopticon, the all-seeing carceral architecture originally conceived by philosopher Jeremy Bentham and later deployed by historian Michel Foucault as a metaphor for modern disciplinary society (see also Dove 2011, 30–31).[19] No one can escape the controlling gaze of these oppressive and all-encompassing pressure points.

Similarities between the spatial layout of the plantation and the garrison extend to their respective inhabitants—oil palm and military men. My companions, for instance, often noted how the endless rows of equidistant oil palms crisscrossing the plantations resembled regiments of obedient soldiers, identical in appearance and size. In garrisons, they explained, soldiers undergo intensive daily trainings that transform and enhance their bodies as objects and targets of power. In plantations, meanwhile, capitalist and agronomic modes of disciplinary power are enacted upon the vegetal body of oil palm in order to optimize its productivity. These forms of disciplinary power include artificial breeding, genetic manipulation, industrial processing, systematic culling, and more. The military carries forward the nationalist doctrine of Pancasila, which lays out the Indonesian State's foundations in religion, civilized humanity, social justice, democracy, and unity. Plantations, too, carry forward visions of progress, modernization, and profit that legitimize the transformation of the forest into a capitalist resource frontier. These resemblances brought my friends to jest as we traveled through plantations about "entering the oil palm commando," "passing the oil palm headquarters," or "saluting the oil palm troops."

Finally, military zones and plantations are comparable in terms of the uncanny tendency of their residents to turn up at unexpected places and times. Undercover soldiers, along with informal spies (intel) and thugs (preman) such as the Javanese men Darius and I encountered on our journey, often pass as traders, plantation workers, and drivers in order to better monitor the activities of local communities. Villagers can only speculate as to who the spies are and how they operate. Strange or suspicious behaviors are telltale signs. For instance, young men by the road are seen texting on their phones in a landscape devoid of telephone signal. Villagers notice the itinerant Javanese meatball seller being dropped off in the city by a company-owned 4WD. Stall owners, like the matron whose kiosk Darius and I visited, are rumored to share information about their customers to military patrols in exchange for a few thousand rupiah. Skillfully navigating the muddy divide between banditry and law enforcement, the collusion of intel with the military and corporations cannot be avoided and their private power cannot be undone. As they move across the landscape, these menacing pressure points spread fear through their unforeseen appearance and uncertain authority.

In this regard, spies bear an uncanny resemblance to oil palm—a plant whose arrival, too, is often both troubling and unexpected. In late February 2019, for instance, villagers from Mirav described their surprise and shock on seeing hundreds of newly planted oil palm seedlings lining the riverbanks where they regularly went to fish. Three months later, a much-anticipated rambutan-gathering trip organized by Bayau villagers was intercepted by a sign declaring that the area was now under the ownership rights of a heretofore unheard-of agribusiness company. Meanwhile, in the neighboring village of Khalaoyam, a sacred agarwood grove had been razed overnight, leaving behind only bull-dozer tread marks and Marlboro butts. Like spies who operate surreptitiously, oil palm encroaches silently yet relentlessly on the last patches of remaining forest, spreading abu-abu in its shadowy trail of death and destruction.

Drawing from Gilles Deleuze and Félix Guattari, Eben Kirksey compares the West Papuan resistance movement to a rhizome that ceaselessly establishes connections, resprouts, and multiplies. Like the rhizome, this underground resistance movement has no fixed center or origin and is therefore difficult for the state to control. Kirksey then contrasts the rhizome to the banyan, the symbol of Indonesia's dominant political party, Golkar. A resilient strangler fig that grows up and around host trees, the banyan, Kirksey suggests, is a power-ful metaphor for the totalizing control of hegemonic projects over spaces and subjects (2012, 55–79). In Merauke, garrisons, roads, and plantations are banyan-like pressure points that invade and occupy the landscape, stifling its inhabitants and restricting their movements. At the same time, spies and oil palm are also slippery rhizomes—an organic form that, as Gilles Deleuze and Félix Guattari note, "includes the best and the worst: potato and couch grass, or the weed" (2005, 7). Like the weed, spies and sawit disperse across space in inexorable yet mysterious ways. In the manner of a parasite, these abu-abu be-ings never stop invading, multiplying, and returning (Serres 1980, 341: author's translation). The areas they occupy have no clear or stable boundaries. Their identities are gray and ambiguous. They are instrumental to, yet exceed, formal state and corporate infrastructures. The source of their power is untraceable. Like lethal fluids, these capillaries of control infiltrate and overflow their sur-roundings in unpredictable ways. Both static and rhizomatic pressure points participate in shaping the topographies of fear of the Upper Bian.

ESCHEWING THE NOTION THAT roots always precede routes in the consti-tution of Indigeneity, James Clifford (1997) identifies movement, travel, and contact as central to the process of becoming that is culture. In the Upper Bian, both roots and relations find spatial expression in the intersecting routes traced by humans and their other-than-human kin. Rather than a static physical backdrop composed of fixed and human-determined sites and boundaries, the landscape is agentive, dynamic, and social as it crosses species lines. It emerges from the relations established between organisms that carry life forward *in* places and *along* paths (Ingold 2007, 2–3). In journeying through the forest, Marind reaffirm their sense of belonging to a landscape that is produced by the fleshy movements, deaths, births, growths, and proliferations of diverse lifeforms. They perpetuate shared pasts through their own moving bodies in the present. Traveling the forest is also a *creative* process. It is pregnant with the possibility of new relations generated by meaningful encounters between human and other-than-human lifeforms (see also Anderson 2011; Leach 2003).

The state and corporate pressure points explored in this chapter exert real yet uncertain violence on the lively topography of the forest, producing what Veena Das calls an ecology of fear (2007, 9). Fear, as feminist scholar Sara Ahmed notes, operates by restricting certain bodies through the movement or expansion of others (2004, 62–81). In the Upper Bian, the expansion of monocrops and their tentacular network of roads and garrisons shrinks the collective body of Marind and their forest kin in spatial and social terms. Plantations in particular erode the landscape's affective materiality by operating negatively toward all that does not sustain them. Obliterating the lifeforms that make place alive, plantations produce spaces of death where humans and forest organisms can neither survive nor thrive.[20]

Yet oil palm plantations are also places of lethal and mysterious life—of abu-abu, in Marind terms. They are home to a foreign and unsolicited plant. This plant entrenches itself within historical landscapes. It proliferates along the banks of ancestral rivers. Its roots propagate through moist, dark undersoil. Its foliage extends uniformly across customary lands and sacred sites. Oil palm does not seem to know or care for relations with the others that it supplants. Its oily fruit feeds and fuels distant and alien communities, who are far removed from the shadow places of capitalist production. Ambivalent and emergent, the ontology of oil palm is not just uncertain, but uncertainty itself.[21]

At the same time, pressure points transform the social topography of the Upper Bian into spaces of absent and impossible relations. Community members read the landscape through what is no longer there—the muted bird of paradise, the lost cassowary, the uprooted sago grove. Absence and death are cast into high relief by a colonizing plant that is strange and estranges beings from each other and their environment. The poetic density of emergent oil palm plantations is further heightened by their indexical relation to preexisting formations—roads and military garrisons. These formations exert pressure on the landscape and its dwellers less through their revelatory failure (Bowker and Star 1999; Star 1999) than through their omnipresent yet often indeterminate force.[22] As (im)moral ecologies of infrastructure, plantations, roads, and garrisons produce growingly arthritic topographies that thwart the movements through which places and persons come to exist meaningfully. Together, they create what literary studies scholar Rob Nixon calls "displacement without moving," or the stripping of place of its life-sustaining attributes and affordances (2011, 19). In a world where landscapes and identities emerge from the flow of interspecies relations along spatial and genealogical paths, the implications of *not* being able to move are deeply unsettling. Fluid territorial boundaries are replaced by militarized checkpoints. A foreign and invasive plant colonizes the land and devours the forest. Cassowaries are forced to choose between flame and famine. Someday, as Miranda mused, the Bian River itself might begin to flow upstream.

My journey across the Upper Bian and the wounds that my companion Darius inflicted on himself embody vividly the disruptive force of the state and the corporate pressure points explored in this chapter. In our encounter with the plantation security guard, Darius's body came to represent what Michel Foucault might term a "dense transfer point of power" (1978, 103)—a material-semiotic fleshiness, or corporeal instrumentality, at once opposed to and oppressed by, hegemonic forms of disciplinary power.[23] My companion's ruptured and bleeding flesh conjured the disfigured topography of the landscape itself as una herida abierta, or an open wound—an expression used by Chicana poet and feminist Gloria Anzaldúa to describe borderlands as sites of asymmetric violence and social suffering (1987, 3).

At the same time, Darius's wound symbolized a visceral *opposition* to pressure points—and the foreign peoples, technologies, and practices associated with them—*from* and *through* his body. My companion tore his hair to affirm his identity and rights as an Indigenous Papuan man. He drew his blood to claim ancestral brotherhood to the sago and dog. He refused compromise to the state and corporate forces incarnated in the armed plantation guard, who denied him entry to his own customary lands. Darius's body thus became the meeting ground of both power and subversion. In harnessing the ripped-apartness of his own flesh, my companion made a visceral statement about the resilience of Indigenous bodies, persons, and relations in the face of dispossessory colonial-capitalist regimes. Darius's unexpected and self-directed actions jolted the hegemony of the state and its corporate allies, in violent and vivid mimesis of the state's own disruptive effects on the landscape and its lifeforms. His wound was one of provocation and resistance to the stultifying control of pressure points over the dynamic topography of the Marind lifeworld.

2. Living Maps

A group of Khalaoyam community members and I gathered to rest in the shade of a mango tree near Bayau village. We had just returned from a three-day fishing trip on the western banks of the Bian River. As they passed around slivers of baked sago pith and coconut meat, my friends started discussing the drought. Several pointed to the shriveled rambutans and jackfruit littering the ground. They compared the withered fruit to the peeling skin of their forearms. The drought, my companions affirmed, was the longest in decades. They vehemently agreed with Claudia, an elderly woman and traditional midwife, that oil palm was to blame. Oil palm, Claudia explained, was grabbing the waters of the Bian, just like it was grabbing the land of Marind. Oil palm was sucking the soil and rivers dry. Soon, she exclaimed, the forest would be just dust and scars. Claudia's impassioned diatribe was interrupted by Paulinus, an energetic young villager, who came running toward us with a bulky cardboard box, hollering, "I found another map! Check it out!"

Paulinus, who was known in Khalaoyam for his sharp wit and savviness, described with satisfaction how he had successfully convinced a representative

Figure 2.1. Yakobus, a Marind mapper. Photo by Sophie Chao.

of the Merauke Board for Planning and Regional Development to lend him the document at a recent oil palm promotion event in the city. The group was excited. Communities in the Upper Bian have limited access to official maps. Those they do possess have been difficult to acquire. Some were stealthily pocketed from the shelves of government offices during cigarette breaks at multistakeholder meetings. Others were photographed covertly during negotiations with agribusiness companies using HD spy cameras in half-frame reading glasses given to community members by technologically savvy Jakartan NGOS. Paulinus carefully pulled out the laminated A3 map. Claudia's husband, Paolo, stood guard in front of the platform where we were seated. He wanted to make sure intel hired by the oil palm companies did not intrude on our gathering. While taking photographs of the map with my phone, I listened to the group's cartographic critique.

Crouched beside me, Claudia used a callused finger to follow the color-coded straight lines carving up the landscape into a series of overlapping geometric shapes: administrative regions; oil palm and timber concessions; military zones; transmigration sites; palm oil mills. The lines on the map segmented vast swaths of forest into nondescript squares and rectangles, classified in the legend according to their productivity type: Permanent Production Forest, Limited Production Forest, Convertible Production Forest, Industrial Plantation Forest, Timber Forest Concession—and many more. Blank patches, where my friends told me "natural forest" (hutan alam) was still standing, were labeled as "degraded land" (tanah terlantar)—a designation under Indonesian law for areas deemed available for industrial development. Over sixty agribusiness company names lined the bottom of the map, each of which owned one or more concessions in the "degraded" zones. A few shaded squares were labeled "protected areas" (kawasan lindung). Claudia told me that Marind were prohibited from entering these areas—including the sacred sago groves and hunting sites within them—because "the government is in control."

The group went silent. Then, Claudia shook her head and said, "There are no straight lines in nature (pada alam itu garis lurus tiada)." The group nodded affirmatively. Beny, Claudia's younger brother, turned to me and explained:

Nothing moves in straight lines. Look at the seasonal flight of the birds. The journey of the cassowary. The monsoonal movement of the fish and prawns. The way the sago reproduces following the curve of the Bian River. The way amai (plant and animal kin) and anim (humans) today follow this movement to feed and protect themselves. Look at the flow of the Bian—swaying right, then left, then right again, it has direction, but it is not rigid. Straight lines exist only on the road, in the army, in oil palm plantations. Nothing grows in straight lines, except oil palm. Oil palm plantations are modern forests.

Beny further contrasted the straight lines of the map to the fluidity of customary boundaries which follow natural markers and the movements of plants, humans, and animals as they grow and multiply. His sister, Claudia, added:

We too have borders between our territories, but these move with the movements of those who inhabit them—the birds, animals, sago, and Marind. Permission is needed to cross these boundaries, but it is not given by the government or companies. Permission is given to us by amai (plant and animal kin). They, too, are landowners. You might not have your KTP (ID card) at hand or any shoes on your feet. But if you show your skin and follow the voices of amai, then the forest and its beings will let you through. This is the way amai and Marind have lived together since time immemorial.

Then, Marcus, a young man from Mirav with whom I frequently went to gather medicinal herbs in the forest, turned to me and asked, "Missus (Miss), can you make this map alive?" Confused, I looked to the others for help, but their rephrasing of the question—"Can you give this map some life? Can you bring it to life?"—failed to make it clear to me. Marcus explained:

> You see, it's like this. This map, I think, is what the government and companies see. Just straight lines and oil palm. But there is something dead about this map. What I mean is that there is no sign of life. This map is not alive (hidup). It is dead. There are no people or birds or animals or plants on it. Just patches of color that don't mean anything. Everything just sort of stands still. Everything is flat. When I look at it this map, I don't know where I am. I don't know who is around me. I cannot hear the birds. Do you know what I mean?

I tried to explain to Marcus that maps could not include everything for reasons of scale and detail and tended to feature fixed locations in the landscape rather than their mobile inhabitants. But Marcus insisted that without that kind of "life," the map was pretty much useless. Borrowing my pen, he started to show me what it meant to make the map alive. Faintly at first, then more pronounced, the weaving line of his hand traced on the map the paths of kindred plants and animals of his clan across forest and marshland. Then, Marcus added the locations where interclan peace-making rituals, warfare, and congregations had taken place, the funerary sites of great warriors of the past, and the routes of the migratory tekle bird along with the places where it rested, fed, and mated. Claudia, Beny, and other members of the group joined in, adding to Marcus's lines a meshwork of other meandering trajectories: the aquatic journey of the angli-angli fish, followed closely by that of its predator, the swooping manhuk bird; the amphibious proliferation of sago suckers weaving along the fluid contours of swamp and mangrove; and the flight of the wallaby during slash-and-burn clearing, followed by its inquisitive return in the wet season to feed on juicy shoots in the forest regrowth. As the lines, arrows, arcs, and scribbles multiplied, administrative boundaries gradually disappeared in a dense overgrowth of intertwined human and other-than-human paths. Taking turns wielding the pen, my friends subverted the state's fixed representation of space by torquing its straight lines into lines of life.

IN THIS CHAPTER, I examine the power and politics of mapping in the Upper Bian. Maps have been historically instrumental to empires and states in affirming their control over territories and peoples.[1] In Merauke, maps have played a major part in the dispossession of Marind of their lands and resources. As Claudia noted, government maps do not represent the customary tenurial rights of Marind clans. They exclude places of cultural and spiritual significance to local communities, such as sacred groves, ancestral sites, and hunting zones. Exemplifying what Black studies scholar and cultural geographer Katherine McKittrick terms "plantocracy logics" (2013, 9), these maps classify the forests in which Marinds' plant and animal kin reside as unused or vacant areas, which are consequently deemed available for corporate development. Keeping Marind and their other-than-human kin off the map thus transforms them into *un*imagined, out-of-the-way communities of diminished political visibility.

At the same time, the marginalization of Marind and their forest kin in spatial representations prevents them from getting *in*-the-way-of government and corporate agendas. For instance, some of my companions interpreted the lack of recognition of their customary rights on paper as a convenient means for the state to exclude Indigenous landowners from negotiations over corporate land leases. Similarly, *not* representing the habitats and locations of rare or endangered species avoids companies losing productive land to excision for biodiversity protection, in line with national environmental laws and corporate sustainability standards. These factors, my friends explained, have facilitated the allocation of vast territories to third parties without the consent of local communities. They have also enabled agribusiness companies to elide their legal responsibilities in terms of both community compensation and environmental conservation. Marind and forest organisms are thus conveniently disregarded or silenced in the state's classificatory vision of space yet never immune to its violence.

For Beny, Marcus, and many other Marind, government maps are dead because they ignore the movements of organisms that make the landscape alive. Instead, the landscape is represented, in philosopher Henri Lefebvre's (1991) terms, as abstract space—a quantifiable commodity, engineered and organized under the logic of technocapitalist modernity. The straight lines of government maps, in Marcus's words, symbolically flatten the liveliness of space by imposing artificial and simplified order on it. Similarly, the expansion of oil palm literally flattens the diversely populated forest to make way for homogeneous plantations, or what Beny called "modern forests." Instrumental to what Ann Stoler calls the "logic of colonial governance," these maps are premised on a rigid classification of space according to productivity (real and aspirational) and scientific rationality (2016, 189–90). Their straight lines exclude Marind from areas designated as conservation zones and corporate concessions. "Dead maps" disregard the fluid boundaries of customary lands. They stultify the dynamic world of the forest. And they dissect the living landscape into regimented zones of agroindustrial exploitation and control.

As Fijian historian Tracy Banivanua-Mar notes, colonial landscapes may appear fixed and hegemonic, but they can nonetheless be stripped back, "making visible the alternative, subversive or repressed narratives of the colonial past and present" (2012, 198). The words and actions of my companions that afternoon under the mango tree in many ways exemplify the form of representational empowerment that Banivanua-Mar describes. By narrating and reinscribing more-than-human lives and relations on a "dead" government map, my friends collectively enacted alternative ways of remembering and reading the storied landscape, along with the memories embedded within it. At the same time, the very fixedness of maps was questioned by my friends from Khalaoyam. This questioning speaks to a broader awareness among Marind that government maps themselves are often inconsistent with one another. For instance, Claudia described how she had identified discrepancies in concession sizes, operators, locations, land-type classifications, and even regency boundaries, both within and across different maps. Some concessions, Beny explained, had figured on maps dating back to 2008 but never materialized. In contrast, other concessions existed on the ground but not on paper. Meanwhile, Marcus added, several maps of MIFEE had been retracted by government agencies shortly after publication due to "incomplete" or "out-of-date" data. The total area covered by MIFEE itself varied wildly across different maps, from 0.2 to 2.8 million hectares. Such variations had direct implications for the effectiveness of Marinds' land rights advocacy. When local activists invoked the higher figures, they were accused by corporations of exaggerating the scale of the oil palm problem. When they cited the lower figures, interest from international news and media outlets waned because the problem simply wasn't big enough.

On the one hand, then, government maps such as the one critiqued by my friends are instrumental in naturalizing the ownership and exploitation of territories and resources by the state and corporations. Indeed, many villagers affirmed that what government maps included and excluded was determined less by the physical landscape itself than by the maps' political contexts. Such contexts included, for instance, jurisdictional rivalries, administrative border disputes, upcoming elections, candidate one-upmanship, and competing adjudications within and across the government's various ministries.[2] Other Marind offer an alternative explanation for the inconsistencies of government maps. One of them is Amanda, a young activist from Khalaoyam who had received extensive map-overlay training from local NGOs. Pointing out discrepancies between the total area of oil palm on two maps produced concurrently by two different government ministries, Amanda exclaimed, "The government intentionally makes maps confusing to keep Marind in the dark. That way, we cannot know what land oil palm will eat up next, which rivers will be drained, or which forests will be razed. The maps make sure everything stays abu-abu." As Amanda suggests, the deceitful fabrication of spatial representations, too, serves a strategic purpose in concealing the state's activities and preempting interference from local communities. Discrepancies in and across maps thus purposefully create grayness, or abu-abu, in order to sustain state and capitalist visions of spaces both imagined and anticipated.

On the other hand, however, the transformational nature of government maps suggests to some Marind an ignorance on the part of the state of what is actually happening on the ground. Responding to the same two maps that Amanda had earlier commented on, Yosefa, a young widow from Bayau and one of the few villagers to have finished high school, reflected, "The state tricks us with false promises of money and jobs. It purposefully confuses us to steal our land. But maybe the state itself is confused. Maybe the state, too, is abu-abu." From Yosefa's perspective, cartographic inconsistencies belie the hegemonic control of the state-military-corporate troika and its omnipresent pressure points. Instead, they reflect the state's fragmented authority over space as an unfinished project—a "confused state," to borrow Yosefa's words. Just as grand schemes never fully colonize the territories on which they are imposed, these spatial representations reveal glimpses of aspiration rather than a state of cohesion already achieved (Tsing 2003, 5102–3). Dubious in their epistemic value and speculative affordances, maps are as volatile and murky as the state itself. When political agendas and actors change, maps in turn transmogrify, exposing conflicting visions of what space should be and for whom. Yet their lines stay straight, even if they shift in abu-abu ways.

SINCE THE RISE OF the Indigenous rights movement in the 1960s, rural communities have increasingly used maps to defend their claims to customary lands and cultural heritage, in what Nancy Peluso (1995) calls *strategies of counter-mapping*. Such grassroots initiatives often take the form of participatory mapping, which is community-led mapping that combines traditional ecological knowledge and modern cartographic technology.[3] At the time of my research, several Marind communities were undertaking participatory mapping as part of their struggle to curb agribusiness expansion. Prompted in large part by the introduction of oil palm into the region, these mapping initiatives were enabled by the technical and financial aid of NGOs and humanitarian branches of the Catholic Church, which first introduced mapping to Marind in 2011 as a means to secure formal recognition of their customary rights to lands and forests. The successful deployment of participatory maps in achieving Indigenous tenurial recognition in other parts of the archipelago constituted an important motivation for Marind villagers to support and participate in mapping activities.[4] Coordinates obtained in the field were then incorporated into *Mata Papua*, an interactive online map produced by a coalition of Indonesian NGOs to monitor deforestation in West Papua (see Yayasan PUSAKA and iLab 2016).

Over the course of my time in the Upper Bian, I participated in five mapping expeditions in the company of villagers from Khalaoyam, Mirav, and Bayau. As with the sago expeditions described later in this book, mapping journeys required little by way of resources from the village. Each evening we fashioned bivouacs from branches, fronds, and fibers harvested throughout the day as we walked the forest. Hunting, foraging, and fishing provided us with our daily source of food—birds, possums, legumes, and sago, which we cooked over small twig fires lit in traditional style by rubbing a stick over a pile of dried moss. Our journeys initially followed the boundaries of the territory of the primary clan leading the expedition, which usually took the form of natural markers in the landscape—the turn of a river, the edge of a sago grove, or the clearings where previous settlements had once stood. Soon, however, our trajectories would be guided by the movements and sounds of birds and mammals encountered in the forest. When customary land boundaries had to be passed, permission was sought from representative clan elders in the group and rituals were performed prior to crossing—a cleansing fire ceremony, an offering of fresh bamboo shoots, or an incantation directed to the ancestral spirits (dema).

The mapping journeys I participated in differed significantly from those undertaken by Marind in the company of NGO representatives. Whereas the latter tended to last only a few days, span smaller geographic distances, and involve at most three to four individuals, the former could last from several weeks to several months and involve from ten to over thirty villagers.

Among them, three or four individuals had received training in mapping techniques from me or from NGO members. But many others joined these expeditions out of curiosity, to forage for kin and friends back in the village, or to pay their respects to the ancestors at sacred sites along the way. For my companions, then, extended cartographic journeys were motivated as much by the strategic ends of advocacy as they were by the pleasure of going to encounter the forest and immersing oneself in its more-than-human liveliness.

On one such journey in late April 2016, a group of Mirav elders and I headed east at the break of dawn to map the customary territory of the Basik-Basik clan. Equipped with two global positioning system (GPS) units that I had gifted to my companions, we started walking. Yakobus, a Basik-Basik elder, led the group, humming tunes of his childhood and regularly calling out our names—a Marind habit that ensures peoples' souls are not captured by evil spirits dwelling in the forest and the swamp. I followed Yakobus, picking insistent leeches off my arms and legs as they undulated up from the moist hummus or dropped down from the overhead foliage. Beneath our feet, the forest floor was a colorful patchwork of rotting leaves, scrambling red ants, and bright blue cassowary plums—the favorite treat of the towering forest ratite. Suddenly, Yakobus stopped in his tracks. He stood in silence, looking up into the lush canopy. Then, he pointed upward and turned to me, saying, "There it is. That's our first point. You can map that." I could not see anything in the foliage and was unsure what Yakobus was trying to show me and what he wanted me to map. He said, "Can't you hear it? Can't you hear the whistle of the khaw (black-crested bulbul) up there? That's one of our amai (animal sibling or grandparent) telling us it knows we are here. It sings for us to continue our journey. This is an important place because it is where we met the khaw amai, so it must be mapped. Put it into the machine."

An NGO friend of mine had set up audio software in my GPS, so I was able to record the shy whistle of the bird whose yellow and black plumage remained invisible to my eyes. Then, Yakobus, who was renowned throughout the Upper Bian for his extensive knowledge of Marind lore, began to tell me the story of the khaw amai: how the khaw sprung feet first and head last from the fertile soils of the Bian, marking its crest with dark blotches; how the bird received the power of flight after drinking from the river, allowing it to circle the sun and catch light under its wings, which from that day on were tainted bright yellow; how the whistle of the khaw guided the ancestors of the Basik-Basik clan as they traveled in search of sago; how the bird sings at dawn to mark sunrise and still, today, protects Marind generations as they walk the forest. Yakobus went silent. Then, slowly and in a low, hoarse voice, he began singing the song of the <u>khaw</u>. The other group members lowered their heads respectfully and joined in. Tears welled in their eyes as they remembered their childhood, their mothers, and the sago groves that once flourished around their villages. The GPS recorded the sound of this place through the stories and songs of human and bird, interrupted by heavy sobs, sighs, and silence. What mattered, I realized, was not so much what my companions could see in the landscape but what they could hear within it and how this made them *feel*—happy, sad, nostalgic. I wondered how a single red dot on the map, a series of digits, could convey the affective textures of this multispecies symphony of a place:

Beep. Whistle. 7°11′18.8″ S 140°43′50.6″ E. <u>Khaw</u>.

After a few minutes, the song of the <u>khaw</u> grew faint. The mapping team advised that we follow it to other "important places." Guided by the flight of the black-crested bulbul, we roamed the forest as my friends pointed out the fruit trees from which the bird feeds, the sago groves in which it nests, the mound where it first came into existence, and the marshlands where it goes to sing its last song. The coordinates of each of these places were recorded on the GPS as part of an intricate <u>khaw</u> life-map, along with the many beings and events that made the bird and its environment "alive."

The names of Marind ancestors.

The location of former Marind settlements.

The sites where peace-making rituals had taken place.

The death of a wise woman from <u>kambara</u> or black magic.

The sago grove where wild pigs come in search of food.

The <u>gakhul</u> trees from which Marind fashion spears and harpoons.

The <u>ngef</u> shrubs whose white-pink inner bark is used as an analgesic.

The resting place of the snake <u>amai</u>.

The shallow sleeping holes of the <u>elekh</u> fish.[5]

Sago-pounding in a nearby grove.

Swaying bamboo shrubs.

The wind.

A child's laughter.

The river.

Thunder.

To these, we added sounds coexisting awkwardly with the familiar acoustics of the forest.

The call to prayer of mosques built for Muslim migrant communities.
The solemn toll of the Bayau church bell.
The xylophonic "Für Elise" chime of the Bayau state primary school.
The Indonesian national anthem, trickling from military garrison megaphones.
Roaring bulldozers.
Distant chainsaws.
Rumbling trucks.
The smoke-chugging palm oil mill.
Pesticide-spraying helicopters.
Crashing trees.
The deathly sonics of the plantation nexus.

At the boundary of the plantation, my friends insisted we map the most deafening sound of all.

YAKOBUS: Listen carefully, child. Tell me now, what do you hear?
ME (*hesitant*): I'm afraid I hear nothing, Elder.
YAKOBUS: Exactly. Put it in the machine.

At the time of writing, *Mata Papua*—the online platform where the bioacoustics coordinates described previously were uploaded—had been indefinitely shut down. The website's closure was prompted by growing pressure from the Indonesian government and systemic reprisals against the NGOs that administered the website and the Indigenous communities involved in its production. What this map offered before its untimely closure was a landscape animated by multiple sounds, songs, and stories, which the viewer could travel through by clicking on various coordinates. Indeed, the centrality of sound in understanding place is what distinguishes Marind maps from the government maps described earlier. Whereas conventional maps identify places primarily though the medium of vision—technological or human—sound is the primary means through which Marind develop what Steven Feld and Keith Basso call a "sense of place," referring to the bodily and affective ways in which place is known, imagined, remembered, and lived (1996, 11). Clicking through the map, one hears the <u>khaw</u> perched high up in the sago palms, the chainsaws gnawing away in the concessions, the grumble of the cassowary in sacred bamboo groves, the national anthem played in a loop by garrison megaphones between shrill crescendos of stubborn static, and the felt-but-not-seen caress of the wind. In hearing these sounds, one experiences not just the acoustic textures of a particular *place*, but also the *process* through which this place was sensed and rendered meaningful by Marind. As we walked the forest, for instance, Marind mappers continually situated themselves in relation to spirits and species of the forest by way of bioacoustic encounters that together produced the landscape as a dynamic, multisensory realm. As sounds travel across space, my fellow mapper Yakobus explained, they connect humans and forest beings as emitters and auditors, even as they are physically removed from, and invisible to, one another. One cannot see or be in multiple places at the same time, but one can hear different sounds in a single place and moment. Knowing what to map, Yakobus continued, thus requires knowing how and what to *hear*—and not just see—within the living landscape.[6]

Place, then, becomes meaningful to Marind as the sites of transitory sensory encounters with beings, like the <u>khaw</u>, who are constantly on the move. In this, too, Marind maps differ from conventional maps. Conventional maps are relatively stable representations of permanent physical sites and objects in the landscape, which exist independently of the mapper. Their legibility relies on legends or map keys that list the symbols and codes used to classify different types of locations. For Marind, in contrast, the referential value of mapped places consists of the stories of the species heard as well as the relations that human mappers entertain with that species. In following the call of the <u>khaw</u>, Yakobus and his kin traced the life-flow of the bird across space and time, and its relations to the many humans that have and continue to be guided by its song. In its cartographic form, this life-flow translates as a dense tangle of crosscutting lines, spreading like capillaries across the forest topography and accompanied by a multitude of sound clips—some lasting a few seconds, others several hours. The living landscape speaks through these sounds. It speaks through the mud, squelching under foot; the screeching parrots, mating in the canopy; and the mounds fashioned by <u>dema</u> at the birthing site of Marind clans. When one sound clip ends, another lights up on the screen. It invites you to travel to the next point in the journey of its mappers. Here, you will hear other stories, interspersed with laughter, silence, whispers, and tears. This acoustic patchwork speaks to Marind mapping as a kinesthetic and affective practice—one that involves tracking *moving* beings across the landscape and becoming emotionally *moved* by what one hears, and in turn, remembers and discovers.

On this map, you will find no cartographic legend to guide you, nor any indication as to what story the next coordinate will offer. You can choose to follow the coordinates as they light up—or you can roam elsewhere entirely. But connections will arise as you travel across this sensory patchwork. You notice a bird that accompanied you in your movement—its cry resounding in one place and backgrounded in another. You recognize the names of <u>dema</u> who walked this landscape before you. You familiarize yourself with the raspy voice of one mapper and the high-pitched song of another. You learn their names, the animals they hunt, and the sago palms they are descended from. Gradually you come to understand that the sounds mapped by Marind evade categorization because different sounds mean different things to individual mappers based on their own life stories and relationships with the species encountered. Unlike the fixed codes of cartographic legends, the meaning of these sounds cannot be described outside the context in which they are heard. Each sound is the crystallization of many other sounds, species, and stories. The protracted narratives that flesh out their dispersed meanings resist reduction to a single symbol or code. Instead, Marind offer you a meshwork of sonic lines and lights to journey through with your eyes and ears.

Within this meshwork, sounds of the forest are juxtaposed with sounds of its destruction. The song of the <u>khaw</u>, for instance, invites the listener to follow the bird as it travels the landscape. Its song is layered over with multiple human voices as mappers recount with fondness their origins and pasts through the cadence of other beings' lifeworlds: the possums they encountered in the forest; the cassowaries they pursued across the landscape; the rivers they canoed across to visit their kin; the rains that nourished the groves of their ancestors. Other sounds, meanwhile, warn you of approaching threats: bulldozers hashing away at the forest; military trucks pulling into the garrisons; helicopters droning overhead.

You can hear my companions gauge the proximity, approach, retreat, and direction of these threatening sounds and their own position in relation to it. At times, their stories will be interrupted by intervening security forces demanding to see ID cards and threatening to confiscate the GPS. Children cry as their parents are brusquely hustled aside for interrogation. Men raise their voices in response to threats from the police. Women whisper anxiously among themselves. Some attempt to redirect the listener's attention away from the tension—to the pitter-patter of the rain, the rustle of nearby vegetation, the fading glow of the sinking sun. In the backdrop, the <u>khaw</u> may continue to whistle. Or it might flee. You may encounter it elsewhere as you continue your journey.

Together, the sonic coordinates of Marind maps constitute powerful sensory embodiments of the radical transformations taking place within the landscape mapped, along with their spatial and affective effects on those who hear them.

Just as important here are the silences that Yakobus insisted on mapping as we traveled through the plantation. This deafening silence participates in the anxious semiotics of landscapes overwhelmingly dominated by humans (cf. Whitehouse 2015). Silence, too, must be recorded, for it represents acoustically the obliteration of life caused by the expansion of oil palm. Sonic absences and dissonances thus powerfully capture the reality of disappearing forests and invading plantations. They produce disturbing cacophonies that mirror the disrupted geographies and lifeworlds of the beings caught in their midst.[7]

IF GPS MAPPING REVEALED to me the centrality of sound in Marind cartography, drone mapping revealed how sight, and specifically perspective, shapes the moral valences of spatial representation among communities of the Upper Bian. During my fieldwork, I managed to sneak a DJI Phantom 3 Professional Quadcopter drone into Merauke. I was convinced that high-resolution drone footage would capture better than any other medium the extent of deforestation underway in the region. A drone would also help overcome the difficulties involved in physically entering oil palm plantations. Where my Marind companions and I could not travel, the drone could. Community members whom I trained to use the drone called it the "plastic bird" (burung plastik). I assumed the name referred to the material the drone was made of and its capacity for flight.

At sunrise, a group of Mirav villagers and I took the plastic bird out for a test run south of Bayau, along the boundaries of the PT AMS oil palm concession. Leading the cohort was Agustinus, a young Marind man who had been one of the most enthusiastic participants in the drone workshops I had organized over the preceding weeks. To avoid drawing attention from passing military patrols and company security staff, Agustinus clambered up a towering strangler fig to seek cover in its dense foliage. A few silky green and orange feathers wafted down to our feet as a flock of parrots burst out of the canopy, squawking at being so rudely awakened. No sooner had the colorful critters vanished than the plastic bird emerged, flying noiselessly out of the bushy treetop. Agustinus hooted from the treetop, signaling that he was safe. My companions hooted back. After hovering briefly in mid-air, the drone disappeared in the glare of the rising sun as it pursued its trajectory toward the PT AMS concession.

At sunset that same day, we gathered around the television in the church precinct to see what the plastic bird had captured. The bumpy footage gradually stabilized, revealing a barren landscape of felled trees and incinerated forest.

Then followed extensive swaths of planted oil palm seedlings, extending as far as the eye could see and cut through by the straight lines of murky irrigation channels. A technical glitch in the display settings on the drone made the horizon appear rounded like the sphere of the earth, further enhancing the visual impact of the devastation captured by the plastic bird.[8] I was speechless and shocked. I was also saddened and worried by the effect this footage would have on the morale of my companions, who had persevered against all odds in their struggle to protect their lands from oil palm. My friends were indeed made despondent by the footage. Some watched with blank expressions. Others turned away from the screen and whispered among themselves. Some shook their heads and pursed their lips. When I asked why, they only said, "We are not used to this." Confused, I packed away the drone and told my friends to use it whenever they wished. They never did.

Several days later, I asked the elders in Mirav whether they believed the drone footage could help them in their land struggle. The elders were hesitant to answer. Eventually, Andreas, the oldest in the group, decided to speak up. Andreas, the elders noted, was of the cassowary clan. He could imitate the sounds of hundreds of bird species. It seemed right for him to have a say on the matter of the plastic bird. Andreas explained: "The drone is a modern tool. It is trusted by the government and companies. For this reason, we are grateful for your gift. But the plastic bird does not show us how humans see the world, or how our plant and animal kin do. The drone travels in straight lines. It is a bird without life. It has no freedom. Real birds swoop and circle—like a river weaving through the forest. Who sees the world like the plastic bird? No one. This is like trying to see like the government."

The problem posed by the plastic bird is best understood in the light of Marind notions of perspective as a multispecies attribute. The forest landscape is the living product of the past and ongoing relations of humans and their other-than-human kin. Each of these species has a perspective that differs depending on its "body," in Eduardo Viveiros de Castro's (1998) sense of the particular physical faculties that shape the habitus of distinctive lifeforms. The capacity of humans and their plant and animal kin to hold a perspective on the world enables them to form relationships with their own and other species. Being-in-the-forest is thus a perpetual encounter between interagentive selves endowed with species-specific perspectival and physical dispositions.

Since the world comes into being through different perspectives, so too the representation of the world *also* depends on the points of view of those who participate in it. For this reason, Marind activists conceive maps as representations of the interspecies relations that produce the landscape, rather than of space as an abstract and static physical entity. Their mapping processes are multi-*sited* in that they follow the movements of meaningful organisms through which space itself accrues meaning.[9] At the same time, community maps in the Upper Bian are multi-*sighted* in that they are shaped by the worldmaking perspectives of different forest beings. For instance, the map guided by the song of the <u>khaw</u> and other animals seeks to represent the forest from the shifting point of view of its various inhabitants. Each coordinate in this map, to torque sociologist Thomas Gieryn's terms, represents, not a singular and authoritative "truth-spot," but a multiple and situated "*truths*-spot," which is rendered meaningful by the intersecting lifeways of itinerant forest and human dwellers.[10] Perspectivally inflected by the organisms who produce and populate them, Marind maps refuse to sit still.

The problem with the plastic bird lies in the point of view it takes on the world. As Andreas points out, the drone is not a living being but rather an inanimate technology. It flies high above the ground, but its movements are controlled by humans who determine its trajectory from predetermined departure and destination points. In this regard, the plastic bird lacks the freedom of real birds, such as the <u>khaw</u>. Furthermore, the drone is oblivious and indifferent to the multiplicity of sentient organisms that animate the landscape beneath it. The footage it captured was unappealing to my companions precisely because it failed to represent the forest as a space enlivened by *diverse ways of seeing* across species lines. Instead, the plastic bird transforms the place- and species-dependent fleshiness of the landscape into a singular and putatively transcendent truth *about* that landscape from a place-less and life-less viewpoint.[11]

The contrast between filming with the plastic bird and mapping with the <u>khaw</u> bird exemplifies Tim Ingold's distinction between the dwelling perspective of animist societies and the building perspective of Western societies. In the former, the perceiver takes a view *in* the world through active engagement with the constituents of the environment. In the latter, the perceiver makes a view *of* the world from a seemingly objective position outside the environment (2011, 9–11). Embodying the *of*- rather than *in*-the-world perspective, the drone footage represents the world as seen, in Andreas's words, by "a bird without life."

The putatively omniscient standpoint of the plastic bird is all the more troubling because of its uncanny resemblance to the cartographic gaze of the state that my friends had critiqued when we sat together at the foot of the mango tree. Both perceive the world from an elevated and supposedly all-encompassing vantage point, whereby the perceiver lies outside the realm that is observed. This elevated vantage point gives the state authority to determine the interests of the places and persons encompassed within its top-down gaze (see Ferguson and Gupta 2002; J. Scott 1998). Similarly, the drone performs what Donna Haraway calls the "god trick of objective science," or "the power to see and not be seen, to represent while escaping representation" (1988, 581; see also Plumwood 2002, 120). In presuming to see more and better than the beings below, this perspective embodies a detachment from the lived world. It also conjures the hubris, in Andreas's words, of trying to "see like the government." Pursuing a mechanical line of flight, the drone is deplorably plain in its monodimensional plane. And like the linear gaze of the state, the perspective of the drone is deceptive, for there is always more in the landscape than meets the (plastic bird's) eye.

ONE AFTERNOON IN OCTOBER 2016, I sat by the Bian River with Ignatius, an elderly man from Mirav with whom I had followed the song of the <u>khaw</u> together with Yakobus and his relatives. Ignatius and I were discussing a meeting due to take place the following day between Basik-Basik landowners and the Merauke Forestry Agency. Since 2010, my friend had played a key role in organizing collective action against oil palm across the Upper Bian. His activism had come at the cost of two years in jail and numerous threats to his family. But Ignatius continued his struggle. I asked him whether he thought the maps his community had produced could help stop the destruction of the forest. Ignatius replied:

> Our maps show the paths of <u>amai</u> and <u>anim</u>. Everywhere in the forest, there are paths like these. But not everyone believes maps are the answer. Not everyone believes the government and the companies will understand our maps. Often, they are deaf. We give them the sound of birds. But they don't listen. We give them the voices of the forest. But they don't listen. We do not know how to make them listen. We are still looking for answers among ourselves. The land, the maps, the plastic birds— everything is abu-abu. And we are still looking for the right path.

Maps that refuse to sit still, such as the one produced by Ignatius and his kin, subvert the straight lines—cartographic and physical—that restrict the movements of forest beings and their human kin. These moving maps counter the putatively singular and top-down point of view of the government by incorporating the viewpoints of those mapping *and* those being mapped. They also challenge the state's anthropocentric vision of space by highlighting the flow of life of more-than-human organisms who, together with humans, cocreate the landscape through their sounds, stories, and movements. While government maps conjure nature as a passive resource available for human exploitation, Marind community maps represent nature in generative motion. These maps operate along a logic of scalarity shaped by the ever-evolving perspectival relations of humans to sentient plants and animals, in what might be described as a process of perpetual, more-than-human différance.[12]

Différance also shapes how participatory maps are interpreted by Marind themselves—albeit in ways that are often highly contested. On the one hand, my fellow cartographers affirmed that maps were crucial in countering the invisibility of Marind and forest species in the spatial classificatory logic of the state. And yet maps that refused to sit still also gave rise to heated debates among my companions. For starters, available mapping technologies struggled to produce maps that were better heard than seen. Sonic coordinates changed continuously as forest beings moved through the landscape. Mapping trips, including the one I undertook with Yakobus and his kin, could never be planned in terms of duration or destination because they followed the trajectories of organisms encountered along the way. When they did not involve visiting NGOs, whose time was often limited to a week at best in the field, these expeditions sometimes lasted for months. We soon ran out of pens for color-coding the increasingly complex legends. Time and resource limitations excluded many plants, animals, and clans from the mapping process. While most villagers agreed that the maps should include all clans' respective amai, at least a dozen maps would have to be produced to represent each species and account for their seasonal migration patterns. Some individuals were adamant that none of their plant and animal kin should be left out, as they would take revenge by inflicting injury and disease on the mappers. For other villagers, a single map could not possibly incorporate all the meanings and movements that make the landscape alive.

The practicalities of mapping were further complicated by the fact that distances and scales were defined less in terms of physical coordinates than in terms of social ties. As a result, amendments had to be made when social relations among Marind shifted or significant encounters with forest beings occurred. For instance, the distance between the territories of the Basik-Basik and Mahuze clans had to be mapped some twenty kilometers further apart than they were on the ground because of a ten-year feud between the two groups. This feud was triggered when a drunk Basik-Basik youth accidentally ran over a dog—the amai, or kindred animal, of the Mahuze clan—with his motorbike. A few weeks after the map was produced, a child from the Mahuze clan was startled by a pig—the amai of the Basik-Basik clan—in the forest. Frightened, the child abandoned a bunch of poisonous mushrooms that he had gathered, which would have killed him had he consumed them. This auspicious event prompted a reconciliation between the two clans. In this and other instances, villagers insisted that the coordinates taken previously were no longer accurate. Distances and scales had to be changed. Customary land boundaries had to be redrawn. Locations had to be added or removed. Maps and their meanings were perpetually differentiated and deferred in light of changing social realities. Meanwhile, oil palm expansion continued unabated.

Contentions surrounding maps further multiplied along gendered and generational lines. For instance, elders refused to map sacred sites because customary law dictated that their location should not be divulged to outsiders, at the risk of punishment from dema (ancestral creator spirits) in the form of illness, death, or drought. Women argued that they should produce maps separately from the men because they entertained different relations to particular species and locations. Similar views were voiced by Marind youth, who often affirmed that the landscape meant something different to them than it did to their parents and grandparents. Some clan heads, meanwhile, worried that fixing land boundaries on paper without consulting neighboring communities would cause conflict between landowners.

The potential benefits and risks of mapping were aptly captured by Perpetua, a mother of two from Bayau who worked part time for the humanitarian branch of the Catholic Church. Reflecting on the numerous cartographic expeditions she had participated in over the preceding five years, Perpetua noted that "maps are a shield for the people, but they can become weapons in the wrong hands." For instance, Perpetua explained, some of her kin expressed concerns than demarcating areas of cultural significance might be opportunistically interpreted by the state to suggest that any areas *outside* these zones were available for development. Some young Marind men and women, Perpetua continued, believed that government and corporate actors should participate alongside Marind in the mapping or ground-truthing process. That way, these actors would better understand how and why Marind conceive the landscape as they do. The parents and grandparents of these younger community members, however, feared that collaboration with state and corporate institutions would be perceived by fellow villagers as collusion. This would not only undermine the validity of the maps produced; it would also jeopardize relations of trust among Marind themselves.[13]

The temporal dimensions of maps, too, gave rise to widespread disagreement among Marind mappers. Some community members affirmed that maps had to incorporate sounds, entities, and places that had been either destroyed or forcefully incorporated into the landscape. As Kristina, a schoolteacher from Bayau in her mid-twenties, explained, "Plantations, razed sago groves, and silence—these must be mapped, because this is the world we live in. This is the destruction we face." Other villagers, in contrast, refused to include in their maps sites and sounds imposed without their consent. For these individuals, excluding such elements from maps is a political, and not just a representative, move. It visually communicates their resistance to the state and corporate topographies of control disciplining the landscape. One such individual was Petrus, an elder from Mirav who had recently joined the local land rights movement. During our cartographic expeditions, Petrus often argued that maps should represent retrospectively what the landscape looked like *before* oil palm arrived because "this is the world we want to restore and protect." Other villagers, meanwhile, affirmed that maps should represent what Marind wanted the landscape to look like *in the future.* Kristina's younger brother, Hugo, for instance, vehemently opposed Petrus in arguing, "We need to show the government what kind of place we want our children and grandchildren to live in. There might be oil palm plantations in that place, but there will also be forest and sago. The government thinks it can control our future, so we need to use maps to show them the future *we* want." Cartographic conundrums over the temporal dimensions of maps thus reveal the imaginative valence of maps as contested representations of worlds at once disappearing, emergent, and yet to come. Abu-abu haunts the multiple temporalities of the landscape—as it was or is, should be or could be.[14]

For these reasons, very few of the maps produced by Marind have been finalized, and as a result, oil palm companies have gone ahead with clearing and planting under the pressure of operational deadlines set by national law. Even where community maps *have* been deployed by Marind as part of their land rights campaign, their effectiveness has been limited. In meetings I attended, for instance, company and government representatives expressed little interest in the sound recordings that constituted the coordinates of these maps or the oral narratives that accompanied them. Changing cartographic scales and distances were taken as evidence that Marind were simply not able or ready to use modern technology. For many government officials, the emphasis on the movements of humans, animals, and plants in community maps also reinforced racially imbued perceptions of Marind as backward tribes, who had yet to evolve from nomadic to sedentary lifestyles.

In this regard, Marind maps, too, come to embody the ambivalent effects of the pressure points described in the first chapter. Like roads, garrisons, and plantations, maps can be at once enabling and disabling. They promise heightened political visibility for Marind communities, but never without the attendant danger of wrongful manipulation or misunderstanding on the part of their audiences—or, indeed, of contestation from among Marind themselves. Maps, landscapes, and plastic birds, together with their perceivers and audiences, thus partake in making and representing a world of abu-abu. The limited effectiveness of maps is of particular concern to those Marind who have seen their bioacoustic representations of the landscape squarely dismissed by state and corporate actors during negotiation meetings. Disagreements have arisen following these meetings over whether visual and sonic coordinates should be combined or the latter discarded in favor of the former. Some villagers have called for the physical demarcation of territorial boundaries on the ground in the form of placards, trenches, and fencing, to complement and bolster their cartographic representations. Yet other villagers have called into question the very usefulness of maps in countering the entrenched power asymmetry at play between Indigenous communities, on the one hand, and state and corporate actors, on the other. Abu-abu thus shapes not only the limited effectiveness of participatory maps, but also what can be done to *counter* their ineffectiveness— and at what cost to their cultural validity.

COMMUNITY-LED PARTICIPATORY MAPPING, SUCH as that undertaken by Upper Bian communities, has given rise to debates among activists and academics over the compatibility of traditional knowledge and modern cartographic practices. For some, Western technology must be radically transformed for Indigenous peoples to achieve representational self-empowerment on their own terms. For others, mapping techniques might be determined by those in power—governments and corporations, but also Western science and technology—but they can still be used by Indigenous peoples in fundamentally alternative ways. However, less attention has been paid to debates on the practical and socioethical dimensions of cartography *among* the peoples directly involved in mapping.

As Ignatius's comment suggests, Marind are acutely cognizant of the problems of validity and effectiveness involved in making maps. The contrasts they perceive between the maps of the state and their own maps reflect their awareness that the landscape is always produced in tension with settler-colonial geographies and their all-pervasive pressure points. Marind spatial representations are thus also perspectival in terms of the *spectators* for whom they are produced and performed.[15] Marind themselves might not like the perspective adopted by plastic birds and the state. But seeing *like* the state is crucial if they are to be seen *by* the state. In reflecting critically on the audiences of their maps, Marind activists undertake "cartography" as the term is used by philosopher Rosi Braidotti (2006), in that they attempt to situate themselves politically within existing architectures of power in order to elaborate effective forms of resistance (see also McKittrick 2006). This suggests that discussions about the compatibility of traditional knowledge and modern cartography must be informed by debates taking place within communities who, in Ignatius's words, are themselves still searching for the "right path."

Cartographic conundrums within Marind communities over how to reconcile the value of maps as cultural resources and as tools for advocacy highlight their struggle to determine what makes maps legitimate, to whom, and with what consequences. Debates over how maps should be made and deployed—both among Marind themselves and in dialogue with supporting NGOs—acquire heightened significance in light of Marinds' conception of the world as shaped by the perspectives of the beholder and the beheld. Upper Bian communities produce maps in order to protect their forest and local livelihoods. But they also fight to reconcile cultural values and strategic activism in the production of these spatial representations. Both struggles find expression in the murkiness of maps as objects and mapping as process.

In this regard, Marind mapping practices complicate the assumed distinction between representation and reality that is central to ontological approaches. Representations are not just passive signals that help us understand reality. Rather, they participate in the production of reality itself. Indeed, cartographic representations are of critical importance in Marinds' attempts to salvage the forest and its diverse inhabitants from obliteration. In this context, cartographic representation, too, constitutes a political act and a medium for ontological self-determination. Marind representational practices further challenge the distinction between knowing and being as separate realms of experience. Instead, knowing, participating in, and representing the landscape constitute mutually imbricated practices of engagement *within* and *of* the world, or what feminist Science and Technology Studies (STS) scholar Karen Barad (2007) calls *onto-epistemology*. Cartographic conundrums among Marind matter precisely because they reveal *conflicting* ways of being in and perceiving the landscape, with all the compromises and contradictions such clashes entail.

Marind mapping practices also bring a more-than-human dimension to existing debates around the meaning of participation in mapping processes.

Such debates have been primarily concerned with the degree and form of participation of Indigenous peoples and other local communities (see Chapin, Lamb, and Threlkeld 2005, 627–29). In other words, the primary issue is which *humans* are involved in the representation of space. Marind mapping practices suggest that the meaning of participation needs to be extended in two senses. First, it should consider how maps affect and are affected by the diverse spectators to whom they are *performed* and the broader power dynamics within which these performances take place. Second, the term participation brings up the question of *other-than-human* perspectives. A distinctive attribute of the living maps produced by the Marind is their attempt to make space—literally and figuratively—for the points of view of sentient organisms whose well-being, like that of the Marind, is jeopardized by agribusiness expansion. These maps are premised on, and represent, multiple vantage points simultaneously in order to generate new ways of seeing and inhabiting the world.[16]

Finally, the centrality of sound and movement in Marind maps highlights how the ocularcentrism of modern cartography can obscure alternative sensory mediums through which humans encounter and interpret the world. As my companions pointed out during our conversation beneath the Bayau mango tree, conventional maps fail to convey the topokinetic nature of knowledge achieved through movement—human and other. Instead, these maps produce the cartographic illusion of geographic fixity.[17] For this reason, Marind activists struggle to produce maps that are deemed legitimate in the eyes (and to the ears) of state and other actors, while also retaining their meaningful sensory and dynamic attributes. The obstacles the Marind face in this regard call for forms of mapping that better accommodate the different cultural modes through which the landscape is lived and perceived. At the time of my research, Upper Bian activists were exploring the possibility of incorporating multisensory simulated-reality systems in their advocacy. Through this medium, they affirmed, users would be able to travel through virtual space and experience the sounds and movements within it. The forest would come alive. The river would retrieve its voice. The khaw would sing again. At the same time, my friends were well aware that such technologies will be challenging to deploy in terms of cost, access, and training. They are also unlikely to overcome the structural power asymmetries that limit the capacity of the Marind, as rural communities in remote areas of a politically volatile region, to undertake effective advocacy. However, new mapping technologies retain their promissory valences. Multimodal and immersive, these technologies may offer a more capacious way to sense and make sense of landscapes, like the Upper Bian, that accrue meaning through movements and sounds rather than sites and sight.

Lost in the Plantation—The Dream of Yustinus Mahuze

I was eaten by oil palm last night. It was the middle of the night. Around me was oil palm, everywhere. I was walking on the scars of trucks and cars and bulldozers. My footprints were so small, and theirs so big. I could hear screaming but I could not tell where it came from. Maybe I was screaming. I looked down and my footprints were no longer anim (human). They were the footprints of the khei (cassowary). I was thunder in the plantation. Around me, oil palm watched in silence. Suddenly, I was watching myself walk as cassowary. I had turned into a buruam (brush cuckoo). I was perched on an oil palm tree, my feathers stained with blood.

The screaming in the air turned into the long howl of I know not what beast. It ripped my khei heart, my buruam heart, my anim heart. Suddenly, I felt heavy and I looked down. My footprints were small and sharp. I did not recognize my skin. I had turned into a basik-basik (wild boar). I was running down a road to nowhere—to abu-abu. The air was filled with the smell of oil and fire and blood. The buruam that I was took off and disappeared in the clouds. As I looked down, I could see oil palm stretching into the horizon. There was only oil palm. I must have walked for days but the sun never rose. I felt like I had not moved at all because everything around me looked the same. All oil palm.

Eventually, I could no longer walk. I lay there, panting and sweating. I saw myself die, slowly, in the middle of the road. I saw my body turn back into anim and watched my skin split and heard my bones crack. I watched my feathers scatter on the road. I watched my tusks turn yellow, then gray. I returned as buruam, perched on top of myself, watching myself dead. The road was like a long, dark river of blood. I turned khei, then anim, then basik-basik, then buruam, dying my death over and over again. Finally, I shrank into a tiny little ball of flesh and bone. Soon, no one would know how many deaths I had died here. Then, I sang as buruam. I sang to the screams that filled the night. My wife tells me I woke up howling. Thus it is to dream in a world of abu-abu.

3. Skin and Wetness

Standing knee-deep in the muddy waters of the grove, Evelina, a young Marind woman from Mirav, held her two-month-old infant up to the trunk of a mature sago palm. Carefully she guided the child's hand down the tree's leathery bark, across its sap-filled cracks, deep fissures, mossy ridges, and rugged internodes. Walking around the tree, Evelina pointed her infant in the direction of the canopy. She described to him the wide, lush fronds bursting from the apex of the palm. Beside her, Evelina's daughter, Marcia, was busy transplanting sago suckers with her father, Gerfacius. She unearthed the delicate shoots one by one, caressed their smooth surface, and then planted them several feet away from the parent palm. "Not too far," her father instructed. Sago children must grow near the the skin of their parents. "Not too close either." Sago children, too, need water and light to develop wetness. A few meters away, Marcia's brother, Bonifacius, was rasping the inner trunk of a felled palm, humming a song about the genesis of sago, back when humans were just shapeless balls of mud floating down the Bian River. Beads of sweat trickled from the adolescent's forearms, mingling with the wet pith that his mother would later leach.

Figure 3.1. Okto, the cassowary man. Photo by Sophie Chao.

Evelina put her infant to sleep in the shade of a sago cluster. Then, she gently pressed her bulging stomach against a nearby sago trunk. That way, she explained to me, her soon-to-be-born child could "know the skin and wetness of sago." In a hushed voice, Evelina described to the baby in her womb the appearance of the palm, the orchids flourishing along its bole, and the suckers emerging from its base. She spoke of the starch that would be obtained from the tree, whose wetness would be absorbed by those who ate it, making them healthy and strong. The sweat of those who had felled and rasped the starch, too, would seep into the food and fortify both them and the sago consumers.

"Being in the forest makes our skin glossy," Evelina told me. "It makes our bodies wet and strong. It allows us to share skin and wetness with each other, and with our <u>amai</u> (plant and animal kin). Thus it is to become <u>anim</u> (human)."

IN THIS CHAPTER, I explore how Marind come into being through their bodily relations to other-than-human organisms in the forest. Like many other Melanesian societies, Marind conceive the body and its transformations as primary sites in the production and evaluation of persons and their constitutive relations. Central to the making and becoming of <u>anim</u>—a term alternately translated by Marind as "human" (manusia) or "person" (orang)—is the cultivation of a good "skin" (<u>igid</u>) and copious "wetness" (<u>dubadub</u>).[1] Skin is qualitatively evaluated—and a good skin is one that is shiny and smooth. Wetness, which manifests in humans as blood, saliva, tears, and sweat, is quantitatively assessed through its degree of abundance.[2] Both features are interrelated—a body becomes glossy when it produces fluids in great quantities. Men, women, and children improve their skin and enhance their wetness by participating in collective physical activities in the forest. In doing so, they absorb each other's wetness along with that of plants, animals, rivers, and soils in the form of sap, meat, pith, mud, grease, water, and more.[3] Diffused across myriad biological and elemental bodies, wetness, along with skin, thus operates within and across species lines.

Procuring and exchanging food are important ways in which bodily wetness and the relations of substance it enables are produced.[4] For instance, rasping and leaching sago generates sweat that makes peoples' skin shiny and moist. Fishing, gathering, and hunting in groups achieves similar ends. As Evelina explained to the baby in her womb, people partake in the skin and wetness of those who obtain and prepare the food that they consume while also imbibing the bodily wetness of the plants and animals from whom the food derives. For instance, those who eat sago take in the moisture of the palm and the wetness of the swamps, soils, and species that sustained its growth. Those who eat meat absorb the sweat, blood, and fat of the game that is captured, along with the wetness of its hunters.[5] People who frequently labor and eat together describe themselves as "sharing the same skin," a statement that is often accompanied by the physical act of rubbing each other's forearms.[6]

Tears are another valued form of wetness among residents of the Upper Bian. For instance, villagers often wept together and rubbed tears onto each other's faces when they recounted stories of the past, met long-separated friends and kin, or encountered plant and animal siblings in the forest. My companions would introduce important sites in the forest with the phrase "we cried here once" and would describe with tenderness the individuals with whom they once wept there. Similarly, when important rituals and meetings end, community members sit on the ground to "share tears with the earth." Collective and public weeping among adults is not just common; it is actively encouraged. It demonstrates peoples' affective investment in the social life of the community. As they seep into the soil, tears mingle with the life-generating wetness of the forest and its organisms, while creating bonds of substance and sentiment between human weepers.

The landscape itself is conceived by Marind as a source of nourishing wetness.[7] On our way to the grove, Evelina's husband, Gerfacius, explained to me how the skin (or surface) of the Upper Bian wetlands, marshes, and swamps allows all manner of vegetation to flourish and animals to thrive. The sleek and smooth skin of the Bian, too, nourishes aquatic flora and fauna, as well as insect and microbe communities that thrive under the soil. Most abundant in wetness, Gerfacius added, are sago groves and mangroves, whose humid climates sustain a wide range of forest organisms. Marind enhance their own wetness and skin by engaging physically with the fertile terrains and tributaries of the landscape through their movements, activities, and labors. For instance, they absorb the wetness of the Bian by bathing in the river (rather than in their homes) and casting water over their heads when traveling down its watercourses.[8] They often bivouac in sago groves rather than sleep in the villages because the clean air, fresh water, and morning dew moisten their skin. When journeys are planned, groups prefer to travel by foot or canoe rather than by car or motorbike. Young Bonifacius piped up to explain that this allows them to "touch the skin" of the earth and rivers—to feel the dampness of the mud, the coolness of the water, the heat of the midday sun, and the relief of rock, bark, leaf, and soil, in which the nourishing moisture of wetlands and wetworlds are inscribed.[9]

Exchanges of substance are instilled into children from an early age so that they, too, develop a shiny and wet skin (see Chao 2021c). Children are not considered fully human, or <u>anim</u>, at birth (dorang belum jadi <u>anim</u>).[10] To initiate this transformation, a child must be delivered in the sago grove in order to come into direct contact with the environment and absorb the wetness of its palms, soils, and streams. Exchanges of skin and wetness between infants and the sentient forest are encouraged even before birth, as when Evelina pressed her womb against the sago trunk so her child could begin to know the forest. Throughout her pregnancy, both Evelina and her husband had actively consumed as wide a range as possible of organisms with whom their child would later share the wetness of the forest—pigs, cassowaries, sago grubs, and more.

After children are born, adults carry or hold them as often as possible, rub their faces and limbs frequently, and sleep with them in tightly nestled groups to transmit to them their bodily wetness. Children are bathed in the river at least twice a day and any wounds are carefully cleaned and treated with a variety of oils and saps to keep the skin moist and prevent scarring. When in the forest, children themselves are encouraged to share skin by holding babies and feeding toddlers, finding food in the forest for each other, and tending each other's scrapes with medicinal herbs and resins. These activities prepare children to participate in the society of adults by initiating them into the exchanges of skin and fluids that are central to becoming anim.

Children come to know the skin of the forest by engaging directly with the environment and its lifeforms, in what might be described as a form of multispecies pedagogy. For instance, during our two-week stay in the forest, young Marcia and Bonifacius were taught by their parents how to identify animal and plant species that can be consumed and those that must be avoided, based on the glossiness of their fur, feathers, tendrils, and leaves. They learned to scrutinize the appearance of the skin (or bark) of sago palms and the wetness of their pith to detect their stage of maturation. Evelina and Gerfacius also taught their children how to protect the skin of the forest—for instance, by encouraging vegetation regrowth through controlled clearing and burning and by maintaining a respectful distance from juvenile animals, whose skin is still frail and whose bodies lack in wetness. In between these practical and meaningful acts of care, Evelina and Gerfacius imparted to their children the sacred myth of Sosom, who lives in the subterranean waters of Mabudauan and fertilizes Marind lands and beings with heavy monsoonal rains. They also recounted the genesis of the first anim, who came into being in the waters of a creek near Kondo. These catfish-like creatures, the story goes, were blind at first, and without limbs or orifices. Eventually, the ancestral creator spirit Aramemb fashioned their arms, legs, eyes, and other body parts from mud, giving them the power of movement and sight.

The collective and kinesthetic labor of making anim brings to mind what James Leach calls the work of "growing people" among Reite of Highland Papua New Guinea. As among Reite, Marind persons emerge within a field of nurture comprising their relationships to other humans and to the environment, which is achieved through various forms of physical work (Leach 2003, 30).[11] Reciprocal exchanges of bodily fluids, both in and with the forest, are central to the production of Marind persons. As a gestational milieu, fluids enable lifeforms to flourish and reproduce—the waters of the Bian, the wetness of the sago palm, the juicy shoots of bamboo clusters, and the blood and fat of game. Each body of water owes and eventually passes on its bodily wetness to others.

As a differentiating force, the flow of wetness connects and distinguishes bodies through their semipermeable membranes, or skin—the bark of the strangler fig, the surface of the river, the skin of infants, and the flesh of the forest. These fluid, transdermal exchanges repeat across time and space as they diffuse through different bodies in different forms—blood, water, sap, mud, grease, tears, and more.[12]

Skin, as the counterpart to wetness, mediates the sensory relationships of human and other-than-human lifeforms. It is an interface inscribed with social memory—a permeable locus of intercorporeal connection that accrues affective materiality through tactile engagements within and across species lines. Together, skin and wetness give rise to bodies as congeries of biological, cultural, and semiotic forces, which exist in their symbiotic becomings with *other* bodies. Together, they produce what philosopher Maurice Merleau-Ponty (1968, 83–84, 169) calls the "flesh of the world"—a meshwork of life in which all organisms are mutually intertwined and implicated. True anim, then, are channeled persons.[13] They come into relational being by opening themselves to literal and figurative flows of shared liquids and life across species lines, in what might be termed a multispecies *skinship*.

FOR MARIND, THE GENERATIVE mingling of wetness, land, and bodies is not about asserting human mastery or control over "nature." Rather, exchanging flesh and fluids with the environment is precisely what enables reciprocal, more-than-human flourishings.[14] At the same time, the labor of making and becoming anim is often described as "hard work." It is contingent on the collaboration of myriad human and other-than-human others. It begins in childhood and never really ends. It requires sustained interactions across species lines. Both good skin and abundant wetness depend on these iterative, intercorporeal engagements. Both can also be lost if such engagements cease.

For instance, individuals who do not participate in collective activities, rarely walk the forest, and demonstrate solitary tendencies are said to develop a dull skin and lack wetness.[15] Individuals may also lose their glossy skin and wet bodies if they spend extended periods of time away from their community and forest or if they travel frequently to the city or plantation by way of the road. These attributes can also be lost if individuals interact frequently with "foreign people," such as non-Papuan settlers, corporate representatives, and government officials, or if they consume vast quantities of "foreign foods," such as rice, cookies, and instant noodles. These individuals are often said to have become abu-abu. They must be actively encouraged by their kin to participate in activities in the forest and to consume abundant volumes of sago and other forest-derived foods.

Only by doing so can they retrieve their depleted <u>anim</u> flesh and fluids. Plant and animal spirits (<u>amai</u>) may also punish <u>anim</u> who fail to share skin and wetness with the forest and its living organisms by making their skin taut, dry, gray, or hard. Such transgressions include neglecting rituals prior to hunting or sago processing, polluting or diverting watercourses, killing female game or their young, or failing to share one's catch with the broader community. Illnesses that provoke dry skin, such as dysentery, malaria, dengue, and more recently, AIDS, as well as climactic conditions such as droughts and floods, are also frequently interpreted by Marind as the consequence of a wrongdoing on the part of the diseased toward fellow <u>anim</u> or <u>amai</u>.[16]

Unhealthy-looking skin and dry bodies thus indicate an imbalance, depletion, or blockage in one's human and more-than-human social relations.[17] The introduction of oil palm over the last decade in particular is widely associated by Marind with a generalized depletion of cosmological wetness—one that Marind community members widely associate with the gray uncertainty, or abu-abu condition, of the world they inhabit today.[18] As Karolina, a young activist and former nurse from Bayau, put it, "Oil palm takes all the wetness of the land, and leaves everyone else thirsty. It wounds the skin of the forest and of <u>anim</u>. It makes everything and everyone abu-abu."[19]

Like Karolina, many Marind interpret the contamination and desiccation of <u>anim</u> and the living environment provoked by the arrival of oil palm and its toxic ecologies in social and moral terms.[20] Oil palm replaces the giving environment of the forest with a rapaciously *taking* environment. It occupies the land, water, and life of forest beings, sapping their fluids and wounding their skin yet giving nothing in return. As oil palm encroaches onto Marind territories, the skin of the forest is stripped of its vegetation through large-scale deforestation and burning. It becomes dry, flaky, and wizened. Trucks carrying fresh bunches of oil palm fruit to the mill, along with bulldozers clearing the forest, dig deep tread marks—or "scars," in Marind terms—on the surface of the soil. Water supplies, too, are depleted as plantation irrigation diverts or arrests the flow of ancestral rivers. The skin of the Bian River, polluted by mill effluents, is no longer shiny and sleek but rather mottled and gray—abu-abu. When a prolonged drought afflicted Merauke in 2015 and 2016, young and old Marind alike affirmed that the oil palm relentlessly sucking up the wetness of the rivers had caused this unprecedented event.

Marind and their forest kin inherit in the flesh the excesses of capitalism and its toxic externalities. Ecological transformations such as the drought, water pollution, biodiversity loss, and deforestation are felt through the fleshy, damp immediacy of their own bodies. As he watched bulldozers crashing through the forest, Antonius, a man in his early twenties and cousin twice-removed of Kristina, pointed to the cracked skin on his calves and commented that oil palm was leaving humans, mangroves, and beasts parched. The wizened skin of the forest, Antonius lamented, could no longer sustain the diverse plants and animals that Marind share relatedness with and depend on for their subsistence. Their bodies, once strong and glossy, become weak and dry, as fertile soils fritter into dust. Meanwhile, the ashes produced from the burning of the forest's skin scatter as sediment across the landscape. These ephemeral traces of once-living kindred species eventually settle in the respiratory tracts of humans and animals in a macabre form of particulate cannibalism.

Meanwhile, chemicals and toxic gases emanating from the plantation and mill penetrate bodies as molecular dispersions of violence both slow and immediate. One of these chemicals is paraquat, a quick-acting, nonselective herbicide that has been banned in many countries due to its toxic effects. Another is glyphosate, a highly carcinogenic, broad-spectrum herbicide and crop desiccant.[21] Very few Marind know the potential long-term health impacts of their exposure to these chemicals. Many voice resignation and powerlessness in the face of the potentially deadly effects of industrial toxins on their bodies and health and those of Marind to come. Rising like a noxious sweat from tropical soils, toxic particle-clouds infiltrate the pores of the poor, inducing nosebleeds, cataracts, sores, dermatitis, abdominal ulcers, cancers, Parkinson's, infertility, and sometimes death (Isenring 2017).

As murky, acidic effluents seep into the splayed bodies of ancestral rivers and soils, pink rashes spread like an insufferable mold on the bodies of infants and children while anthracite blotches of eutrophied algae multiply like tenacious ink stains on the greasy sheen of streams and springs. Chemicals dumped into watercourses, Antonius explained, contaminate the forest animals who drink from them and corrode the flesh of fish, turtles, and crocodiles living within them. "Drunk" (mabuk) on toxins, these and other aquatic critters float to the surface by the hundreds. Similarly, people's skin, once glossy and soft, turns rough and irritated when they drink from and bathe in streams polluted by pesticides, fertilizers, and sludge. The lively permeability of human and other-than-human bodies is subverted by lethal transfections in this murky, abu-abu, milieu, where invisible toxins gnaw at the skin and fluids of humans and their forest kin.

The polluted wetness of rivers and soils in turn jeopardizes the living bodies and relations, present and future, mediated by their flow. This contamination thwarts the intergenerational continuity and shared survival of the forest's diverse communities, in a life-negating process that Deborah Bird Rose calls *double death* (2004, 175–76). Chemicals absorbed by humans and other lifeforms are unwittingly transmitted to those who touch or consume them—through the breastmilk of mothers, the sweat of hunters, the blood and grease of game, the flesh and scales of fish, and the shoots of budding sago palms. Those who are exposed to these harmful fluids partake in new and troubling chemosocialities that are produced and inherited across kin and kind.[22] Infiltrating soils and bodies, noxious substances emitted from oil palm plantations and mills produce the *chemical sublime*, Nick Shapiro's term for the condensation of vaporous displeasures caused by the corrosive effects of industrial progress.[23] In the face of disquieting yet omnipresent toxicities, indistinct harms crystallize into a generalized vulnerability distributed across the diverse species and substances inhabiting the chemical geographies of abu-abu (Shapiro 2015, 369).

Within these toxic geographies, Marind and their forest kin inhabit a condition of lethal suspension. In the state of suspension, humans, other-than-humans, and elements together become participants, vessels, and mediums of chemical diffusion in a shared atmospheric environment. Such forms of toxic suspension extend indiscriminately across species, time, and space, but they are always unevenly distributed. For instance, chemical assemblages serve to sustain oil palm's growth and productivity to feed a global community of consumers. Yet these same chemical assemblages enfeeble the bodies exposed to them through contaminated atmospheres, rivers, foods, and wombs.

Plantation toxicity thus disrupts certain levels and kinds of socioecological orders in order to maintain others. It, too, constitutes a pharmakonic force that alternately enables, enhances, restricts, or terminates different forms of life and their constitutive relations, enhancing some worlds and bodies at the expense of others. Nonlinear yet inescapable in its effects, toxic exposure reveals the visceral ways in which the bodies of Marind and their other-than-human kin are connected, and rendered vulnerable to, the structural violence of the capitalistic world system.[24] Mass-produced for its oil—a lucrative yet lethal form of wetness in itself—oil palm severs the chains of substance exchange that nourish forest lifeforms. The plant and its chemical ecologies rupture Marinds' self-constitutive affective, bodily, and moral relations to the landscape and its diverse beings. In doing so, oil palm jeopardizes the capacity of Marind to become and remain, anim, or human.

IN CHAPTER 1, I DESCRIBED how Darius proved to the plantation guard his identity as Marind and his skinship with dog and sago through the bodily mediums of lacerated flesh, dripping blood, and ripped-out hair. In this and many other, less violent, everyday acts, skin and wetness among Marind come to constitute what Merleau-Ponty calls the *thickness of the flesh*, or the phenomenological interface that enables selves, human and other, to establish and communicate the nature and networks of their mutual relations (1968, 135). The centrality of skin and wetness as idioms of morality is heightened by the fact that these fleshly substances and surfaces reveal what cannot be known directly of the mind—a dimension of the self that many Marind also describe as abu-abu.

Like other societies in the Pacific region, Marind insist that the thoughts and feelings of others are difficult, if not impossible, to know.[25] For instance, while my companions commented openly and frequently about the condition of each other's skin and wetness, they were generally reluctant to speculate about or interpret fellow community members' desires, intentions, and feelings. When pressed to speak on behalf of others, they were visibly wary and uneasy and responded that they could not do so or simply "do not know." Actions, I was often told, mattered more than words and promises because only actions could manifest peoples' genuine intentions and commitments. Insisting on finding out what others think was often described as "disrespectful" or "rude."

Among Marind, then, respecting the opacity of others' minds is both the consequence of the impossibility of knowing others' minds *and* an actively upheld moral stance. It constitutes, as Webb Keane puts it, "an assertion of the right to be the first person of one's own thoughts, and an acknowledgment of others' right to be the first person of theirs" (2008, 478). The secrecy of mind also extends to plants and animals. How, or what, plant and animal amai think or feel consequently matters less than what they reveal themselves to *be* through their bodies and behaviors—for instance, the way they move, reproduce, grow, and sustain themselves; the condition of their skin and wetness; and the nature of their encounters with humans and other-than-humans in the forest. It is these bodily, interactive affordances that determine the capacity of organisms to affect and be affected, or what philosopher Brian Massumi aptly calls a *bodying* (2015, 144, 203). Consequently, nothing can be known about a body, human or other, other than what that body can *do*. Points of view, in other words, exist not as thoughts but as actions, and species exist not as nouns but as fleshly verbs. However, it is not only the thoughts of others that are unknowable, but also one's *own*. Marius, an elder from Bayau, explained this to me as we sat in the shade of a sago palm watching a passel of possums jostling in the canopy. Marius turned to me and said: "When I walk the forest, I encounter my amai and those of my kin. I spend hours in their company, watching them, hunting them, whistling to them, listening to them. When I return home, I carry amai in my body, my smell, my blood, my sweat. I have shared skin with amai. Maybe I have changed. Maybe I am no longer anim. I cannot know for sure. This is why the forest is a dangerous place for anim."

On the one hand, the forest is the best place to become anim because there, humans and their plant and animal kin can achieve intercorporeal relations by sharing skin and wetness. In this sense, the forest is a space enlivened by fleshy animacies, affective and material practices that subvert binary systems of difference and invite new kinds of communalisms and intimacies across species lines (cf. Chen 2012, 3, 10–11). And yet as Marius suggests, the forest is also a dangerous place. Anim who spend protracted periods of time within it may unknowingly begin to behave like the teeming living beings who inhabit it. Becoming anim in the bewildering multitude of the forest is thus always also a struggle to *retain* one's anim self and resist perspectival capture by other-than-human entities one becomes-with. The pharmakonic nature of more-than-human becomings is particularly marked during skin-changing, a practice I discovered through the troubling fate of my friend, Oktavianus Ndiken.

Oktavianus—or Okto, as he was called by friends and family—was a middle-aged man from Bayau village. In his youth, he had been a notorious trouble-maker who frequently got involved in village brawls and caused conflicts within and beyond his clan. Okto was also renowned throughout the region for his capacity to "change skin" (ganti kulit), the Indonesian translation of a Marind expression that refers to the temporary and willed transformation of men into animal form.[26] During our many trips to the forest, my companion often told me that Okto particularly enjoyed wearing the skin of the cassowary because he could frighten away other creatures with his towering height and muscular legs—much as he reveled in intimidating his human rivals in the village. Okto could also explore the forest at his will—like the cassowary that travels alone rather than in a flock—and gorge himself on delicious forest berries that no human could eat. Wearing the cassowary's skin, my friend would exclaim, gave him "freedom."

Over the years, Okto began to develop peculiar physical traits. A thyroid follicular adenoma, a benign tumor of the thyroid gland, developed under his chin. The small excrescence grew until eventually it hung limply from his neck, burgundy and mottled like the fleshy wattle of the cassowary. Okto's nose, once flat and wide, became singularly aquiline. It stood out against his tight, thin jawline like a cassowary's beak. People described Okto's gaze as sharp or piercing, like that of the tall forest ratite. At night, Okto did not snore like other men but grumbled in a cassowary-like timbre. The soles of his feet, too, became unusually callused. His toes were splayed and elongated and his nails hard and dark, like cassowary claws. Many of his friends and relatives took these slow transformations as troubling evidence of a gradual process of cassowary-becoming.

Okto's bodily transformations were accompanied by behavioral changes. For instance, he was no longer the outspoken and aggressive man of his younger days. He had grown to dislike the taste of sopi, the traditional Marind coconut palm wine, for which he once had had a distinctive (and destructive) penchant. Okto also appeared reluctant to share skin and wetness with his kin and friends, avoided most social occasions, and preferred to eat alone. To evade the company of others, he often left the village well before sunrise to forage. Whereas other community members brought back game, fish, and plants from the forest, Okto infuriated his wife and sisters by returning with snails, small rodents, rotting fruit, and sometimes soil, which cassowaries are known to consume when food is scarce. When he was not hunting, Okto would stay at home and care for his children rather than sit on the porch with friends or in the men's hut. The villagers said he was behaving like the male cassowary, who incubates and rears its young while the female moves away to breed with other mates. Widespread rumors of an affair between Okto's wife and a younger man from Mirav seemed to confirm this. And just like the male cassowary, who retreats when in a standoff with a female, Okto rarely stood up for himself when his wife and sisters publicly reprimanded him from spending so much time away from the village in the forest. Instead, he suffered their browbeating in silence.

On a Sunday afternoon in July 2016, Okto was found dead in his bed. He had returned to the village at the break of dawn after three days of absence. Community members believed Okto had been in the mangrove forest because his feet and thighs were covered with thick mud stains. Like the smell of decomposing detritus and anaerobic bacteria in mangroves, his body reeked of rotten eggs. Some community members claimed to have heard his low grunt in the mangroves where they had fished the previous day and later recognized him from the distinctive color of his wattle. That evening, village members gathered in the customary hut to deliberate on how to handle Okto's death. The meeting was standard practice. Like many other Melanesian societies, Marind regard any human death other than that of the very elderly or the very young as an act of killing, either by another human, by amai (plant and animal kin), or by dema (ancestral creator spirits) (see, for instance, Hau'ofa 1981; Lipset 2016; Stasch 2001; and Stewart and Strathern 2000). Consequently, extensive consultations are required to figure out why the individual was killed, who bears responsibility for the death, and what must be done to assuage the anger of the killer and compensate the family of the victim.

However, Okto's death posed a problem. To begin with, it was unclear who was responsible. The man's belligerent temperament meant he certainly did not lack for enemies. More worryingly, it was uncertain who Okto *himself* was at the time of his death. Everyone agreed that Okto was no longer fully human, but neither was he fully cassowary. Community opinion was divided over how to deal with Okto's split identity. Were they dealing with a cassowary-like human or a human-like cassowary? If indeed a villager had killed Okto, was the crime that of killing a human or killing an animal? How might this in turn determine how the killer should be punished? Who would the community grieve in the required forty-day mourning period following Okto's passing—a bird or a man? Should the mourners also include Okto's cassowary kin? At the end of the day, was this not a matter for cassowaries rather than humans to decide? Two days later, Okto was buried in the local graveyard. However, the dilemma posed by his abu-abu identity haunted Bayau for several months and remained a subject of conversation for much longer. No one hunted cassowary that season for fear of attacks by the grieving birds. Cassowary eggs were respectfully left in the forest where they were found, along with offerings of betel nuts and nipa fronds. When I left the field three months after Okto's death, the case of *Oktavianus-as-anim* v. *Oktavianus-as-cassowary* remained unresolved.

LIKE SOCIETIES THAT PRACTICE shape-shifting in Amazonia and elsewhere, Upper Bian Marind change skin to harness the physical abilities of their kindred plant and animal amai and thereby improve their hunting outcomes. These capacities include, for instance, the mighty tusks of the wild boar, the piercing eyesight of the bird of paradise, and the stealth of the green tree python. They encompass the agility of the tree kangaroo, the speed of the cassowary, and the lightness of the heron. Yet hunting and its outcomes, while important, are not the main reasons why people take on animal forms. Rather, and as Okto himself had described to me days before his passing, people change skin primarily for the pleasure of experiencing the world from a body endowed with faculties that anim do not possess. For this reason, many Marind men describe skin-changing as an enthralling experience of "freedom."

However, and echoing the ambiguous qualities of the pressure points and community maps examined previously, skin-changing is also dangerous and risky. In part, the danger relates to the trade-offs of strength and weakness inherent in different bodily forms. Later during our journey through the forest, Marius explained to me that people who adopt a cassowary skin might be able to run with great speed but also suffer from short-sightedness. The skin of the monitor lizard, he continued, allows one to climb trees with remarkable agility but it also forces one to stay close to water and shade. Similarly, the powerful arboreal leaps of those who have become tree kangaroos are offset by slow and awkward locomotion on the ground. The greatest risk, however, lies not in these physical limitations but rather in the possibility that one may not be able or willing to retrieve an _anim_ skin because one has "lost oneself" (hilang diri) to the skin one has adopted.[27]

Encounters with humans in the forest can help prevent shape-shifters from losing the capacity to free themselves from their animal form. For instance, before slaying their catch, hunters repeatedly address the animal by the names of different kin and friends. If the animal is a skin-changer, it may react to its name or be reminded of its human relatives, thereby prompting it to return to its human form and family. When they walk the forest, community members also leave behind artefacts they have made, such as sago frond huts, machetes, bows, and arrows, to remind shape-shifters who encounter these objects of their once-human condition. However, these efforts can fail. For instance, Okto, who had once been able to transition back into human form easily, unwittingly came to resemble and act like a cassowary in his everyday life. He began to care for his children like a cassowary father, roamed the forest alone and for days on end, and avoided the company of his human kin. Okto also sought to extend his cassowary habits to his wife and children by bringing them cassowary foods to eat, such as snails, rotting fruit, and soil. In the worst cases, skin-changers may try to reproduce in the forest with the species in whose form they are trapped. They may also attack _anim_ in the forest whom they no longer recognize as their own kin. The actions of these individuals become increasingly unpredictable and potentially harmful to they themselves and to others. Neither fully _anim_ nor fully _amai_, these abu-abu beings are trapped somewhere between a human skin that they cannot retrieve and an animal skin that they cannot shed.

The potentially disruptive transformations entailed in Marind skin-changing practices reveal the human body to be less a static biotic entity than a malleable locus of ontic contestation. On the one hand, skin-changing is pleasurable because it allows humans to inhabit the lifeworld of other organisms and become perceivers and agents in and of the environment in radically different ways. But skin-changing also entails the risk of becoming confined to the perspective of the skin one wears, in what might be described as a form of perspectival (dis)possession. The opacity of mind and the threat of perspectival dispossession thus pose limits to the capacity of Marind to make _anim_ of themselves and others. In everyday life, you cannot know _what_ others—or even you yourself—think. In skin-changing, you cannot know *how* others, or even you yourself, think—as <u>anim</u>, as <u>amai</u>, or somewhere precariously in between.

Here, the body itself constitutes a corporeal pressure point—one whose transformations may be alternately beneficial or lethal. As Marius's words convey, shape-shifting offers the possibility of bodily transformation and perspectival accrual—but never without the attendant danger of ontic abduction. In the hazardous metamorphosis of humans into animals, the animal other both signals and subverts the self of the human who is gazing and gazed on.[28] Becoming-with <u>amai</u> may thus also involve *un*becoming <u>anim</u>. The sentient ecology of the forest, as Philippe Descola artfully puts it, "obliges [humans] to play by rules they do not always control" (1986, 221). In this diversely populated realm, multispecies ontic conversations can easily veer into perspectival conversions. For this reason, changing skin is a perilous endeavor that only the most courageous can attempt. As a Marind proverb goes, "The brave man head-hunts. The braver man changes skin."

IN THE UPPER BIAN, interspecies relations simultaneously sustain *and* jeopardize the identity of Marind as <u>anim</u>, giving rise to a world populated by precarious beings. Fleshy exchanges of wetness and skin enable life-sustaining reciprocal exchanges between humans and their environment, but they also expose humans to contamination, contagion, and the possible dissolution of their <u>anim</u> self. As the ambiguous personhood of Okto reveals, transmutations of the body and the body's consequent capacity to affect and be affected can never be fully predicted. The permeability of bodies that do not end, but rather emerge from and through the skin, can produce both lively relations and lethal transformations. The potential for such transmutations is heightened in the intersubjective space of the forest, where the boundaries between <u>anim</u> and animal, seer and seen, and object and subject are troublingly fluid and indeterminate. The human condition, then, is neither permanent nor finite, but rather susceptible to dangerously irreversible, albeit experientially exhilarating, processes of becoming-animal. In the difficult work of making <u>anim</u>, becoming-with forest organisms—the *other others*, in Jacques Derrida's (1996, 68–69) terms—reveals itself as a vital yet potentially perilous necessity.

The bodily dimensions of making <u>anim</u> fleshed out in this chapter resonate with the characterization of Melanesian persons as arising from, and being shaped by, the embodied relationships they form with each other, or what Marilyn Strathern (1988) calls *dividuals*.[29] Bodily fluids such as tears, sweat, and saliva, come to stand for the persons who exchange them. They are at once *produced* through cumulative social interactions and *produce* new social relations when imparted on others. As such, <u>anim</u> bodies emerge *from*, and *as*, the composites of intercorporeal relationships achieved processually and through the flesh.

At the same time, the poetics and politics of multispecies skinship in the Upper Bian expand Strathern's notion of the dividual and its attendant critiques by highlighting how persons are constituted through relations that extend *beyond* the category of the human.[30] Among Marind, becoming <u>anim</u> entails corporeal cobecomings with bodies human and other-than-human, riparian and estuarine, geological and meteorological. Resonating with the etymology of the term *skin* itself, which originally referred alternately to animal hide, fruit skin, or tree bark, walking the forest enables Marind to partake in more-than-human skinships that enfold myriad organisms who together produce, exchange, and circulate life-sustaining cosmological wetness. This skinship also extends beyond the animate to include abiotic bodies such as rivers, rocks, and soils. <u>Anim</u>ism, so to speak, involves the processual generation of meaningful and agentive bodies in relation to other bodies—only some of which are human and not all of which are benign. Marind ways of becoming <u>anim</u> thus suggest that the problem with the dividual as an analytical category may have less to do with its assumption that Melanesian persons are radically different to other kinds of persons, or that relational persons consequently lack individual agency. Instead, Marind ways of becoming <u>anim</u> point to the need to push the concepts of agency and alterity *further* to encompass the array of other-than-human elements and entities whom Marind persons more or less actively—and willingly—become-with.

At the same time, the secrecy of individual minds complicates the more-than-human relationalities that together produce Marind persons. Edward LiPuma's (2000) theorization of Melanesian personhood is useful in untangling this relationship. Drawing from his fieldwork among the Maring of Highland Papua New Guinea, LiPuma describes how modernity brought to the fore individual aspects of Maring personhood that were traditionally marginalized in local valuations of the social self in favor of relational modes of sociality. Persons, whether Melanesian or other, are thus neither fully dividual nor individual, according to LiPuma. Rather, persons are constituted in and through the tension between differentially valued aspects of the self that are always already present but alternately foregrounded or backgrounded in different social and historical contexts.

A similar dynamic shapes the relationship between the legible body and the secret mind among Upper Bian Marind. On one hand, physical interactions in the form of shared labor, physical contact, and the exchange of skin and wetness foreground the *dividual* dimensions of organisms constituted through their bodily interactions. In the absence of access to the interiority of others, encounters through the skin-as-surface become the central modality of intra- and interspecies relations. On the other hand, respect for the secrecy of others' thoughts and intentions foregrounds the mind as the *individual* dimension of living beings as autonomous entities—even as the mind may be inaccessible to the self itself. Building fleshly relations with other beings as dividuals thus requires scrupulously refraining from undermining their psychic agency as in- dividuals, in what might be called an ethos of *restrained care*.

Yet as shape-shifting highlights, the relationship of the body to the mind as the dividual and individual dimensions of the self is inherently ambiguous. The mind of skin-changers may be captured unwittingly by the skin that they adopt such that they gradually lose the capacity to interact with other anim as anim. Shape-shifters like Okto are no longer individuals in the sense of independent and self-reliant agents of free choice. This is because the self they rely on is no longer fully human nor fully under their control. In other words, the bodiliness in which dividuality is contained both enables and en- dangers the personhood of those who journey across species boundaries at their pleasure and peril.

The cultivation of bodily relations, the respect for secret minds, and the risks of shape-shifting among Marind also demonstrate that the contextual fore- grounding and backgrounding of aspects of the self theorized by LiPuma does not necessarily require or arise solely in the context of intervening external factors, such as globalization, colonialism, and capitalism (see, for instance, Dwyer and Minnegal 2007; Hess 2006; Mosko 2013; Robbins 2004). Among Marind, the challenge of finding a balance between affective engagement and respectful disengagement in the hard work of making anim stems directly from the tension between the awkwardly coexistent dividual and individual dimen- sions inherent in all sentient beings from the outset. Otherness *as* a relation operates *within* the self because the capacity of anim to control and ascertain their own mental lives is inherently tenuous. At the same time, this tension is heightened in the forest, where anim constitute just one of many *kinds* of agen- tive entities. As the lethal lure of shape-shifting starkly reveals, making anim is never just a human affair, but rather always a precarious, abu-abu negotiation distributed across species, skin, and substance.

Deforestation and agribusiness expansion pose a further peril to the hard work of becoming anim in a world where selves come into being through their relations to human and nonhuman others. Fleshly communions achieved through exchanges of skin and wetness transform into deadly contagions with the arrival of oil palm and its cloud of chemical companions. Infiltrating the all-too-porous bodies of rivers, soils, and living organisms, these capitalist nomads challenge the capacity of Marind to retain their humanity in an increasingly posthuman ecology, engineered and exploited by deadly biocapitalist regimes. Marind thus experience the troubled waters of global capitalism as a visceral crisis of wetness at once embodied and environmental, symbolic and material, immediate and latent, dehumanizing *and* posthuman. Technoscientific and agroindustrial flows, as incarnated in the figure of oil palm and its noxious biochemical allies, subvert life-generating transfers of substance in and across species lines. These lethal flows replace interspecies connectivities and lively topographies with relations and landscapes of radical disjuncture. They erode the elements and beings that are fundamental to Marind ways of becoming human with, and through, myriad more-than-human others. These dehumanizing effects are exacerbated by the abu-abu provenance, affordances, and relations of oil palm itself—a plant whose opaque ontology I explore in the chapters to come. As Okto himself, in uncanny mimesis of his own transformation, told me just a few days before his passing, "The problem with oil palm is that we do not know its skin."

4. The Plastic Cassowary

The village of Khalaoyam was home to a juvenile cassowary. In August 2015, a group of women discovered the remnants of its nest with three eggs inside in a newly dug irrigation ditch, amid the smoldering remains of a forest that had been leveled and burned to make way for an 18,000-hectare oil palm plantation. Certain that the hatchlings would die with no father cassowary to nest or brood the eggs, the women carried the eggs back to the village, and incubated them in a vat of raw rice. Two rotted away but one held on. Seven weeks later, a tiny, scraggly chick squirmed out of the pale blue-green shell. The villagers named him Ruben.

Ruben was a shy creature, who was gangly on his skinny legs. He had a brown-and-cream striped plumage, a tiny beige wattle that always seemed somewhat of a nuisance, and a barely noticeable casque under the tufts of light feathers crowning his angular head. Ruben often roamed around the settlement, his fragile whistle wafting in the mid-afternoon air as a melodic backdrop to villagers' conversations. Community members created a resting place for the young cassowary in the back yard of the local church, where the Dayak priest from

Figure 4.1. Ruben, the plastic cassowary. Photo by Sophie Chao.

Kalimantan kept chickens and doves. Marind describe these birds as "settlers" (pendatang) because, unlike forest animals, they come from outside Papua and have never lived in the wild. The chickens often chased angrily after Ruben when he tried to eat the food scattered by villagers. The cockerel would peck at him viciously when he got too close to the hens or sometimes for no apparent reason at all. Although he towered over his avian yard-mates, Ruben never attempted to defend himself. Instead, he ran and hid between his keepers' legs or sought shelter in a cardboard box that once held instant noodles, where he waited patiently for the cockerel to calm down.

At first, the community tried to feed Ruben forest foods, such as fruit, berries, and small invertebrates. When Ruben refused to eat, the villagers tried to feed him sago, their staple starch. Eventually, they realized that Ruben would only accept purchased foods such as raw rice, cookies, and instant noodles. Community members frequently coaxed the bird to the edge of the forest, hoping he would return to the wild—back to his kin and home. But Ruben always retreated to the yard or crouched behind the stone well, his feathers ruffled and his meager frame trembling slightly. Every day, I fed Ruben leftovers and bathed him in a plastic bucket. I found him rather sweet in his gawkiness and felt sorry for the many attacks he suffered in the yard. I considered him a pet-like creature because he was domesticated and made dependent on humans. I believed Ruben deserved particular care because he had lost his forest and kin to the rampant devastation wrought by oil palm expansion. Seeing my friends face similar threats to their lands and livelihoods, I assumed they felt the same way about him.

One evening, I was sitting in the back porch of the church with a group of men and women after a hearty meal of sago jelly and fish soup. Everyone was relaxed, smoking clove cigarettes, and chatting animatedly. During a momentary lull in the conversation, Ruben's shy whistle echoed through the night. I smiled and commented on how sweet his song was and how lucky we were to have such a cute pet among us. I glanced around at my friends and, to my surprise, faces that only an instant ago had been happy and smiling were now somber. Elena, an old woman employed by the priest to tend the chickens, raised her head and sighed deeply, her eyes moist. She said: "This is no song, namuk (sister). This is a weeping. This is the cry of the cassowary. Can you not hear the sadness, anak (child)? Does it not rip through your heart with the speed of a hardwood ngef (Arenga pinnata) arrow? We hear only a weeping, a lament. We feel the grief of the khei (cassowary) as it seeps through our skin and bone. We hear death and mourning in its call. No longer wild (liar) or free (bebas), the cassowary has become plastic."

Sitting beside her, old Marcus began to sing, swaying side to side, his head bent low:

> Without a father, you were born
> Without a forest, you were found
> Khei, khei, wither your kin, wither your home?
> Khei, khei, your call, once proud and glorious
> Its rumble was the earth shaking
> Today, you are an orphan
> For food you beg, for water you beg
> Khei, when will you sing again?
> When will you roam the forest free again?
> When will your call boom across the land?
> When will your solitary journeys resume?
> Wild, you were, strong you were
> Khei, khei, orphan of the land you are
> Without kin, and without home you are
> Now, you eat raskin (poor peoples' rice) and Indomie (instant noodles)
> You take baths and know the sound of HP (mobile phones)
> Like us, you turn plastic
> Like us, you have become modern
> A pendatang (settler) far from home

Khei, your cry fills my old bones with grief
Without a father, you were born
Without a forest, you were found
Khei, khei, wither your kin, wither your home?[1]

In this chapter, I examine the ambivalent ontology of animals who once thrived in the forest and now dwell in the village. Ruben is one of several creatures to have found refuge in the settlements of the Upper Bian following the destruction of his native habitat to make way for monocrop oil palm plantations. In Bayau, for instance, I encountered a pair of New Guinea crocodiles confined in an unused outdoor water trough and a lame deer tied to the foot of a mango tree. In Mirav, a bright-eyed spotted possum was housed in a makeshift cage hanging precariously from a gnarled jackfruit tree. In Khalaoyam, Ruben the cassowary lived alongside three young tree kangaroos, who always wandered around the village together. With some two million hectares of land in Merauke slated for conversion to agribusiness concessions, today many more creatures approach the settlements in search of shelter and subsistence.

As Elena's words and Marcus's song poignantly convey, salvaged animals provoke sadness and pity among their keepers. These creatures and the humans who tend them suffer from the effects of environmental degradation in ways that transcend their differences as species. Together, they form what Deborah Bird Rose calls "multispecies communities of fate" (2011, 91, 142). Yet Upper Bian Marind are deeply reluctant to cultivate relations of care with domesticates and find their tame and human-like behavior unsettling. For instance, villagers actively avoid forming affective attachments with animals in the village, frequently mock them, and constantly encourage them to return to their native habitats. Many of these creatures were not actively sought out by community members, but rather approached the village seemingly of their own accord and in unpredictable numbers. For instance, the deer entered Bayau one morning after a nearby plantation operator started burning the forest to make way for oil palm. The crocodiles crawled into the village square when toxic effluents were released into the river by an adjacent palm oil refinery. While Ruben differed in that he was rescued from the forest, the women who brought him to Khalaoyam affirmed they did so only because leaving him in the forest meant certain death. Yaref, a young man from Mirav, explained the position of Marind vis-à-vis the village animals as follows: "We are forced to care for animals that come to our village. We must save them from the forest, because the forest is disappearing. We must feed them, bathe them, and keep them in the village. But we have no choice. If they go into the plantations, they will die of hunger. But Ruben and the others—they don't belong in the village." For Yaref and many others, creatures who take up residence in the village impose themselves in a space where they do not belong. Far from claiming to own these animals, community members emphasize that animals come to *them* for help and that Marind have "no choice" but to accept them. Villagers often said they did not know how to interpret these animals' behavior or deal with their needs, which they often ignored. For instance, on one occasion Ruben suffered a deep wound after a particularly ferocious attack from the cockerel. I was concerned for Ruben's health, but the community members seemed indifferent to his plight. When the cut later became infected, I purchased antiseptics from the local clinic and visited Ruben every morning and evening to clean his wound. The community observed my actions with dubious expressions, saying that it did not seem right to heal a cassowary's skin with medicine intended for humans.

People worried when domesticates showed signs of dependence on humans for food and attention and even more so when they refused to return to the forest. They also actively discouraged physical interactions and expressions of emotional attachment toward village animals. For instance, adults admonished children for playing with, cuddling, or stroking the deer and the possum. Villagers avoided the tree kangaroos unless they had to feed them. Yaref and his relatives often kicked Ruben away when he approached them for protection from the cockerel's attacks. Village animals, my friends affirmed, do not merit care in the way domesticated pigs and dogs do because they serve no practical function. They cannot be eaten because they were not hunted, and therefore the rituals required to consume them have not been carried out. Furthermore, these critters display strangely un-animal-like behaviors. For instance, they take baths, eat cookies, and in Ruben's case, even weep.

The similarities that Marind identify between the comportment of animals and that of humans make the idea of eating them incongruous and repulsive. As Geraldina, a young woman from Mirav, put it, "These animals no longer have animal skin or wetness. We cannot eat them because we would not know what we were eating. A possum? A human? It is not clear. These creatures, they are abu-abu." At the same time, villagers worried that keeping animals in the village would adversely affect their hunting and foraging activities in the forest. For instance, Khalaoyam men feared Ruben's wild kin would attack or injure them as a punishment for taking one of their young. They consequently refrained from hunting cassowary for several months following Ruben's hatching. Similar avoidances were maintained by villagers in Bayau and Mirav, where the crocodiles, deer, and possum are kept.

If, as Geraldina suggested, domesticates are not food, then neither are they friends. Marind do not consider village animals pets, and indeed no Marind equivalent exists for the English word "pet." Instead, they refer to them as settlers (pendatang), refugees (pengungsi), or animals that are no longer wild, no longer free, and have "become plastic." When I explained to Elena and Marcus the Western notion of pets as friendly domestic companions, they were perplexed and found the idea of taming animals strange. Like Geraldina, many community members described village animals themselves as abu-abu. "Creatures like Ruben," my host sister Elizabeth told me, "are confusing. They are not pets. They are a problem." Ruben was a particular problem for Khalaoyam villagers who feared he would become increasingly violent as he matured. Unlike other New Guinean societies, Marind do not traditionally rear cassowaries because they are associated with the deep forest (hutan dalam) and are considered dangerous. The bird's large size and lethal kick make it difficult to tame, and indeed to hunt. Unlike pigs and dogs, cassowaries are unpredictable in their behavior, do not respond well to human care, and cannot be allowed to wander the village freely. They do not breed in captivity and they die young. These elements suggest to Marind that cassowaries are best left in the forest where they rightfully belong.

The discomfort of Marind around domesticates arises from the way these creatures' comportment—and the relations they demand of humans—disrupt the capacity of forest organisms to behave and interact as relational yet autonomous beings. Forest animals and plants participate in Marind becoming-withs through sustained and mutual exchanges of skin and wetness. These animals and plants are considered to be agentive entities who are endowed with attributes and capacities that are specific to their bodily form as a species. When in the state of "wildness" (liar), animals can harness their species-specific abilities to roam the forest, hunt, forage, and reproduce. For instance, cassowaries use their mighty speed to travel the forest in search of fruit. The sinewy limbs of crocodiles enable them to meander effortlessly down the serpentine flow of the Bian River, while the dexterity of possums allows them to find shelter in the dense canopy, where they mate and nurture their young. Being "wild" thus entails being "free" (bebas) to live out one's species-specific aptitudes, and Marind use both terms interchangeably to describe animals in the forest.

Humans must respect the wildness of animals as free and autonomous beings. This is the element of restraint that interspecies care demands. To this end, Marind actively refrain from overly influencing or controlling the movements, growth, and reproduction of forest species. Instead, community members seek to enhance the *environment* of plants and animals in ways that are conducive to these organisms' autonomous thriving, or what archeologist Les Groube (1989, 300–301) calls *strategies of minimal manipulation*. For instance, villagers clear pathways for pigs and deer to travel to water catchments, leave fruit and nuts behind when foraging for cassowaries to feed on, and avoid disturbing the canopy during birds' mating season. These minimal manipulations make the forest a realm of natural abundance achieved through a range of indirect human actions that are situated within a wider web of past and other-than-human activities. The material intimacies of human-animal relatedness thus go hand in hand with a need to respect their different *modalities* of being. In this light, being wild means being able to live autonomously yet in relation to others as part of multispecies social and ecological networks, or what I call an ethos of "restrained care."[2]

In line with the principle of restrained care, Upper Bian Marind rarely rear animals (or for that matter, cultivate plants) because, in the words of Arthur, an elder, "it takes away their freedom." For instance, and in notable contrast to many other New Guinean societies, Upper Bian Marind do not engage in gardening nor tend to vegetable or fruit plots.[3] The only exception is kava (*Piper methysticum*)—an otherwise widely popular Pacific root crop—which Marind only cultivate occasionally and primarily for the purposes of feasts and clan-wide congregations, where it serves as a ritual drink. Over the last decade, agribusiness corporations have funded and initiated several small-scale horticultural projects—sweet potatoes, spinach, and various fruit trees—in the villages as part of their Corporate Social Responsibility schemes (see Amo 2014). These plots, however, are by and large neglected because cultivation is said to violate the autonomy of plant organisms and is seen as an inadequate substitute for wild forest foods and the cultural practices entailed in their procurement, preparation, and exchange.[4]

Similarly, Upper Bian Marind practice very limited pig rearing compared to their coastal Marind neighbors and other societies across Melanesia.[5] Pigs are captured in the forest primarily to use in rituals and castrated rather than bred to increase their size and weight. As pigs belong to the wild, the number of individual pigs kept in the village and the amount of time they spend there are subject to stringent customary laws. Furthermore, numerous rituals must be undertaken before and after their slaughter to avoid violent retribution from their wild porcine kin. Other domesticates in the Upper Bian include dogs, which are used as hunting aids, and chickens, which are kept by the local priest for consumption during church events. Chickens do not figure in Marind cosmology and are considered settlers because they come from Java and have never lived in the forest. Dogs are said to belong in the village and are petted by adults and children alike. They are not owned by specific individuals or families, and emotional attachments with them are sustained only so long as they perform their role in hunting activities. Practical forms of care enacted toward domesticated pigs, dogs, and chickens, such as feeding, petting, and housing them, are justified because they serve a specific *function*—as food for special events or as hunting companions. However, strong affective ties with these creatures are generally discouraged because they can become hindrances when pigs and chickens must eventually be slaughtered or exchanged or when dogs must be punished for failing to make themselves useful during hunts. Again—this is care, but with restraint.

No longer wild or free, animals like Ruben lose their autonomous agency and species-specific capacities when they become dependent on the humans who surround and control them. They begin to eat and behave like humans and acquire human names—Ruben the cassowary, Mina the deer, and Theo the possum, for instance. Severed from their kin and home, these critters make their *domus*, or home—the etymological root of *domestication*—in the village, and find shelter in manufactured instant noodle boxes, courtyards, wooden crates, water troughs, and wells. Their bodies are frail and weak because they no longer hunt, forage, or eat forest foods. For instance, community members commented that the glossy fur of the deer had turned dull and patchy since it took up residence in Bayau. The possum's eyesight had become so poor it barely reacted to the movements of humans around it. The tree kangaroos who once leaped across the canopy now hopped with a slight limp along dusty village paths. As Elena and Marcus lamented, Ruben would likely never roam the forest freely again. Instead, he knew only the taste of rice and the sound of mobile phones. Unlike his kin in the wild, whose resounding call boomed across the forest, Ruben was skinny and awkward. He had forgotten how to sing and instead begun to weep—like a human.

These transformations are not celebrated by Marind as an achievement of human mastery over nature. Tamed creatures provide no emotional or material gratification to their keepers. Their human-like comportment—a prevalent attribute of pets in the West—gives rise to anxiety rather than heightened intimacy. Neither is domestication seen to benefit the animals, who are instead pitied and reluctantly cared for. The trouble with domestication, to invert the phrase of historian William Cronon (1996), is that it violates the ontic alterity of forest beings. In an uncanny reversal of the perspectival capture of shapeshifters by their animal skins described in the previous chapter, creatures like Ruben are no longer able to self-identify and interact *as* animals with their own kin and other forest beings. Relations of reciprocal capture are superseded by relations of one-sided *captivity* as animals become physically trapped in the space of the human village and ontically trapped in the habitus of human-like domesticates.[6] Rather than petting and nurturing them as human-like kin, Marind cultivate what Juno Parreñas (2018, 33–59) calls "relations of rejection" with animals *as domesticates* because domestication subverts the moral relations of humans to animals *as autonomous selves* in the forest.

The tragic fate of domesticates has political implications for their human keepers. Many of my companions, for instance, compared the loss of wildness of forest creatures to the freedoms denied to them as West Papuans under Indonesian rule. Like animals constrained to live in the village under the control of humans, West Papuans were forcefully incorporated into the Indonesian nation-state following the controversial Act of Free Choice of 1969 and have had little say in the numerous top-down projects that have undermined their livelihoods, culture, and environment. Just as Ruben can no longer roam the forest freely, Marinds' movements across the landscape are constrained by the tentacular control emanating from omnipresent state and corporate pressure points—plantations, military garrisons, and roads. Much as village animals are reduced to begging for food and water from their keepers, rural Marind are increasingly dependent on financial aid and food handouts from the government and from agribusiness companies to whom they have ceded their customary lands. Meanwhile, deforestation and agroindustrial development have prompted a radical shift from traditional forest-based modes of subsistence to a diet of commodified and processed foods—which is experienced by both Marind villagers and domesticated animals. The threat of extinction faced by forest organisms in the context of oil palm expansion and the incapacity of village organisms to reproduce with their own kind also resonate with the self-positioning of many Marind as victims of a government-endorsed genocide. Growing population dilution, forced sterilization, and cultural assimilation are often cited by villagers as evidence that the Indonesian government is on a mission to eliminate Indigenous Papuans. And with state-endorsed agribusiness projects multiplying relentlessly, both humans and animals face the dire prospect of a rapidly disappearing forest to live in and from. As Gerfacius, an elder from Bayau, put it, "Ruben, me, Marind people—we are all the same. We are all victims of the government and of oil palm."

For many Marind like Gerfacius, village animals are powerful icons of self-determination denied both within and across species lines. The forced displacement and arrested autonomy of these creatures evoke to Marind the violation of their own right to sovereignty as coerced subjects of political taming on the part of a paternalistic state, whose developmental and nationalistic rhetoric and practices recast the wildness of Marind as archaic primitivism in need of eradication and that of their forests as dormant resourcefulness awaiting exploitation. Together, animals and humans partake in interspecies memoria passionis, a term used by West Papuans to denote their collective history of suffering and oppression under Indonesian rule (see Hernawan and van den Broek 1999).[7] In this light, Ruben's wounding cry can be conceived as a lament for the wildness lost by animals *and* a poignant evocation of the fate of their keepers, whose freedoms are denied by a state determined to control their future. In sharing the bird's grief, Elena and her companions were mourning not just *for* domesticates, but also *with* them (van Dooren 2014, 143). By refusing to pet village creatures and exhorting them to return to the forest, community members thus seek to reverse the troubling transformation of animals from autonomous agents to captive dependents, while also expressing in a veiled medium their own desire for political and cultural emancipation.[8]

DURING LUNCH BREAKS AT the primary school where I volunteered as an English instructor, my colleague Barnabus would often joke that Ruben the cassowary was "born in rice and lives off rice. Ruben," Barnabus would continue in jest, "is the first modern cassowary to walk Marind soil. He doesn't know how to sing like a cassowary or speak cassowary tongue. Who knows, someday, this pendatang (settler) might start teaching us Javanese!" As Barnabus's comments suggest, the troublingly human-like attributes of village animals are further heightened by the particular *kind* of human they resemble—pendatang, or settlers. The term pendatang, which literally means "arrival" or "comer," is used in Indonesia primarily to refer to non-Papuan migrants who have relocated to the lesser-populated regions of the archipelago through government-sponsored programs or spontaneous migration.[9] Like many West Papuans, Marind resent the presence of settlers and describe their relations with them in terms of mutual dislike and avoidance. Pendatang live in settlements established on customary lands without the consent of local landowners. They increasingly outnumber Papuans in Merauke, monopolize formal employment in the villages, and own most kiosks and public transport. For this reason, many Marind associate settlers with money, capitalism, and business, and more specifically with oil palm monocrops, which have prompted a renewed influx of migrants into Merauke in the last decade to work as plantation laborers.

Many Upper Bian Marind consider settlers deeply untrustworthy because they work for or collude with the police and military—state organs that routinely harass, intimidate, interrogate, and incarcerate Marind community members. Cultural differences also shape Marinds' characterization of pendatang. While Marind are predominantly Catholic or Protestant, settlers are primarily Muslim, a creed that many like Barnabus associate with Java, the heart of political and economic power in Indonesia, and with Javanese people, Merauke's single largest non-Papuan ethnic group. In contrast to the ancestral relationships, care, and respect that Marind entertain with forest species, settlers are said to poach game and fell precious woods indiscriminately, without asking permission from Marind clans or undertaking the required rituals. My companions also affirmed that settlers consider them backward because of their forest-based livelihoods, lack of formal education, and dark skin. Some said pendatang avoid coming into physical contact with them because they assume Papuans are unhygienic, ridden with tropical diseases, or infected with HIV. Several villagers reported having been refused goods and treatment at the local kiosks and clinics for these reasons.

For many of my companions, similarities between non-Papuan settlers and settler-like animals express their common status as aliens. For instance, neither are native to the environments they inhabit—Indonesian settlers in West Papua and animals in the human world of the village. More and more pendatang are moving into Merauke and more and more animals are turning into domesticates. Village creatures eat rice, instant noodles, and cookies—commodities sold and consumed primarily by pendatang. Ruben himself, as Barnabus joked, was born in a vat of rice—the staple starch and agricultural crop cultivated by settlers. His inability to sing like a cassowary suggested to Barnabus that the bird would rather speak Javanese—the language and culture Marind associate with pendatang. And just as Indonesian settlers disrupt Marind ecologies and ways of life by appropriating their lands and resources, domesticates invade the village and exhibit alien habits that are deeply unsettling to their keepers. Endowed with a foreign ethnic identity as pendatang, domesticates thus constitute living vectors of the social differences and disparities underlying Marinds' antagonistic relations with migrant populations. The troubling parallels between domesticates and settlers are further exacerbated by their shared association with oil palm. The proliferation of monocrops drives animals to the villages and brings settlers to the Upper Bian. The plant was introduced by national Indonesian corporations and is cultivated by non-Papuan workers. Foreign and domesticated, oil palm too is an uninvited settler in Merauke.[10]

At the same time, and much like the indexicalities at play in the transformation of beings from wild to domesticated, animals' behaviors give rise to morally imbued reflections among community members about their *own* transformations from native to settler. Geronimo, a widower from Mirav, once told me, "Ruben used to be native. Now, he has turned into a settler. There are few native cassowaries left in the forest. There are few native Marind left in Merauke. All of us are becoming pendatang." Just as Ruben has forgotten how to sing, children forget how to speak Marind and respond to their parents in Indonesian. These settler-like children refuse to eat sago and instead want instant noodles and rice, which make their bodies weak. Like Ruben, who was born in a vat of rice, Marind children are born in state clinics rather than in the forest, as custom requires. Resonating with Martinican political philosopher Franz Fanon's (2008, 4) notion of "epidermalization," or the desire to possess and inhabit the skin of putatively superior others, young Marind men and women increasingly use urban cosmetics and whitening products to "beautify" themselves (jadi ganteng, jadi cantik, or "becoming handsome, becoming pretty"). In doing so, they lose their authentic Papuan traits of curly hair and dark skin and begin to resemble Javanese settlers. Similarly, community members become pendatang when they sell their land or seek employment on the plantations because, like village creatures, they forget how to live in the forest in the company of their plant and animal kin. By taking on pendatang habits, Marind lose their native identity as West Papuans and acquire an Indonesian identity. Consequently, as Geronimo suggested, few organisms today—human or other—can claim to be truly wild and native. "One day," he speculated, "we will no longer know wildness. We will no longer know anything but settler life. We will forget what freedom tastes like. The whole world is becoming plastic."

Geronimo's comment highlights the classificatory logic of contrasts that operates in the simultaneous transformation of animals and humans from wild to domesticated and from native to alien. These concomitant transformations are associated by Marind with modernity, a condition described by Geronimo and others through the idiom of "plastic." As suggested earlier by my companions' critique of the lifeless and hegemonic perspective of drones, or "plastic birds," plastic things and beings more generally are often associated by Upper Bian Marind with deception. For instance, modern foods like instant noodles and candy come in glossy plastic packaging but are never satisfying. Government and corporate representatives, wrapped in fashionable clothes, pledge to support community development, but fail to deliver on their promises. Villagers adopt modern habits and become fixated on material wealth, even as the items they purchase fail to gratify them. Plastic birds offer an all-encompassing perspective from above ground but they are ultimately blind to the myriad goings-on of interspecies life across soil, water, and air.

Upper Bian Marind also associate plastic things and beings with uncontrolled proliferation. For instance, more and more plastic goods enter the villages, more and more pendatang move into Merauke, and more and more young Marind turn to a modern way of life. State officials and company staff intervene incessantly in village affairs to introduce yet more agribusiness projects, which they promote as key to the modernization of West Papua and its "primitive tribes" (suku terlantar). As Geronimo suggested, Upper Bian community members can do little to stop the flow of plastic things and beings into their world. In a world of abu-abu, plastic, as STS scholar Jody Roberts puts it, incarnates "the futility of resistance and the inevitability of accommodation" to alien yet imposed modernities (2010, 104).

Domesticates represent and participate in a world dominated by plastic. For instance, Marcus sings about Ruben eating modern foods wrapped in plastic, such as instant noodles and cookies, bathing in plastic buckets, and recognizing the sound of modern products like mobile phones. Domesticates are also deceptive. They may look like crocodiles, deer, possums, tree kangaroos, or cassowaries in form, but they behave like human pendatang. Like plastic, domesticates are inedible because they were not hunted and therefore cannot be consumed. Just as the accumulation of plastic defies the logic of decay, regrowth, and transformation, humans and tamed creatures no longer participate in the cosmic economy of sharing that animates the forest realm through life-sustaining exchanges of bodily flesh and fluids. The sterility-inducing effects of plastic also resonate ominously with the fate of village animals who are confined to human settlements and therefore cannot perpetuate their kind in the forest. As deforestation continues, domesticates multiply in the villages. Like plastic wrappers that will not biodegrade, no one knows what to do with these alien beings who refuse to go away.

Marind theories of plastic and its local manifestations also bring to mind broader ecological concerns with the physical accumulation of plastic at a planetary scale. Plastic is an omnipresent material of almost inconceivable spatial and temporal dimensions, or what philosopher Timothy Morton (2013) calls a hyperobject. With more than eight billion tons produced so far and twelve billion tons expected by 2050, no place on earth today can claim to be plastic free. Indeed, the ubiquity and impact of plastic have earned it the status of a possible geological indicator of the Anthropocene, or what journalist and author Christina Reed (2015) terms the *Plasticene*. The whole world, as Geronimo put it, really *is* becoming plastic.

In the Upper Bian, plastic pendatang, plantations, products, and pets relentlessly invade and multiply, much like plastic substances with as-yet-unknown effects penetrate the planet's soils, oceans, and atmosphere. Like settlers and the Indonesian State, who refuse to disappear from Marind lands, and domesticates, who impose themselves in the village, plastic does not biodegrade but rather breaks down into imperceptible but troublingly resilient fragments. It also contaminates the flesh and fluids that enliven and connect forest organisms—their blood, urine, fat, breastmilk, and organic tissues. In describing modern beings and objects as plastic, Marind thus recognize a more widespread problem with the materiality of modernity—its troubling permanence and proliferation. And just as ambiguity surrounds how we can and should live in a world of planetary plastic cobecomings, so too uncertainty, or abu-abu, surrounds cassowaries like Ruben, who not only burden Marind with the responsibility of their survival, but also remind Marind of their own fraught efforts to engage with modernity and its plastic manifestations.

SARA, A YOUNG WOMAN from Khalaoyam, watched Ruben as he struggled to untangle his bony feet from strands of the instant noodles she had scattered for him. The young bird eventually tripped up and collapsed onto his side. The incident drew peals of laughter from nearby community members, who jeered at the plastic cassowary for not being able to stand on his own two legs. But Ruben seemed unfazed by his fall and the villagers' mockery. Instead, he began to chew voraciously at the mushy noodles around him. Sara turned to me and said anxiously, "Maybe there is no hope for merdeka (independence), no matter how hard we struggle. Maybe Ruben will never return to the forest. Maybe some of us have already given up. We try to make Ruben return to the forest and become wild again. But what can we do if Ruben himself does not want to be free?" Many community members express pity, sadness, and compassion toward animals who have been alienated from their native environments and demoted from wild to domesticated beings because their plight resonates with Marinds' own experiences of subjection to state control. In this light, the villagers' efforts to return animals to the forest can be conceived as an attempt to liberate animals from their dependence on humans and as a symbolic enactment of Marinds' own unfulfilled dreams of sovereignty. However, animal-human relations in the village also give rise to widespread frustration and anxiety, which stem from ambivalence surrounding the *intentions* of domesticates and their keepers.

On the one hand, Marind say animals are forced to seek human help because their forest is being destroyed. Yet many villagers also note that domesticates seem to relish being in the village and living a modern, settler way of life. For instance, community members' efforts to coax Ruben back to the wild were met with resistance. He chose human food over cassowary food and flapped around excitedly when it was time for his warm-water bath. Rather than retaliating against the attacks of the cockerel with his strong legs, the young bird invariably turned to humans for protection. Similarly, when Bayau villagers carefully transported the crocodiles to the river to set them free, the reptiles refused stubbornly to set foot in the water. Meanwhile, women in Mirav described how the possum seemed to prefer dozing in their laps rather than in the treetops. These behaviors bring villagers to speculate at length over whether animals are truly forced into subjection or whether they in fact enjoy being domesticated. Both possibilities are equally disturbing. Coercion suggests there is no hope for animals to become wild again, while compliance suggests that domesticates are resigned to, or even embrace, their subjection. As Sara suggests, it is not just than animals have been robbed of their freedom but also that they appear to no longer *want* to be free. In other words, the village animals' bodies and behaviors are just as abu-abu as their intentions.

Again, both interpretations have ominous implications for Marind themselves in the context of West Papua's struggle for political independence. Up to the year 2000, Marind were actively involved in the pursuit of Papuan autonomy, including as members of OPM, the Free Papua Movement. This involvement culminated with the "Bleeding of Merauke" (Merauke berdarah), a violent attack by the military on civilians in Merauke City on December 3, 2000, during which twenty people were killed or severely wounded. Since then, Marind engagement in political activism has dwindled and OPM presence in the region is practically nonexistent. Village elders in the Upper Bian often contrasted the political apathy of contemporary Marind with the resilient resistance movements of Highland Papuans and attributed this to young Marinds' desire for material wealth rather than political autonomy. As Nikolaus, an elder from Bayau, complained: "The young generations are too lazy to fight for merdeka. They just want money and mobile phones. And those are easier to obtain than merdeka. Merdeka is not like rice or instant noodles. You cannot buy merdeka. You must bleed for it. But now, Marind don't fight for more freedom. They fight for more money." Nikolaus and Sara's statements communicate a sense of futility or hopelessness concerning West Papuan autonomy and self-determination. Just like village animals who refuse to return to the forest, many Marind have given up the fight for political freedom and have resigned to their ensnarement within the Indonesian State—much as Ruben symbolically surrenders to the mess of instant noodles trapping his bony legs. In suggesting that villagers have succumbed to the lure of money and material comfort, Nikolaus also highlights Marinds' awkward relationship to modernity as a way of life at once fetishized and reviled (cf. Rutherford 2000, 2003; Stasch 2016). On the one hand, community members condemn the shiny new world promised by modernization projects and their plastic manifestations, because it creates an impression of abundance that is illusory and short-lived. Yet just as village animals appear to enjoy novel experiences like taking baths and the sound of mobile phone ringtones, modern life is also an object of ambivalent longing for many villagers.

For instance, some of my companions told me they envied modern pendatang for their cars, their relative wealth, and their knowledge of the world outside Merauke. Like Ruben who appears to enjoy devouring the instant noodles tangled around his feet, many Marind are drawn to manufactured and foreign products entering the villages—instant soft-drink mixes, laundry powder, and Tiger Balm salve. Some community members are tempted by the prospect of cash income from employment in the plantations, which will allow them to send their children to better schools in Jayapura or Jakarta. For young Marind keen to leave the villages and make a career in the cities, modern things, technologies, and practices become particularly potent sources of agency and objects of desire. As Mira, a girl in her twenties who hoped someday to train as a nurse, told me, "I want progress. I want to keep up with the rest of the world. We cannot be left behind. Look—even Ruben knows what a mobile phone is, and what it sounds like, even though he is just a cassowary. Even Ruben must become modern to survive."

At the same time, many of my friends pointed out that animals' attempts to behave in modern ways were fraught with failure and violence. For instance, villagers frequently noted that Ruben seemed unable to integrate into the avian community of the yard. Watching the bird suffer yet another vicious pecking by the cockerel, Darius, a Khalaoyam elder, said sadly:

> Ruben wants to be like other pendatang. He tries to eat and walk like them. But still he is not accepted, and they won't share their food with him. Ruben will never be a real pendatang, no matter how much rice he eats. He was born to be in the forest—not the village. He tries so hard to be like the chickens, who are real pendatang because they come from outside Papua and have never lived in the wild. But Ruben is different. He should be in the forest. Ruben is a nuisance to them. And he is a nuisance to us, too.

Darius's words point to the ambiguous merging of settler and Papuan positions in the figure of the domesticate. On the one hand, animals like Ruben are a nuisance to Marind because, like settlers, they impose themselves in a space they do not belong to, and behave in an alien, modern manner. Yet Marind also empathize with animals who are forcefully displaced from the forest in light of their own experiences of dispossession and minoritization as a result of migrant influx. The contradictory amalgamation of Papuan and settler identities in domesticates is further problematized by the fact that many Marind *aspire* to a settler-like way of life—even as it remains largely beyond their reach. Indeed, like the violent rebuff that Ruben suffers in attempting to bond with the hens and the cockerel, Marind too face rejection in their attempts to survive, in Mira's words, by partaking in projects or activities associated with modernity. For instance, their efforts to interact with pendatang face limited success because pendatang widely deem them savage and uncivilized. Marind who attempt to participate in formal employment suffer entrenched discrimination from Indonesian settlers, who recast their dark skin, glossy wetness, and forest-based livelihoods as symptoms of racial inferiority and cultural primitivism. Indigenous landowners surrender their customary territories in the hope of social welfare schemes and compensation that turn out to be short-lived or unfulfilled. Like Ruben, who was born in a vat of broken, subsidized rice—known locally and officially as "poor peoples' rice," or beras miskin—Marind face poverty and food insecurity in remote villages where government funds rarely materialize. Just as the chickens Darius observed refuse to accept Ruben as a messmate, Marind remain excluded from the sites and circuits of wealth and opportunity promised by modernity and its deceptively plastic promises. As such, it is not just the aspirations of humans and animals that are abu-abu, but also their actual capacity to fulfil them.

A final layer of ambivalence arises in relation to Marind as animal keepers and as subjects of the state. Alfons, a young man from Khalaoyam, once compared these two roles as follows: "We control Ruben—what he eats, where he goes, where he sleeps. So, when we control Ruben, we become like the government. We don't like the government. But we control animals, just like the government controls us." Alfons's statement can be understood in terms of a relation between oppressed and oppressor. On the one hand, Marind and domesticates are victims, as Gerfacius put it, in the face of ecological and cultural transformations imposed by the state in the role of oppressor. These shared experiences of subjugation give rise to sentiments of pity and compassion toward animals on the part of the keepers. At the same time, domesticates are problematic because they impose themselves in the village and force Marind to become their caregivers. Human control is necessary for these animals to survive, but it also violates the ethos of restrained care that enables animals to retain their wildness. By domesticating animals, Marind thus participate, to return to Arthur's words, in taking away their freedom. From this perspective, Marind and domesticates no longer stand together in metonymic relation as the oppressed in the face of the state as oppressor. Rather, they stand in disturbing *opposition* to each other in a relation of asymmetrical domination. This contrapuntal dynamic in turn produces a paradoxical and ominous equivalence between Marind and the state, as captured in Alfons's statement. By controlling the lives of animals, Marind end up replicating in disquieting ways the role of the state in their own political disempowerment. In sum, domesticate-human relations are profoundly disconcerting because they position Marind simultaneously within two opposed categories—the oppressors *and* the oppressed.

AS EMIC CATEGORIES, WILD/DOMESTICATED and native/nonnative are central to understanding the ambivalent ontology of domesticates and their keepers in Merauke. The ambiguous status of domesticates in particular is enhanced by their concomitant position at the gray intersection of other morally charged sliding scales of difference—food and not food, human and animal, forest and village, person and possession, and oppressor and oppressed. As relational, contextual, and polyvalent categories of *trans*species valence, wild/domesticated and native/ nonnative distinctions also reveal how interspecies relatedness is shaped, not just by "species" as generic categories of life, but also by the actual, real liveliness of particular organisms, like Ruben, that matter precisely because they belong to, yet do *not* adequately or fully represent, their species' ways and wants. As such, the tensions at play in the villages of the Upper Bian are not just between humans and animals as different species, but also between their shifting individual and collective indexicalities as wild or domesticated, native or alien, or somewhere in the abu-abu interstices of established categorial orders.

In the West Papuan context, such transspecies indexicalities also bring to light the abu-abu entanglements of ecology with questions of politics, race, and difference. In particular, they foreground, to borrow political scientist Claire Jean Kim's words, how the question of human domination *over* animals intersects ambiguously with the question of racial or cultural domination *among* humans (2015, 11; see also Moore, Pandian, and Kosek 2003). Like animals deprived of their forest habitats and ways of life, Marind suffer imposed civilizational and developmental projects as reluctant members of the Indonesian State. As native species populations dwindle, Marind too become a minority on their own lands due to the ongoing influx of non-Papuan migrants. And yet many Marind aspire to the modern lifestyle promoted by the government and embodied in the settler population whom they at once resent and envy. Caught between a waning forest-based way of life and visions of modernity that are at once alien and alienating, Marind, like their metonymic animal counterparts, are becoming matter out of place.[11] And if liminal persons are not responsible for their marginal condition, the problem with domesticates is that they appear to *enjoy* their anomalous situation—as do some of their human keepers.

The conflicting affects provoked by problematic "pets" among Marind invite greater attention to how radical socioenvironmental transformations *reconfigure* the content and moral significance of preexisting and transspecies classificatory vectors such as *wild* and *domesticated*, and as *native* and *invasive*. In Merauke, plastic critters are the byproducts of historically entrenched colonialist ecologies and emergent agroindustrial assemblages that sever them from the relations that make them animal within the more-than-human forest lifeworld. These "postanimals" matter precisely because they constitute a problematic breach of the categorial order that distinguishes species and shapes their relations with humans. Furthermore, as domesticated animals are a relatively new phenomenon in the Upper Bian, no cultural mechanism, ritualistic or other, exists to resolve the condition of these disconcerting tricksters who are abu-abu because they are no longer fully animal yet also not entirely human. Unwanted in the settlements yet struggling to survive in the wild, these anomalous creatures find themselves caught between the clashing ontoecologies of plantation, forest, and village.

By ignoring, condemning, and mocking domesticates, Marind seek to distance themselves literally and figuratively from beings that are out of place in the village, while simultaneously affirming the categories to which they do not conform. These tactics of avoidance, however, are complicated by uncannily similar transformations that Marind perceive among their own kind and uncertainty over whether these transformations are imposed or embraced. In this light, animals who refuse to return to the forest become disquieting symbols of resignation to the hegemony of powerful others. Their desire to stay in the village suggests a willingness to adopt alien habits dictated by encroaching settlers and an occupying state. The ambiguous intentions of village animals thus reflect the ethical and existential quandaries faced by West Papuans themselves as they attempt to exceed the conditions of their exploitation yet also retain their sense of humanity in an increasingly uncertain world of modern things, technologies, and peoples.[12] Conflicting emotions, speculations, and interpretations coalesce around these abu-abu creatures who reveal troubling truths about their keepers' own desires, destinies, and dilemmas.

ONE YEAR AFTER I last saw Ruben, I received a flurry of WhatsApp and Facebook messages from Khalaoyam community members who were visiting Merauke City. Ruben had left the village. When I returned to the field one month later, Ruben's disappearance was still the object of much speculation. Some said Ruben had returned to the forest of his own free will. Others believed Ruben's cassowary kin had finally come to rescue him. If so, this meant Ruben had managed to retain his native and wild cassowary self, retrieve his place in the community of his own species, and finally find his freedom in the forest. Others were less optimistic about Ruben's fate, suggesting he had been captured by settlers on the prowl for wildlife to sell or eat. Perhaps Ruben had entered the oil palm plantations, where he would starve to death. Maybe he was in one of the neighboring military garrisons, where soldiers kept dogs and cats.

My friends imagined Ruben wandering around the barricades or roaming the forest. They envisioned Ruben's feathers sold as key rings in Merauke tourist shops. Some spoke of the bird's body being displayed as a stuffed mount in a museum in Jayapura or even Jakarta. No one was saddened by Ruben's departure or missed his presence. I asked Marcus, who had sung poignantly about Ruben's fate over a year ago, what he thought had happened to the plastic cassowary. He responded: "Maybe Ruben has returned to the forest. Maybe he is wild again. Maybe he does not want to be plastic anymore. I don't know. It is still abu-abu. People say Ruben has found his freedom. I hope this is true. I *must* hope."

Eben Kirksey (2012) describes the West Papuan independence movement as a pursuit of freedom in entangled worlds, in which activists collaborated with changing figures of hope in the anticipation of dramatic transformations on the horizon. In Merauke, the entanglements of animal and human worlds and their complex relations of captivity and domination resonate with the political subjection and quest for autonomy of West Papuans. For some of my companions, Ruben's disappearance was a somber omen of the fate of West Papuans facing ongoing political oppression, cultural assimilation, and environmental crisis. For others, the event materialized the hopeful transformation they had so long encouraged—the recovery of animals' freedom. Many people, like Marcus, were ambivalent about Ruben's fate. However, they must sustain the *possibility* that Ruben has regained his freedom, for this in turn sustains the possibility of freedom for West Papuans themselves—somewhere on the abu-abu horizon.

Metamorphosis—The Dream of Yosefus Samkakai

I tried to get out of bed, but my limbs wouldn't move, and my skin was burning. I looked at my arms and saw that I had scratched off a lot of skin. It was the same with my legs and my chest. I thought it might be <u>kambara</u> (black magic), but then it must have been a new kind of <u>kambara</u>, because I had never seen anything like this before. My wife thought I had cheated on her when I went to the city and that I had caught the four-letter disease (AIDS). She was very quiet and angry. She brought me food and water without speaking. Finally, I began to feel better. My children came to see me. They were no longer afraid because my skin was beginning to look normal again.

We chatted. Then, everyone went very quiet. My children were staring at my face and arms. They seemed afraid. I looked down at my forearms and saw several lumps had formed under the skin—round and hard. Maybe it was the four-letter disease after all? I ran my hand across my face and felt more hard lumps, bulging under the skin, across my forehead and temples. My wife asked me, 'What is happening to you?' I swore I had been faithful. Later, the lumps grew bigger and harder, and more grew across my entire body, many, many, many. The lumps were orange and yellow under my skin, pushing hard against my flesh, like they were about to burst out. It was like they were filled with pus. Some were black and red, like rotten flesh and blood. It was disgusting. I was afraid of myself for what I had become. But I did not know what I had become.

The pain grew as the lumps grew. Then, a terrible pain hit me, like a machete in the middle of my chest. I screamed—*aiiii!* I saw that a huge sharp spine had burst out of my flesh, in between two lumps. It was hard and brown. I thought I had gone mad. One after the other, spines shot out of my body. Blood and white water burst from my skin. It smelled of oil. Burned meat. Broken bones. I had become oil palm. I am a brave man, but I screamed from the pain. My wife screamed too in the corner where she was crouched, watching. Then, I woke up.

5. Sago Encounters

Ayo, kita pigi kenal sagu! A strident voice sliced through the muggy air of dawn. Soon other voices joined in: the guttural singing of men; the bubbly chitchat of women; the languid whimper of infants; the startled cry of a barking owl. Every so often, the boom of a ritual <u>kanda</u> drum reverberated through the landscape as the villagers made their way from Bayau village to the sago grove. It was the end of the dry season. The heavy smoke and glow of fires in nearby oil palm concessions reflected in muted pastels off the placid skin of the Bian River. But this ominous sight, signaling almost two weeks of unremitting forest burning, did not stop the group in their steps. Not today—today, they were going to know sago.

Following my Marind companions, we now leave the space of the village and plantation and head to the forest and grove. This foray offers a moment of respite for the reader. Here, we step away from the destruction, grayness, and murk that has haunted much of the first half of this story. We venture out of the world of abu-abu—the topographies of fear and violence, the rumble of bulldozers and chainsaws, the toxic ecologies of monocrop concessions, and the

Figure 5.1. Processing sago in the grove. Photo by Sophie Chao.

disturbing fates of plastic cassowaries. This is a chance to catch your breath— to con-spire with the grove and plot with the forest.[1] Like my friends that early morning, we go to meet a plant in whose company, as Marind put it, one can finally—if only temporarily—forget about oil palm.

Going to know sago (kabekhat dakh kmakh uyu or pigi kenal sagu) was one of the most popular topics of conversation among Marind of the Upper Bian. For instance, my companions in the field talked at length about where they visited sago groves, in whose company, how they processed sago pith and whom they shared it with, and what events took place during their journeys. Echoing Marcus Gebze's song in the opening of this book, my friends' narratives often celebrated sago (dakh) for bringing life and growth to share, as well as food and water, shade and rest. The starch obtained from its trunk, they explained, is the most tasty, filling, and nutritious of foods. It makes Marinds' bodies strong, wet, and shiny.[2] In the grove, Marind learn to sing, sweat, and work with a plant that "knows how to live with amai and with anim" and whose sounds and growth "make the forest alive." True Marind, I was told repeatedly, are Marind who eat, encounter, and know sago.

I once asked Andreana, who was sitting on the front porch of her house in Khalaoyam village, to describe the taste of this deeply valued and nourishing palm. An imposing matron in her late fifties, Andreana was well known in the village for her unparalleled storytelling skills and in particular, for her capacity to craft everyday events and scenes into exquisitely detailed narratives. On this occasion, however, my question was met with hesitation. For several minutes, Andreana stared in silence at the half-braided sago bags strewn haphazardly at our feet, which she had painstakingly been teaching me to weave. Then, her furrowed brow relaxed into a beaming grin. Throwing her head back, Andreana laughed and replied:

> I can't come up with a good story for you, child. It is difficult to say what sago tastes like. Let's see . . . Sago tastes of many things. Sago tastes of water. Sago tastes of land. Sago tastes of forest. Sago tastes of amai (plant and animal kin). Sago tastes of anim (humans). In the grove, we walk, we watch, we tell stories, we learn, we share, we eat, we listen, we sing. If you want to know why sago tastes good, then stories in the village simply won't do. If you want to know why sago tastes good, then you must go to the forest. You must go to know sago.

As Andreana suggests, going to know sago encompasses a wide range of activities that take place in and around sago palms. These include walking to the grove, processing and consuming sago pith, foraging, fishing, and hunting.

Going to know sago also entails morally and affectively charged labors of care. For instance, Marind express and affirm their social relations to each other by sharing food produced in the grove and engaging their bodies in wetness-inducing collective hard work. They pay acute attention to the relations of sago to its diverse companion species and carefully modify the physical environment to support their mutual growth and well-being. Marind also immerse and participate in the bioacoustics of the grove, including through songs that celebrate the lifeforms, events, and stories associated with sago. Going to know sago thus constitutes a total social fact that condenses seemingly disparate aspects of culture in a single phenomenon or practice. Alongside humans, sago and its symbiotes, too, participate in enlivening the grove through their own distinctive yet related lifeworlds, producing more-than-human socialities that are distributed across people and palms.

Inspired by Martin Heidegger's concept of being-in-the-world, I examine the multiple dimensions of pigi kenal sagu, which coalesce in "being-in-the-grove." Being-in-the-grove involves a range of embodied and affective interactions between humans, sago, and other organisms as sympoietic collaborators who collectively bring forth the world, and themselves within it, through their fleshly relations.[3] As they walk, rasp, pound, eat, sing, listen, and observe their surroundings, community members attune to their being-with-others across species lines. They engage with the movements, sounds, and behaviors of the bodies and things around them, such that their being becomes indissociable from the world in which it is grounded—or perhaps in this case, rooted.[4] Together, the social, affective, and moral dimensions of going to know sago, as a multispecies process, are what endow sago pith with its distinctive social taste.

SAGO PALMS ARE USUALLY found in the vicinity of mangrove and nipa forests along riverbanks and in lowland forests where sago stands can find partial shade from taller dicotyledonous trees. The densest clusters tend to be located at the site of former Marind settlements established during the colonial and precolonial periods.[5] Community members set off to the grove at the break of dawn, traveling across swampland and grassland by foot in the dry season and by canoe in the wet season. Songs and stories accompany the cadence of peoples' footsteps as they pass sacred sites, former villages, scenes of past wars and reconciliations, and places where significant amai encounters once took place. Along the way, the groups stop frequently to rest, smoke betel, and catch fish and birds, which they grill on twig fires. Mothers feed their infants under the shade of juniper bushes and bamboo clusters, while children forage for roots, nuts, and fruit.

Once in the grove, group members look for a <u>dakh</u> <u>kupsan</u>, a sago palm at the verge of inflorescence and therefore containing the greatest amount of pith.[6] Such trees are identified by their shorter and greener fronds, opening and downward-bending leaf midribs, and widening trunk girth. Community members chip at the bark to reach the inner pith, which turns white when chewed if the palm is mature. They cut or burn vegetation around the foot of the tree to get rid of weeds, lianas, wasp nests, and lurking snakes. The group then performs prayers and ritual songs to seek permission from their plant and animal kin and ancestral creator spirits to fell the tree and to ensure the pith they collect is abundant, wet, and dense. Men and women hack away at the trunk, aiming as close as possible to its base to avoid wasting pith. A creaking groan rising in crescendo signals the imminent fall of the stand, accompanied by strident cries warning children to stay clear. The resounding crash of the palm hitting the ground sends vigorous tremors through the soil and bodies of those present.

When the dust has settled, the group walks around the collapsed tree, examining its bark and thorns, clearing the trunk of fungi and epiphytes, and noting any deformations, diseases, or parasitic growths along the ridges of the bole. The fronds are chopped off and used to build bivouacs and filters for the sago-processing structure (<u>dakh</u> <u>sekuka</u>), a canoe-like construction erected close to a water source or dug-out waterhole. The bole is chopped open and a group of individuals sit alongside each other to rasp the inner starch into shreds with an axe or adze. While men hack away at the trunk in synchronous movement and song, children bring the shreds of pith to the women working the <u>sekuka</u>, who add vast quantities of water to it and pound it vigorously with a wooden rod to extract the edible sago starch. The sago starch and water mixture trickles down the toboggan-like funnel at the top of the <u>sekuka</u> and then into a sago frond trough erected at its base. Over time, the edible flour settles at the bottom of the trough. The water is either poured out and reused or left to evaporate. Once dry, the flour is usually cooked as <u>dakh</u> <u>kakiva</u>, a chewy mixture of sago, coconut flesh, sago grub larvae, or meat, cooked inside bamboo stalks over a fire.[7] In another variant, <u>dakh</u> <u>sep</u>, the ingredients are wrapped in banana or coconut leaves and cooked on hot stones covered with sago fronds or eucalyptus leaves, which enhances their aroma and flavor.

In the weeks leading up to the sago expeditions, an almost tangible sense of excitement permeates the villages of the Upper Bian. People travel far and wide to invite their kin and friends to participate and school classrooms are practically empty. Expeditions may involve as few as five individuals or as many as fifty and last between a few days and three months. While a sago trunk usually requires around six to twelve days to process, my friends affirmed that a truly enjoyable trip is one in which the group bivouacs for several weeks in the grove. The pleasure Marind derive from sharing time with sago (bagi waktu sama sagu) motivates trips even when sago procurement is not the primary objective. For instance, villagers regularly visited groves to observe the growth of sago stands and of recently transplanted suckers; to note the state of surrounding vegetation; to detect evidence of animal passage in the form of droppings, half-eaten fruit, or nests; or simply to "walk around" (jalan jalan).

A considerable amount of coordination goes into planning sago journeys. For instance, groups must decide who will participate, undertake different tasks in the grove, and provide tools, food, and other necessities. They must account for upcoming events for which extra sago flour may be needed, such as feasts, clan-wide congregations, weddings, and births. Women make note of community members who are unable to join and need their sago stocks replenished, such as parents observing ritual food taboos after the birth of a child, a family in mourning, an ailing patient, or an aging widow. Those who participate in this meticulous and protracted planning demonstrate that they know and care about the wider community's needs.

At the same time, many Marind affirm that an enjoyable trip depends as much on their own prior organization as on the collaboration of forest amai and elements. For instance, access to the grove is conditional on weather conditions and river levels and currents. Community members look to recent encounters with plants and animals, and notable events that have taken place in the forest, to determine the right time to travel, the route to take, and the most propitious sago grove to visit. Much like the mapping processes described previously, groups follow the sounds and movements of plant and animal amai as they weave their way across the landscape toward the grove. These physical and acoustic encounters prompt frequent detours, such that journeys can end up lasting weeks longer than originally forecast. Whether and how one experiences the joy of "sharing time with sago" is thus always a matter of *hap* (chance)— the etymological root, Sara Ahmed reminds us, of the word happiness itself (2010, 208, 222–23). Where people walk to, from, and through, and in whose more-than-human company, bestows each journey with unique and often unanticipated affective textures.

Like other Melanesian societies who associate sago production and consumption with the creation of social relations, Upper Bian Marind often say that "sharing sago makes kin" (bagi sagu jadi keluarga). The making of kin through sago consumption extends beyond the Marind clans to encompass members of Wambon, Kei, and Muyu communities who have settled in Merauke, and who are said to "eat sago from the same grove and share women of the same blood." It also includes other Papuan ethnic groups across the island of New Guinea who, like Marind, derive their subsistence from sago. Together, sago eaters partake across space in a shared gastro-identity that encompasses the ecological assemblages, cultural values, and social relations embodied in the food they consume.[8] The environment of the grove itself, according to Matthias, a Khalaoyam elder and ritual healer, is conducive to the affirmation of social ties. Unlike the village, road, and plantation, the grove is calm, peaceful, and cool. It is, as Matthias explained, "where kin find each other again, and where one remembers what it is to be anim."

Like Matthias, many of my companions describe the grove as a place of safety and freedom. Here, Marind can at least temporarily avoid the military, police, government, agribusiness corporations, and intel (spies), whose presence pervades everyday life in the villages. Morally valued forms of social behavior, such as collective work, sharing food, joking, teasing, singing, and storytelling, serve to enhance the sociality of the grove. Community members recount these experiences in minute detail after returning to the village, such that the relations established in going to know sago are sustained long after the expedition itself has ended. These conversations often continue late into the night. Hushed murmurs diffuse across houses and outdoor platforms, while thick veils of cigarette smoke shroud the supine bodies of those bound by their shared experience of encountering sago.

Sago also reaffirms connections between Marind across time, as community members are reminded in the grove of those who produced and consumed sago before them. For instance, they recall the lives of long or recently deceased family members and friends with whom they once journeyed to the grove. They recount the story of sago's creation by dema, or ancestral creator spirits, and how they came to be enskilled by dema to extract its nourishing starch. They describe the exploits of former headhunters, whose peregrinations across the landscape were enabled by the abundance of sago growing along the banks of the Bian. Naturally plentiful, easily stored, and requiring no cultivation, sago could sustain warriors for months on end when cached in large war canoes as whole logs. Marinds' collective social memory thus crystallizes in the shared experience of knowing and eating sago.

Marind affirm that sago processing and consumption "make the body strong" (sagu deh bikin badan kuat).[9] Marind who process and eat sago can toil until the fall of night without feeling tired or hungry. Sago allows men to hunt for days on end and women to bear many offspring. It enables children to grow tall and healthy. It produces a glossy skin and abundant wetness in the bodies of all those who go to know sago together. The social relations achieved through eating sago and the taste of sago itself are further enhanced by the production of bodily wetness among those who process its pith. As Petrarchus, a school-teacher from Khalaoyam, explained during one of our visits to the grove, "The harder you work, the more you sweat, and the better the sago tastes." As they labor, villagers comment approvingly on each other's perspiration, compare with pride their glistening foreheads, and rub their forearms to encourage each other to work harder. Sago pounders set up their <u>sekuka</u>, or pounding structures, in a close semicircle so they can chat, sing, sweat, and be lively together. Those who eat sago, Petrarchus continued, take in not only the feeding substance of the plant itself, but also the skin and wetness of the people who obtained and prepared it. For this reason, sago made by others is always tastier and more nutritious than sago procured by oneself.

The abundant wetness of the sago palm and its ecology, too, participates in increasing the health and vitality of sago pounders and eaters. As a hydrophilic plant, sago flourishes in and around large bodies of water, such as swamps, marshes, and mangroves. Various plant species that grow in close proximity to sago clusters sustain the wetness of the palm by offering it the shade and coolness of their lush foliage. Just as wetness enhances and communicates the moral and social well-being of <u>anim</u>, so too many Marind consider the wetness of sago to be one of its most valued attributes and the primary source of its nutritious "good taste" (rasa enak).[10]

For this reason, Marind fell palms at the point of inflorescence, when the fluids in the bole are most concentrated and the tree is ready to share its wetness. Boles whose starch is particularly moist are a source of great satisfaction and excitement. People gather around these trunks, gently stroke them, and then massage their pith with both hands to draw out its moisture. The sago heart, or cabbage, which Marind consider the wettest part of the tree, is a much-treasured treat that is consumed by the group before undertaking the hard labor of sago processing. The pounding of sago itself requires vast quantities of water, which children carry from the river or waterholes. Every so often, adults pause in their labors to smear moist pith onto the children's faces so they learn to know the wetness of the palm. The vigorous thrashing of the starch sends innumerable particles of sago flying into the air. These wet fragments land on the sweating faces and limbs of those who are hard at work, cooling and moistening their bodies as they stand knee-high in the muddy waters of the grove. Sago thus imparts its nourishing fluids to those who labor and consume its flesh, along with the wetness of all the places, times, and organisms inscribed in its bodily matter.

Finally, Marind frequently invoke sago to distinguish themselves as sago-eating Papuans from rice-eating Indonesians (see Chao 2021d, 2021e).[11] Although rice has become an increasingly prevalent foodstuff in the Upper Bian as a result of the loss of forest-derived foods, rice is associated by many of my companions with agricultural projects implemented without their consent by the Indonesian government, and prior to that, by the Dutch colonial administration. Indeed, rice featured centrally in the original design of MIFEE under then president Susilo Bambang Yudhoyono, who spoke of transforming Merauke into a "national rice barn" (lumbung padi nasional). Rice cultivation also contravenes the ethos of restrained care because it entails the control of humans over plants rather than respect on the part of humans for plants' self-driven processes of growth. Furthermore, rice is the staple food crop of the non-Papuan migrants who are increasingly taking over Marinds' lands and who frequently treat them as racially and culturally inferior. Rice itself was introduced from Java, the political heart of the Indonesian State, where decisions about West Papua's future are made without West Papuan consultation or consent. Finally, rice is often criticized by Marind as a "plastic" food that fails to provide satiety and well-being to those who consume it. The divergent composites of humans, relations, and ecologies that coalesce in rice and sago respectively in turn give rise to what Arjun Appadurai (1981) might term *gastro-politics*, which is the morally imbued expression of cultural values through food choices and transactions. By choosing to eat sago, Marind symbolically oppose the alien and invasive capitalist-colonialist forces incarnated in rice and its human producers and consumers.[12]

THE NURTURING ENVIRONMENT AND pith of sago accrue heightened significance in light of the plant's association with women in their role as mothers. This association was brought to my attention by Evelina, the pregnant mother of three whom we encountered in chapter 3. As she hacked away at a freshly felled sago palm, Evelina explained, "The sago grove is where <u>anim</u> children and sago children (<u>dakh</u> <u>izmi</u>, or sago suckers) grow together—strong and wet. It is where Marind become mothers. It is a realm of life. The sago grove is the realm of women."[13]

Many other Marind also characterize sago groves as the "realm of women." Such statements are often supported by direct reference to the shared morphological attributes of sago palms and women, which together give rise to their nourishing qualities and lifegiving capacities. In the grove, for instance, Evelina pointed out the swollen boles of mature palms, which she said resembled the protruding belly of pregnant women. Other women in our group compared sago pith that turns dark red, milky pink, or light orange after pounding and leaching to menstrual and postpartum blood, and white sago pith to breastmilk.[14] The reddish and wet lumps of inedible fiber left over after sago processing, Evelina added, resemble the human placenta, which is buried in the forest after the child is born, while the filtered flour forms lumps of white starch that are "soft and wet, like a baby's body." If stored when still damp, sago ferments and releases lactic acid, which Marind say smells like the breastmilk of women who have undertaken vigorous physical exercise. Indeed, sago flour mixed with water, which is referred to by villagers as "sago milk" (susu sagu), is often substituted for breastmilk for infants whose mothers have passed away or are too weak to breastfeed. In the grove, Evelina and her female kin enhanced their bodily contact with sago palms by carefully kneading the bole interior to induce greater starch yield, in the same way they massaged each other's stomachs to encourage contractions and alleviate labor pains. The women also shared skin with sago by wrapping fronds around their torsos and bending the fronds into a semicylindrical shape when fashioning sago filters. Molding the fronds to the contours of their own bodies, they gently coaxed the plant into yielding its nourishing starch while simultaneously enhancing their own capacity to bear many healthy and strong children.

Many Marind women take great pride in their affinities to sago and the hard work they invest in processing its starch. During breaks between pounding, for instance, women inspect with satisfaction each other's worn palms and flowing sweat and congratulate each other on these visible testimonies of their hard work.[15] The most nutritious and flavorsome flour, they often told me, is that produced by women because both the plant and its pounders are bearers of sustenance and fertility—the one of edible starch, the other of human offspring.

Gendered distinctions also inflect the "freedom" that many community members associate with the grove. Selli, a mother of five from Bayau who played a leading role in organizing sago-processing expeditions, explained, "In the grove, women are freer than in the village. Our husbands follow us to be fed, just like our children follow us for milk." Rosina, Selli's younger sister, giggled at her words and added enthusiastically: "In the sago grove, it is the women who do the hard work! So, it is the women who tell the men what to do. In the village, men are the boss, but in the grove, women become the bosses!" As Rosina and Selli's comments suggest, women's distinctive relationship with sago gives them license to direct the group's activities in the grove and also to tease, joke about, and make fun of men with greater audacity than would be acceptable in other settings.

On one occasion, for instance, a group of women with whom I was pounding sago complained that the men accompanying us were not working hard enough. They walked over to the fire around which the men had been chatting and chewing betel for the last three hours and virulently stamped it out. Poking the panicked men's backsides with blazing embers, the women chased them into the forest, to the great amusement of all those watching. The men eventually hobbled back from the forest, bearing impressive second-degree burns on their bodies. However, none among them complained or blamed the women for their injuries. Instead, they smiled and laughed, while many reminisced with fondness how they were once beaten in a similar way by their mothers when they were naughty as children. The affective dimensions of being-in-the-grove are thus inflected along gendered lines. Marind women associate the pleasurable taste of sago with the affirmation of their status as maternal caregivers. Meanwhile, Marind men associate the grove with tender memories of their childhood. They are neither angry nor ashamed when they are teased or bossed around by women in the grove. After all, it is in the company of sago that, as Selli put it, "women become mothers, and men become children."[16]

The association of sago palms with human mothers is accompanied by correlations that Marind identify between sago processing and child rearing as analogous processes of growth. For instance, vast amounts of shared wetness are needed to make anim out of children, just as abundant quantities of water are required for sago to thrive and taste good. Leaching, rasping, and pounding sago demand skill, time, and dedication on the part of all the members of the group, much as raising children necessitates the affective and physical investment and care of the whole community. Correlations between the stages of maturation, reproduction, and senescence of palms and humans are commemorated in various ways.[17] For instance, palms are named after Marind children whose birth occurred concurrently to those of their suckers. Palms that mature after fifteen to sixteen years are given the names of pubescent girls who have begun to menstruate. The shared growth of humans and sago is fostered by placing newborns in a sago bag (kabuh) made of fronds from palms that produced suckers at the time of their birth. Children are carried in a succession of sago bags made of the same palm as they grow, learn to crawl, and eventually walk. In this way, Rosina explained, "Sago and anim follow each other's lives."[18]

The correlations between the growth of palms and people differ in form and duration according to the mode of reproduction of sago, which can be either sexual or vegetative. In sexual reproduction, seeds are produced through fertilization and disperse to form new plants upon germination. In vegetative reproduction, new plants grow directly from parts of the parent plant, such as its root, stem, leaf, or bud. Plants that reproduce through their shoots and suckers are known as soboliferous plants. While sexual and vegetative reproduction occur naturally in sago, vegetative propagation is more frequent and is actively encouraged by Upper Bian Marind through sucker transplantation and the felling of trees prior to inflorescence.

The sexual reproduction of sago involves four stages. In the rosette stage, which lasts three to six years, the sago is weak and unripe, like an infant, unable to communicate or act autonomously. The second stage, bole formation, lasts four to seven years. During this time, the trunk elongates and increases in starch content while the fronds develop one leaf every month. Marind community members compared this stage to children between the ages of seven to thirteen, who learn to speak properly and participate in social activities in the home and grove. The third stage of sexual reproduction is inflorescence. Palms in this period of the life cycle are gakhum, or "pregnant." Like young women's bodies during gestation, the "womb" of the palm—its dense, starch-filled, and nutrient-rich bole—swells laterally for seven to ten months. The starch moves up the bole to feed the flower and fruit until, eventually, the trunk becomes saturated with starch and the inflorescence bursts out—first as bracts, which my informants likened to a baby's spindly arms and legs, and then as a massive flower bearing anywhere from 250,000 to 850,000 fruits. The starch content of the palm decreases following the fruiting stage and its regenerative capacity declines rapidly. The parent palm dies shortly afterward, as sago is a hapaxanthic plant, meaning it flowers once and then senesces. Community members often compared the senescence of sago to that of women whose bodies become tired and dry from bearing children. Just as the palm dies after its profusive flowering, women's reproductive capacities cease once they reach menopause.

In vegetative reproduction, sago palms produce suckers and stolons, which Marind call dakh izmi, or "sago children." These suckers and stolons emerge in regular succession from adventitious buds on the underground roots or lower part of the trunk. My friends often compared these sucker clusters to groups of young children sitting at the feet of their parents as they are fed, cajoled, and otherwise cared for. Vegetative reproduction in the grove also enables the formation of clonal colonies, dense clusters of genetically connected and morphologically similar plants. Like human children among their kin, suckers and parent palms grow in close proximity; resemble one another; live off the same nutrients, soil, and water; and entertain symbiotic relations with the same organisms. Parents and children, plant and human, share roots with each other and with their circle of relatives in the grove and the village, respectively.

Community members actively support sago palms' production of sucker children through transplanting. This practice, which involves pulling out sago suckers and replanting them a few meters from the parent palm, demands a careful attunement to the condition of the young suckers and their environment. For instance, my companions affirmed that transplanting should only take place when the suckers are several months to a year old. Before that time they have no root system and rely on their parent for photosynthetic products, nutrients, and shade. Just like human infants, these sago children do not have enough wetness to live and their skin is weak. To move them away from their parent would kill them. When suckers begin to develop their own root system, the parent palm can no longer provide them with enough sustenance and its shade begins to hinder their growth. The sago children are ready to "find their own space" (cari tempat diri). Transplanting helps the sucker finds its own space and access more nutrients, water, and light. Its maturation accelerates and its wetness, or starch content, increases rapidly. Like human children who learn to speak, walk the forest, and procure food for themselves, the sucker learns to survive more autonomously. Transplanting also benefits the parent palm by allowing it to regain its regenerative capacity and utilize its nutrients to produce more sucker children. Throughout this process, both the sucker and its parent remain organically related and physically close to each other and to their kindred stands in the grove.

Community members also encourage vegetative reproduction in sago by selectively felling palms that are dakh kupsan, or just about to flower. As in transplanting, felling prior to flowering redirects the nutrients of the palm toward the production of suckers, who will perpetuate the life of their parent as they spread and multiply. Felling also opens a space in the canopy for suckers to grow into, which allows them to gain full exposure to sunlight, develop an optimal number of leaves, and achieve their full trunk circumference and height. This in turn prevents an overdensity of sago clusters and growth-stunting competition for nutrients between the parent and suckers. Felling palms and transplanting suckers can thus be conceived as forms of generative destruction that sustain the collective and intergenerational continuity of both plants and humans.[19] The pith of palms feeds and strengthens children so that they can grow into adults, nurture future suckers, and bear offspring, who will in turn support sago maturation and reproduction through their labors in the grove.

Meanwhile, the felled palm lives on through the sucker children, who flourish in abundance within its close proximity.

OKTAVIUS, A MIDDLE-AGED MAN from Khalaoyam, was mending his bow beside me in the grove. At his feet, lumps of freshly leached sago were grilling over the woodfire, the crackle of the flames drowned out by the chirrup of cicadas in the canopy. Oktavius paused, cocked his head, and listened, smiling. After several minutes, he turned to me and said: "The grove is full of life because sago knows how to share space with others. The sawfish rests in the rivers between its roots. Birds nest at the tip of its trunk. Insects sing with the wind in its fronds. Sago is a tree of many stories. Sago is a tree of many lives."

If "sagoscapes" are conducive to the affirmation of gendered and intergenerational relations among Marind, and between Marind and other "sago people" across New Guinea, they are also realms enlivened with more-than-human socialities. A complex ecosystem develops around the sago palm, whose "many lives," in Oktavius's words, encompass myriad organisms that thrive in and from the palm and its environment. In this multispecies assemblage, diverse species adapt to and construct ecological niches out of each other to sustain their shared growth and propagation. In doing so, these organisms produce themselves *as* environments for others to flourish in and from, in mutual ontogenesis.

Just as Marind people come into existence through their bodily relations, so too the sago palm as a symbiotic multiplicity emerges from its fluid and fleshly alliances with different lifeforms at different scales, in a process of perpetual transspecies becoming. Resonating with Deborah Battaglia's (2017) description of the reciprocal enlistment of yams and humans in the Trobriands, sago engages the lives and labors of human and other-than-human beings in a dynamic and mutual sympoiesis. Going to know sago consequently involves attending to how the palm's Umwelt, or lifeworld, is shaped by multiple lifeforms who together generate the plant as a community of growths rather than a singular and bounded entity. As they encounter sago, Marind attune to the development and decay of sago's relationships with diverse forest beings in the grove as a diverse space of multispecies sociality.

For Marind, going to know sago involves observing and attuning to the movements and growth of forest beings who share space with sago. For instance, during our journeys, community members tracked pigs attracted by the pith of damaged or deliberately felled palms and the cool waters of the sago swamps. They pointed out agile tree kangaroos and sugar gliders leaping across the canopy and noticed solitary spotted possums nesting between tightly bunched sago fronds. Children identified various avian species drawn to the grove's springs and fruit trees, where they drink and feed, and to the safety of the crown of the sago tree, where they roost and mate. These critters, the children explained, include inquisitive, red-billed brush turkeys, emeraldine buff-faced pygmy parrots, swooping white-breasted wood swallows, silver-crowned friarbirds, sharp-eyed gray goshawks, and more. Sago's companion plants, too, abound in the grove. They include wild sugarcane, campnosperma, swamp oak, pandanus, *Alstonia scholaris*, Bishop wood, bur tree, *Planchonia papuana*, nipa palm, and various species of the genera *Syzygium* and *Neonauclea*. Juicy ferns in circinate vernation surround the feet of sago trunks while plump paddy straw mushrooms flourish along mature boles and in piles of rotting sago pith. Other mycorrhizal fungi grow in symbiotic association with the roots of the palm, receiving organic nutrients from it while enhancing its surface capacity to better absorb water and minerals.

Even in dense groves where little sunlight reaches the ground, the undergrowth flourishes with a diversity of flora and fauna. Various insects, for instance, make a host and home of sago. Termites congregate and nest on the palm in the wet season, while sago palm weevil grubs and larvae incubate in its stumps and pith. Borers and beetles abound in the grove alongside skipper butterflies and bagworms. The humid air is alive with the buzz of pollinating insects, including stingless bees, honeybees, and various wasp species. After it dies, sago continues to feed forest organisms, nourish the soil, and sustain diverse microbial, bacterial, and fungal communities with its rotting pith. The plant perpetuates its afterlife in the diverse bodies sustained by its generative decay.

The mutual nourishment provided by sago and its other-than-human *companions*—a term that, as Haraway (2008, 208) reminds us, originally meant "those who break bread together"—multiply the sources of sustenance available to Marind in the grove. For instance, community members obtain larvae of the sago palm weevil—a delectable and important source of protein—from rotting sago stumps, lesions in living stands, heaps of waste pith, and the soft tissue of frond sheaths. Women and children gather nuts, seeds, fruit, tubers, and edible leaves in the grove. People set up fishing nets or fish traps fashioned from dry and spiny sago fronds in nearby rivers. Men hunt in bands for game, which is plentiful in the grove, and build sago frond huts near ponds, in which they wait for birds and other game to approach the water source at the break of dawn. Pigs, attracted by the smell of sago pith, become trapped in hollowed-out trunks lying around the grove. As my companion Oktavius explained, these pigs provide an important source of meat for community members who are unable to hunt, such as widows, ailing individuals, and persons subject to hunting restrictions during mourning.

Marind themselves sustain the lifeworld of sago's symbiotes through their presence and activities. For instance, children are tasked with burning leftover fronds and other waste around the working area at the end of the day. This stimulates plant growth, attracts animals who feed off the shoots of regenerated vegetation, and over time, increases the level of nutrients in the soil. Animals forage for grubs and insects in leftover piles of pith and other byproducts of the leaching process. Maleo fowl build nesting mounds out of these mulchy remnants, in which they bury their eggs, which incubate as the pith decomposes. Eventually, the rotting mounds transform into hummus which nourishes the soil and its diverse plant and insect communities. Marinds' own blood provides sustenance to hematophagous critters, including hungry mosquitos and plump leeches, and then returns to the groves in which these organisms feed, reproduce, and eventually decay. Reciprocal relations of eating and being eaten thus connect Marind to sago palms as mutual sources of nourishment.

Community members support the growth of sago itself through various means. As exemplified by the apprenticeship of young Marcia by her father, Gerfacius, transplanting suckers a little further away from the parent palm allows the stands to access more sunlight and nutrients. Earlier that day, Gerfacius had spent several hours clearing senescent branches off the bole of the parent plant, weeding its base and trunk, and judiciously pruning its fronds. These activities, he explained, helped increase leaf formation and starch accumulation, and thereby made the palm and its sucker family strong and healthy. Meanwhile, other members of the group busied themselves with canopy thinning, ring-barking, and controlled burning, which are also said to help the sago thrive. Those who carried out these tasks described themselves as "sago guardians" (penjaga sagu). "Sago guardians," as village elder Marianus explained, "help sago grow. But sago guardians do not tell sago what to do because sago is wild. Sago is free." Indeed, Upper Bian Marind widely affirm that they do not plant sago and are not sago cultivators. Rather, they understand the wildness of sago palms to be as much the product of multiple acts of restrained care on the part of "sago guardians" as the outcome of sago's own self-willed modes of development and interspecies relations. In this light, Marind practices of environmental care are premised on a fundamental respect for the relational yet autonomous growth of the palms. In exercising restrained care toward sago palms, sago guardians eschew human mastery. Instead, they cultivate vegetal wildness.[20]

Fine-tuned observation of sago palm morphologies is central to knowing and understanding the multispecies sociality of the grove. Indeed, time devoted to such observation is deemed just as important by Marind as time spent procuring, preparing, and consuming sago flour. For instance, my companions on sago expeditions would scrutinize sago palms at length and in great detail for evidence of plant growth, indicated by the height of boles, the budding of internodes, and the changing color and size of fronds. They noted the activities of birds and insects in and around palms. They also detected signs of disease that may have stunted the palms' maturation. Many conversations in the grove, too, revolved around sago's physical characteristics—unusually sharp thorns, extended pneumatophores, or the contorted growth of a sucker—which Marind discover through minute inspection and physical contact, or what STS scholar María Puig de la Bellacasa calls "vision-as-touch" (2009, 309).[21] Etched into the flesh of the palm, each of these traits incarnates the more-than-human intimacies that sago sustains with other lifeforms, and which have shaped the palm's growth and senescence—or in Marind terms, its "story."

Echoing Oktavius' extolment of sago as a "tree of many stories," my friends in the Upper Bian often told me that the best way to know sago was through its cerita (stories). As Eben, a young man from Bayau who enskilled me in the arduous labor of sago rasping, once explained, "When you go to know sago, you discover its story in its bark, fronds, and suckers. When you go to know sago, your own story changes. Sago becomes part of your story and you become part of sago's skin. That is how we make stories together." Eben's description of the interweaving of human and sago stories achieved through fleshly interactions in the grove resonates with the words of Kabi, a Gimi man and father, friend, and protector to Paige West during her fieldwork in Maimafu, Papua New Guinea. On being asked in what way he knew the forest, Kabi responded, "My eyes, my ears, my nose, my mouth, my teeth, my skin, my father, my bones" (West 2006a, 237). To echo Kabi's words, knowing the story of sago for Marind is a multisensory, phenomenological, and embodied experience—one that reaffirms their identity *as* a relation to a larger world of animacy (see also West 2016, 128–29). And just as one cannot touch without being touched in the mingling of bodies from which the capacity to affect and be affected rises, humans, plants, and their shared stories accrue meaning through the flesh-to-flesh materiality of their interactions in the grove, whose traces are written in their storied skin.

As Eben's words suggest, sago stories (cerita sagu) are relational in terms of *what* they designate—namely, the relations of sago to other organisms, both human and nonhuman. Sago stories are also relational in terms of *how* they can be known—namely, by establishing a relation with sago through being-in-the-grove. Much like knowing the taste of sago, as village storyteller Andreana explained, requires that one travel the forest to the palm's encounter, so too knowing and narrating sago's story requires that one be in the fleshly presence of the palm or grove whose story is being told. Only by doing so can community members scrutinize the morphology, textures, and transformations in which the plant's life events are inscribed. As they immerse themselves in this literal and lively form of palm reading, plants and people partake in producing the grove as an ongoingly storied and sensory multispecies world.

Marind read the storied animacy of sago palms from their physical attributes. As we waded across dense mangrove swamps, for instance, my companion Oktavius pointed out to me plants with particularly broad fronds that were cared for by families who, in the image of the palms' abundant foliage, had borne many offspring and spread widely across Marind territory. Holding my hand tightly as she guided me across the deeper swamps, Evelina explained how changes in the wetness and yield of palms caused by seasonal change, human management, and the condition of the Bian River could be read from the color of the back of the petiole and rachis—for instance, green without bands, dark gray bands, or brown bands, with or without tinge. Along with their width and hardness, the color of the bole and leaf—brown, cream, deep or pale green—indicates whether the stand was a prolific parent. Thicker spines on the leaflets, Evelina's husband Gerfacius noted, often feature in the history of palms that had to defend themselves from parasitic insects or ravaging forest fires lit by invading tribes in times past. Trees whose growth occurred in times of strife, warfare, and conflict show a darker degree of redness at the base of the leaflet. Conversely, Gerfacius added, palms that proliferated during times of peacemaking and intertribal reconciliation are distinguished by the wider spacing and smaller number of spines running along the petiole and rachis of their suckers. As we gradually made our way from swamp to grove, battles and alliances, ancestors and relatives, dema and amai, myth and history, and births and deaths multiplied, the story of one palm interweaving with that of another in a vibrant flourish of lives, words, and deeds.

In some instances, the story of palms arises from their memorable encounters with particular animals whom they eventually came to resemble, which live on through the palms' anatomies and nomenclatures. As we reached the outskirts of the sago grove, for instance, Oktavius introduced me to the "dog sago" (dakh mahu), whose welcome shade had once spared a pack of canines and their hunters from severe dehydration and whose inwardly curled bole sheath was aptly reminiscent of a dog's floppy ears. Later that day, Gerfacius took me to visit the "cassowary sago" (<u>dakh</u> <u>khei</u>). This stand and its suckers, Gerfacius explained, were known to produce smooth and curvate cabbages resembling the casque of the male cassowaries who roost in their proximity at the wane of each monsoon season. Other palms named by way of sago-animal compounds and identified to me by my companions in the grove included the "scrubfowl sago" (<u>dakh</u> <u>mizur</u>), which had distinctively claw-like and elongated basal thorns; the "kangaroo sago" (<u>dakh</u> <u>walef</u>), whose thick, beige-brown, and pubescent suckers evoked the muscular tails of forest macropods; and the "rainbow fish sago" (<u>dakh</u> <u>uzaim</u>), which is recognizable from afar by the variegated hues and colors reflecting off its sun-soaked foliage, conjuring up the iridescent scales of native riverine critters.

The texture of sago pith, too, holds secrets to the story of the palm. Where the pith of felled palms revealed itself to be particularly wet, soft, and dense, for instance, my companions reveled in enumerating the names of the various birds, insects, and mammals who had collectively sustained the growth of the stand. The multispecies story of palms continues to propagate through the growth of organisms who community members bring back to the village after visiting the grove. For instance, villagers may salvage epiphytes from the boles of felled trees, such as orchids and ferns, and plant them in other groves or offer them to their kin. Often these plants share the name of the grove whose story they once shared. As they visit each other's plants, people retell with fondness the events that took place when the epiphytes were found and the story of their host palm. These stories flourish in meaning as the plants propagate in their new homes. Eventually, the epiphytes outlive the host palm and continue its story through their own thriving. Years after the plants were moved from the grove, community members continue to describe their story in conjunction with more recent events in the lives of the humans who collected and exchanged them, such as marriages, births, and deaths. In this intergenerational process, plants salvaged from the grove become living indexes of the webs of more-than-human actors and relations invested in their growth (cf. Fajans 1997, 69–70; Tammisto 2018b, 39–43).

Finally, the movements and growths encompassed in the cerita, or stories, of sago motivate new forms of relatedness among humans. For instance, suckers who spread naturally on lands owned by other clans signal an invitation to form closer social ties between their respective members, such as through inter-clan marriage, child adoption, or shared land-use arrangements. Conversely, a sucker who rots or fails to mature on the land of another clan is an omen of impending conflict between the two groups. The sucker is allowed to grow but carefully monitored to ensure decay occurs fully. People burn the plant when decay is nearly final to preempt the social strife foretold by the stunted growth. The living flesh of sago stands thus participates in the making and unmaking of human relations through their idiosyncratic patterns of growth and senescence.

Sago cerita thus speak to the relational ontology of sago palms as organisms that are *generated* by the field of relationships set up through their presence and activity. As the morphology, doings, and encounters of sago palms, as individuals and collectives, transform over time, so too their stories change and multiply as information continuously *in*-formation. Each organism, event, and relation becomes a part of the palm's evolving *dermography*, a term used by Gísli Pálsson (2016, 102–3) to refer to the course of life or biography of organisms as it is embodied in their skin (see also Howes 2005). The textures of the palm stories generated from this taxonomic bricolage thus constitute a perpetual and participatory multispecies "growing"—the Greek term for which, coincidentally, happens to be *phuton*, the etymological root of "plant."

AS I DESCRIBED EARLIER in the context of mapping in chapter 2, sound is the primary medium through which many Marind perceive and engage with their environment. Similarly, going to know sago involves becoming attentive to the sounds—or in Marind terms, the "voices" (suara)—of the grove and its animate beings. In the grove, for instance, people commented extensively on the sounds they heard, their possible provenance, and their cause. Community members would interrupt lively conversations and activities to listen to the sounds of wind, water, and birds or other animals. Every so often, they congregated on nearby sago stumps or budding suckers or under the shade of dried-out sago fronds. Together, and for up to an hour or more, my companions would look around the grove, up into the foliage, sitting still and listening in silence. Children, too, were encouraged to listen to the forest as they carried out their tasks. I myself was often reprimanded by my companions for asking questions incessantly when in the grove. "Stop talking," they implored me; "Start listening!"

The voices of the grove include sounds produced by human activities, conversations, and songs. They also encompass the sounds of other-than-human beings and elements. Together, human and other-than-human voices give phenomenal texture to the grove as a bioacoustic assemblage of dynamic lifeworlds. Conversely, angry insults, arguments, and harmful gossip, along with screaming, shouting, and other kinds of "loud noise" (suara keras), are strictly prohibited during going to know sago because they disturb birds and other animals and cause them to flee.[22] Marind also affirm that loud noises distract people from investing themselves emotionally and physically in their encounters with, and attunement to, sago and its living environment. Such disturbances, in turn, affect the taste and texture of the sago pith obtained, which becomes dry and bland, rather than nourishing and tasty.

The sonic mélange of the grove is transposed and harmonized across manifold substances, species, and scales. Villagers from Bayau, with whom I traveled to the grove that early dawn in the wake of the dry season, taught me how to notice and distinguish the diverse voices of forest: the crackling of bamboo shoots in the midday heat; the deafening mating calls of cicadas; the susurration of rippling rivers; and the clumsy flopping of fish in nearby streams. Around me, men and women panted to the cadence of syncopated sago pounding and the soft gurgle of pouring water. Trunks and fronds creaked and swayed to the patter of raindrops and the caress of the wind. As fires crackled on the forest soil, crowned pigeons sung high up in the foliage in chorus with the soporific drone of insects. Every so often, the sounds of the grove were silenced by a crashing frond, and then resumed in gradual and synchronized crescendo. Crystalline laughter melded with the pattering of small feet as children chased one other across muddy creeks and shallow swamps. Comic shrieks and splashes occasionally interrupted melodic songs and animated conversations as sago pounders noticed stray leeches crawling up their legs. At night, the melancholic hoot of perched owls, the crackling of dying fires, and the low hum of human voices filled the darkness, interrupted at the break of dawn by the whistle of <u>menggep</u>, <u>ofa</u>, <u>khi</u> <u>auba</u>, and other familiar nesters of the canopy.

Each stage in the sago-collection process, too, yields a different kind of sound—felling, rasping, pounding, washing, kneading, straining, leaching, and more. Some sounds are intensified by the smells that accompany or precede them, in a spontaneous form of sylvan synesthesia. For instance, thunderous downpours follow the release of petrichor from rocks and soils when humidity levels rise. Juniper twigs crackle and release a fresh, balsamic scent as they burn. Damar wood resins hiss and fill the air with a molasses-like fragrance as they melt. The drone of flies attracted by mounds of leftover pith mingles with the sickly-sweet smell of damp rotting fiber.

Songs sung in the grove participate in the sensory experience of going to know sago. These songs often incorporate well-known verses and tunes that celebrate sago as the staple food of Marind and Papuans across space and time. They recount the myths of sago's creator spirit (<u>dema</u>), narrate the discovery of sago by humans (<u>anim</u>), and describe how the wetness of sago made Marind strong and fearless warriors. Other songs are more personal and refer to singers' childhood experiences in the grove in the company of their parents and grandparents. In between bouts of rasping, for instance, Gerfacius sang nostalgically about his first sago expedition—what he learned, saw, heard, felt, and in whose company, human and other. He described how his predecessors sustained the growth of sago by transplanting suckers, thinning the canopy, and burning undergrowth and the pride he took in passing on this knowledge to his daughter and son. Some songs celebrate the diverse uses of the sago palm. For instance, Oktavius's songs often revolved around the textures of the sago bark from which villagers fashion machete handles and the pliable sago fibers they weave into ritual pig cages and ceremonial cases. Oktavius also described the sago sap used by men to bind arrow tips and fix canoes, whose resilience enables them to travel and hunt far and wide. The women, including Evelina and Andreana, meanwhile, sang about carrying leftover sago bark to the village as fuel for the hearth and scattering leftover sago pith in the grove as bait for pheasants. They also described the sago fronds we slept in when bivouacking in the grove, whose form folded slightly over our bodies and protected us from insects while also keeping us cool.

Other songs I heard in the grove narrated the particularities of the sago tree being felled—the color and shape of its spines, the texture and wetness of its pith, the epiphytes growing along its rugged bole, the size and regularity of its fronds, and the scars left behind by parasitic insects and molds. My companions also sang about the plants and the insects and other animals that live on and from the sago's bole, fronds, fruit, and flower. These songs, Evelina explained, pay tribute to the multiply populated lifeworld of sago and remind the palms themselves of their relations to diverse human and other-than-human organisms. The longer people sing, Gerfacius added, the more the palms' core softens. The pith swells and the bole expands. The trees gradually start to yield their pith and wetness, which in turn enhances the quantity and flavor of their flour.

When people take turns felling, chopping, rasping, pounding, and leaching, they pick up the song and beat of those whose labor they continue. By the end of the working day, everyone has participated in the collective chain of singing and processing some dozen times or more. If inspiration runs dry, community members improvise a new song and theme. They commend each other on unexpected innovations, additions, and variations on familiar tunes. They tease each other about their singing abilities. They also sometimes incorporate comical events and anecdotes from daily life in their lyrics, which draw peals of explosive laughter from their companions. Songs thus become auditory chains of affect, to which different community members contribute their experiences and memories through the creative verses they perform.

These living ribbons of braided words, thoughts, and feelings embody the shared identity of those who know how to encounter sago and accrue aesthetic and affective value in symphony with other-than-human voices in the forest. As anim, amai, and sago grow, reproduce, and die, their voices interlace, mingle, and transform, enlivening the realm of the forest-in-song.

The most extraordinary moment, my friends often told me, occurs when song, sounds, and the bodies producing them fall into perfect harmony. I first experienced this moment late in the afternoon of my eighth day in the grove with my companions from Bayau. Around us, swaying sago fronds rustled to the rhythm of children's giggles. Mud squelched underfoot to the snapping of fronds pulled or chopped off the bole. The regular rasping of the bole interior synchronized with the shy whistle of a khaw and the high-pitched song of a child. Multiple voices came together in an echoing choir of tonalities and vibrations distributed across species lines. In this moment, my friends described with awe and excitement how they could feel their amai (plant and animal kin) and dema (creator spirits) in and through their bodies, moving their arms and legs, as they did at the beginning of time, when anim were enskilled by dema in the arts of sago processing. Stefanus, an elder from Bayau and cousin twice removed of Gerfacius, described this experience as follows: "It happens after many, many times repeating the same movement. At first, you are tired, your joints hurt, you feel hot and sweaty. But then, the sago becomes soft under your hand. You hear the voices of the forest. Then, when all the sounds come together, the tiredness vanishes, as if you had been pounding sago since you were born. Suddenly, you feel like someone else is moving your body and singing with your voice—dema, amai, sago itself. Suddenly, anim become one with sago." Stefanus's description of the apogee of sensory immersion in the grove resonates with the spiritual force of listening (ho'olono) in Hawaiian epistemology, as described by Hawaiian elder, scholar, and activist Manulani Aluli Meyer. To really listen, Aluli Meyer (2001, 132) writes, is to actively become "lifted beyond the mundane"—to become intimately tied to human and more-than-human others through conjoined processes of doing, feeling, and knowing. The experience recounted by Stefanus also brings to mind the spiritual phenomenology of bir'yun, or "shimmer," among Yolngu in Aboriginal Australia. As they dance and sing together throughout the night, Yolngu become acutely aware of and attuned to the multispecies relations and pulses surrounding and entering their bodies (Rose 2017, G53–55). Just as bir'yun suffuses Yolngu when they connect to the earth and ancestral power through the medium of performance and music, Marind in the grove become one with amai, dema, and elements through their shared movements and sounds.

As they work, sing, and share skin and wetness with the grove, community members attune to the lively pulsations of its lifeforms, spirits, and elements. In kinesthetic and acoustic unison, they experience a heightened awareness of the sentient beings surrounding and penetrating their working flesh and fluids, whose presence coalesces in the movements and sounds generated by sago pounders themselves. This produces a powerful and affectively charged energy across species and spirit lines—one akin to the boost of extra-being, which philosopher Brian Massumi suggests arises during communal events that surpass the individual self (2015, 199–202). In these moments of collective effervescence, Marind *become*, as Rupert Stasch eloquently puts it, "the trees they fell, the sago they pound, the streams they bail, the fish they get from those streams, the holes they dig, and the paths they walk along" (2009, 146). Sago that thrives in this rich bioacoustic environment is not only good to eat but also good to hear.

ACCOUNTS OF HUMAN-SAGO ECOLOGIES have often focused on adaptation and nutrition in exploring the significance of sago palms to their consumers (see, for instance, Ellen 2011; Ohtsuka 1983; Ulijaszek 2002). Such accounts, however, do not account for the cultural and affective significance of palms outside their economic and ecological functions. Certainly, sago is of central significance to the majority of Marind as their staple source of food and other daily materials. But the affective and sensory dimensions of being-in-the-grove reveal that sago-human relations extend far beyond the pragmatics of subsistence. In the Upper Bian, sagoscapes arise from interspecies minglings that together produce the grove as a perpetually shifting and relational realm of more-than-human liveliness—a dynamic semiotic-ecological system in which people and plants communicate with each other through trophic flows of matter and meaning. By immersing themselves in the grove and ingesting the nourishing flesh of sago, Marind "take in the world" (Bakhtin 1984, 281) as it is produced by, and with, a multitude of other organisms.

Plants, as philosopher Michael Marder notes, may not engage in deliberation or dialogue in the way humans do, but they nonetheless "get involved with their environments, one another, and other species in sophisticated and responsive ways" (2016, 45). Originally derived from the Latin *involvere* meaning "to roll into," the notion of "getting involved with" conveys the fleshy, sensual ways in which species become imbricated in each other's lifeworlds. For Upper Bian Marind, getting involved with sago entails walking, noticing, listening, singing, and eating as crucial elements in the perceptual and pedagogical immersion of humans within the phenomenal lifeworld of the grove. In doing so, community members fashion their own identity as <u>anim</u>, or humans, in relation to sago and its diverse companion species. Being-in-the-grove thus enfolds the lifeworlds of the diverse beings who sustain, and are sustained by, the sago palm. It also encompasses practices of care that take the form of practical labors, affective investments, and ethically imbued engagements across species lines (cf. Puig de la Bellacasa 2012). Care in the grove, to borrow Haraway's (2011) words, is wet, emotional, and messy. Water, mud, sweat, and other nourishing substances imbue this space with a cosmological wetness distributed across the skin of myriad organisms who are involved and invested in each other's lifeways.

As Andreana's statement in the opening of this chapter suggests, the taste of sago is difficult to express in words because it incorporates the countless things and beings that endow sago palms and sago pith with life and meaning—water, land, forest, <u>amai</u>, <u>dema</u>, and <u>anim</u>. This taste can only be understood by engaging with the sticky fleshiness and fluids of the grove, in a process I described in a phenomenological vein as being-in-the-grove. By participating in this kinesthetic and multisensory experience, people immerse themselves in the storied lifeways of sago while at the same time enhancing their own relations with humans and other-than-humans. Finally, eating and knowing sago are also politically imbued actions. By choosing to eat sago, Marind affirm their gastro-cultural connections to other Papuan communities as "sago people," in opposition to non-Papuan "rice people," whose presence and influence erode the material and affective ecologies that enable sago palms and sago people to flourish.

6. Oil Palm Counterpoint

In March 2016, I traveled to Merauke City with a group of Marind villagers to attend a sosialisasi, or awareness-raising meeting, concerning the implementation of the Merauke Integrated Food and Energy Estate (MIFEE) project. Marind from across the regency had been invited to the event, which was hosted by the Ministry of Forestry and the National Land Agency as well as a dozen or so representatives of oil palm companies operating or intending to operate in the region. As is usually the case in sosialisasi, the meeting was dominated by formal presentations from the government and corporations. The speakers described the expansion of oil palm as key to building the capacity of Papua's isolated tribes. Oil palm would provide communities with the money and resources they lacked and the development opportunities they needed. Marind would thus free themselves from a life of poverty in the forest. They would no longer depend on the land and rivers for their subsistence. Instead, they would be able to take up proper jobs, buy modern foods, and build a better future. Their children would no longer be naive, dirty, and stupid. It was time for Marind to join the world of progress. Oil

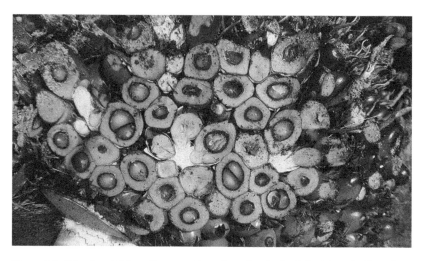

Figure 6.1. Oil palm children. Cross-section of an oil palm fresh fruit bunch. Photo by Sophie Chao.

palm, which the presenters called "the tree of hope" (pohon harapan), would make that happen.[1]

At one point, Agus, the human resources officer of the agribusiness conglomerate KORINDO, displayed a PowerPoint slide showing a dish of papeda, or stewed sago jelly, and a bottle of palm oil next to a packet of instant noodles. Agus commented: "Marind eat sago every day. Sago, sago, sago—always sago. If you work with the companies, you can eat instant noodles every day. Instant noodles are made with palm oil, and everyone in Indonesia eats instant noodles. Have you not heard the advert for Indomie? 'From Sabang to Merauke, Indonesia, my homeland, Indonesia, my taste.' All Indonesian citizens eat Indomie! Sago is poor man's food. You can't eat sago forever."[2] The Marind participants, who had so far remained silent throughout the speeches, stirred in their seats. Many looked angry. Some started whispering to each other, then grumbling, louder and louder. The meeting facilitator's call for order prompted a resounding ripple of protest from the audience. Then, Agustinus, a clan elder from Bayau towering at over six feet tall, marched over to the podium, and grabbed the microphone off Agus. His hoarse voice trembled with rage as he spoke: "You oil palm bosses and government officials have given many speeches. But don't you dare tell us not to eat sago. You know nothing about sago. You think we want instant noodles? You think we want palm oil? Marind are strong because they eat sago. Now, our children eat instant noodles and palm oil, and their bodies are small and weak. Don't ever take away our sago. Nobody wants oil palm. Oil palm kills the sago (sawit bunuh sagu)."

Agustinus's words were met with vehement applause. Several community members rose to their feet, chanting, "Oil palm kills the sago." The commotion in the room was extraordinary. Agus, the corporate human resources officer, hurriedly stepped off the podium and retreated to the back porch for a cigarette. He had triggered the visceral anger of my companions by attacking a plant with deep-seated and affectively charged significance—one whose existence is directly and increasingly threatened by the expansion of oil palm. Repeating "Oil palm kills the sago" in unison, all Marind in the room stood up and walked out, with Agustinus in the lead. My friend paused in his stride when they reached the front gate. He glanced back, tensing his muscles, and running his fingers repeatedly through his thick, matted hair. Kicking the dust at his feet, Agustinus muttered to me, "You know, they almost had me with all that oil palm talk. Their promises and commitments and all. But then they said those things about sago. And that I cannot accept. That I cannot accept because oil palm kills the sago."

IN THIS CHAPTER, I examine the ontology of oil palm and its contrapuntal relation to the sago palms we visited earlier in the grove. Like Agustinus, many Upper Bian Marind frame the socioenvironmental transformations wrought by agribusiness through a series of contrasts between oil palm and sago palm, as social actors of contrasting attributes, needs, and affects. The relation between these vegetal protagonists, and between everything they represent, is primarily antagonistic. Recall here the song of Marcus Gebze that opened this book: "Oil palm killed the sago, oil palm killed our kin. Oil palm choked our rivers, oil palm bled our land." The expression "oil palm kills the sago," invoked by Augustinus and echoed by the meeting participants, was one that I heard frequently throughout my fieldwork. When I would ask my companions what they believed to be the most devastating impact of agribusiness development, almost all responded with this same statement. Similarly, my friends' speculations about where oil palm comes from, what it wants, and how it grows were invariably set in counterpoint to the attributes of the sago palm. Oil palm's opaque and threatening ontology was invariably cast into high relief against the affective companionship of sago and its giving environment in Marind discourse. As Ruben, a young Mirav villager, told me during one of our many land rights advocacy meetings, "If you want to talk about oil palm, then you must talk about sago."

Originally derived from the combination of different colored threads in quilt-stitching, the term *counterpoint* is used in music to describe the combination of two or more independent instrumental or vocal melodies. The counterpoint works through contrast to highlight harmony produced by the interplay of different cadences, colors, and concepts. This synergetic interplay, German biologist Jacob von Uexküll suggests, lies at the heart of the composition of nature itself. Just as at least two tones are needed to produce a harmony in music, so too nature arises from the harmonious relationship between a meaning-utilizing subject and the meaning-carrying objects encompassed within the subject's Umwelt, or perceptual lifeworld. Nature, as such, arises from the contrapuntal relation of living organisms and their environment or milieu (von Uexküll 1982, 52–54).[3] By the same token, species, as Marisol de la Cadena puts it, are complexly and relationally multiple—"they are *with* what they are not" (2019, 480). The counterpoint, as an analytical and empirical optic, thus foregrounds how things and beings accrue meaning through their differential yet relational juxtaposition to *other* things and beings. Difference—organismic and ontological—is the central operative of the counterpoint. But more important yet is the difference that this difference makes for the lives, futures, and well-being of the various parties involved in the counterpoint-as-relation.

The approach to the counterpoint deployed by Cuban novelist Antonio Benítez-Rojo in his analysis of Caribbean plantation histories is useful to think with here. As Benítez-Rojo notes, the emphasis on homeostasis and symbiosis underlying von Uexküll's theory of nature-as-counterpoint is not the whole picture. Counterpoints are not always or necessarily harmonious (Benítez-Rojo 1996, 150–76). Some produce melodic congruence. Others generate cacophonic dissonance. The parties to a counterpoint, to return to von Uexküll's words, are not always "made for each other" in a positive or mutually beneficial way (1982, 54).

Note that we have already encountered several ambiguous counterpoints in our story so far: tortured landscapes and moving maps; forest birds and plastic drones; humans-turned-cassowary and cassowaries-turned-human; dividual bodies and individual minds. Like these and other awkward couplings, the counterpoint of oil palm and sago palm stressed by my friends Agustinus and Ruben is far from harmonious. The relentless proliferation of the former results in the violent destruction of the latter, along with that of the forest and the diverse storied selves that animate it—human, other-than-human, and elemental. Entangled in a weave of fraught consequence, sago and oil palm acquire heightened significance in relation to each other's distinctive lifeways as two extremes of a symbolically charged moral-vegetal spectrum—one that nonetheless remains shot through with abu-abu.[4]

In the Upper Bian, human and other-than-human organisms collectively produce the living landscape through their intersecting trajectories of movement and growth. Organisms within this sentient topography accrue value through their skin-to-skin, wetness-to-wetness interactions with other meaning-making lifeforms. The sago grove in particular is cherished by Marind as a realm of sociality animated by the many lives of organisms sustained by the palm's growth. In contrast, oil palm thrives in heavily guarded concessions that, alongside other state and corporate pressure points, erode the dynamic topography of the forest and rupture the flowing paths of plants, animals, and humans. The plant's expansion fragments the sympoietic lifeworlds of human and other-than-human beings and dislocates them from their constitutive relations and environments. In the monocrop ecologies replacing the diversely populated forest realm, few species other than oil palm can survive. Oil palm, Anna, a young activist and former schoolteacher from Mirav, told me, refuses to coexist with other lifeforms because it "does not like to live with others" and "prefers to be alone." Similarly, Selly, a middle-aged widow from Bayau and cousin twice removed of Anna, described oil palm as a selfish plant that "has few friends" and that "does not want to share space with others." The plant's relentless proliferation is also associated by some Marind with insatiable greed. "When oil palm is hungry," Selly continued, "it eats everything, just like the bulldozers and the government. It eats the land and it eats the water. It kills <u>amai</u> and it kills <u>anim</u>. Oil palm eats everything. It leaves only scraps for the others."

In the sago grove, the voices and songs of myriad elements and beings mingle in harmonious polyphony, giving rise to a lively multispecies sensorium. As agribusiness expands, this realm of "enchantment"—a word itself derived from the verb "to sing"—is replaced by the *dis*enchantment of the plantation. Here, an uncanny silence presides, punctured only by the sounds of destruction—trucks, bulldozers, chainsaws, and fume-chugging mills. Songs in the grove celebrate the abundance of life and nourishment derived from sago and its environment. In contrast, songs such as that of Marcus Gebze in the opening of this book, describe the threat posed by oil palm to sago, humans, and other organisms, as they are displaced or uprooted from their land and kin. Much like the songs of Muyu refugees in Papua New Guinea, whose exile is accompanied by a deep yearning for sago and for the memories and social relations embodied within it (Glazebrook 2008, 95–102), oil palm songs speak of sorrow, loss, and death. Anna and Selly's songs, for instance, often compared monocrop plantations to the Freeport McMoRan mine in highland Papua, the largest gold and copper mine in the world. Freeport, they explained to me, is like the agribusiness concessions in that it operates through a business-state-military nexus. It pollutes waterways and undermines the livelihoods and rights of Indigenous Papuans. The riverine disposal of tailings and waste rock by the mine also contaminates and destroys the sago stands of downstream Kamoro communities. Like MIFEE, my friends sang, "Freeport kills the sago."

In the grove, sago, <u>amai</u>, and <u>anim</u> exchange bodily wetness and nourish each other's skin in ways that communicate and enhance their relations of substance. These interactions are rooted in an ethos of restrained care, whereby Marind establish reciprocal relations with other beings while also respecting their fundamental autonomy. Unlike sago, oil palm does not know how to share skin and wetness with other beings. It subverts the multidirectional traffic of bodily flesh and fluids that enables organisms to form meaningful relations with others. As the tributaries of the Bian River are diverted to irrigate plantations and forests are cleared to make way for monocrops, the skin of land and river bodies becomes scarred, mottled, and muddied by tread marks, rubble, and chemicals. A deadly and generalized cosmological desiccation disrupts the aqueous skinships binding humans to forest lifeforms and elements. Reciprocal flows of substance between sago and humans are hindered by a plant-being that enacts unrestrained violence—rather than restrained care—on the organisms and ecologies it undermines or destroys.

The social taste of sago arises from the collective labor and activities entailed in the much-enjoyed practice of going to know sago. As sago groves diminish, people are no longer able to generate or share bodily wetness through communal hard work. Instead, they increasingly consume processed goods purchased or received from agribusiness companies, such as rice, cookies, and instant noodles like those displayed on Agus's PowerPoint. These foods, Ruben affirmed, do not taste good because they "do not taste of the grove." They make bodies frail because they are devoid of nourishing wetness.[5] As we explored in the previous chapter, sago's valued taste arises from the physical and affective investment of those who process its pith. In contrast, no one knows who produces processed commodities or where they come from. More than this, processed foods are said to exacerbate the hunger of those who consume them. Parents, for instance, complained of their children clamoring for more food within hours of eating instant noodles. Women including Anna and Selly described how they would snack on cookies throughout the day but always crave more. Ruben and other young men, meanwhile, talked of having become addicted to rice, which they would eat in copious amounts without feeling full. In contrast to the satiety offered by the nourishing foods of the grove, processed commodities give rise to new and different kinds of hunger.[6]

Foods like rice and instant noodles are also associated by Marind with the oppressive and unsolicited presence and influence of non-Papuan migrants and the Indonesian state. As commodified foodways invade the villages, the gastro-identity of Marind as sago people gives way to "gastro-anomie," sociologist Claude Fischler's (1979) term for the anxiety caused by the breakdown of traditional food systems and the social relations they encompass. This gastro-anomie is, in turn, symptomatic of what Chamorro poet, scholar, and activist Craig Santos Perez (2013) calls "gastrocolonialism," referring to imposed alimentary regimes that perpetuate the violence of colonization and neoimperialism through the medium of food (see also Chao 2021e). And yet ironically, many of the commodities replacing sago and other traditional foods contain the very ingredient responsible for the destruction of the forest—palm oil. The fact that the oil is usually labeled as generic vegetable oil, or as one of more than two hundred different organic compounds, adds elusiveness to ubiquity. Maria, a teacher from Mirav, commented on this while we waited in line at the village kiosk to purchase cookies for the students' mid-morning tea. "Oil palm is everywhere," she muttered, "but it keeps its secrets well."

Sago creates bonds of substance between those who produce and consume its nourishing pith. Conversely, the arrival of oil palm creates conflict within and across communities who are increasingly divided over the benefits promised by the agribusiness sector. As Viktorinus, an elder from Khalaoyam explained, "Now, the clans no longer share sago. Instead, the clans eat each other up." While many villagers like himself actively oppose oil palm projects, the lure of compensation has given rise to widespread elite cooptation and collusion. Some community members, Viktorinus continued, have ceded their lands (and those of others) in exchange for money. Others now act as informal middlemen for the companies. These individuals forget how to eat sago and instead "eat money," "eat from oil palm," or "eat from the companies."[7] Many have moved out of the village to live in the company precincts for fear of retribution from the kin they have betrayed—including three of Viktorinus's own nephews. Clans that were once united through their shared experiences of being-in-the-grove now turn against each other over cash and contracts.[8]

The decimation of sago groves has particular implications for women. Specifically, it robs them of the freedom and pride they derive from their hard work in the grove and its associations with the labor of making anim out of children.[9] As we walked through a recently razed patch of forest to fetch water from the Bian River, Selly spoke of how the serenity and sociality of the grove were disappearing as plantations multiplied. Instead, the landscape was becoming a place of anger, sadness, and suffering. The vulnerability of Marind in the face of these transformations takes viscerally corporeal forms for young women and girls. As rivers become polluted and access to the riverbanks is restricted by plantation boundaries, women must walk great distances to obtain fresh water. In doing so, many have suffered sexual abuse by plantation workers and military men.

Finally, and as the speech from Agus the corporate representative exemplified, the promotion of agribusiness expansion frequently denigrates the sago-based livelihoods from which my Marind companions derive their sense of collective identity and belonging as primitive and backward.[10] For instance, MIFEE is often framed in government rhetoric as an opportunity for Marind to progress from a sago-based livelihood to formal employment within the agribusiness sector. Echoing Paige West's critique of development, extraction, and conservation processes in Papua New Guinea, discourses of *capacity building* in West Papua are premised on, and perpetuate, the notion that development's subjects inhabit an archaic and retrograde world. Framed through the idiom of "lack" and consequently addressed through technical and external interventions, this developmental logic both elides and undermines Indigenous ways of seeing and being in the world (cf. West 2016, 65–66). Oil palm plantations and the modernization discourse surrounding them thus come to represent to many Marind the botanical arm of a long-standing and racialized process of colonization in West Papua—one that is rooted in the forceful incorporation of the region and its sago-eating communities into the Republic of Indonesia, and which entails their subjection to the growing influence and presence of rice-eating Indonesians. Like the state, settlers, and soldiers, taking over Marind lands, oil palm is both alien and invasive. It participates in the region's ongoing domination as a biological ally of the state and its hegemonic architectures of violence and power.[11]

THE MARIND EXPRESSION "OIL palm kills the sago" crystallizes the contrapuntal relationship at play between these two vegetal beings—one that is characterized by violence and asymmetry, rather than harmony and complementarity. The expression brings to mind the phrase "the soy kills" (la soja mata) deployed by peasants in the context of the Paraguayan soy boom.[12] This formulation, Kregg Hetherington notes, suggests that Paraguayan farmers attribute the environmental destruction wrought by agribusiness expansion less to the human actors and institutions driving it than to the being of soy itself

(2013, 66–68). Similarly, many Upper Bian Marind attribute the destructive effects of monocrop expansion to the voracious and selfish disposition of oil palm—a disposition heightened by the cash crop's contrapuntal relation to the nourishing attributes of the sago palm.[13]

Oil palm's solitary and selfish mode of existence is also what distinguishes it from other plant and animal species that were introduced into New Guinea in prehistorical and historical time—for instance coconut, rambutan, durian, mangosteen, jackfruit, pigs, and deer. Alba, a mother of four, explained that oil palm grows alone in industrial monocrops, destroys forest ecologies, and does not procure food or resources for Marind. Introduced plants and animals in the forest, in contrast, have learned to "live with each other like friends and family" (tahu hidup bersama-sama macam kawan, macam keluarga). Their symbiotic relations to one other, and also to native species, speaks to their ability and willingness to make a home for themselves in the company of others—a trait that oil palm appears to lack. This capacity is in turn what explains the recognition of some introduced species as kin, or <u>amai</u>, of particular Marind clans—for instance, the dog for the Mahuze, the coconut for the Gebze, and the pig for the Basik-Basik. While the precise origins of introduced yet familiar species are said to be unknown, numerous stories and songs recount how they arrived on the shores of New Guinea aboard canoes guided by the voices of ancestral spirits, or <u>dema</u>, and how they gradually spread out across the landscape following the migratory flows of consecutive Marind generations. Echoing the contingent identities of animals and humans across forest and village examined in chapter 4, the significance of these species for Marind is thus shaped less by temporally and geographically fixed referents of origin than by the nature of the *relations* that they entertain with humans and with other forest lifeforms.

Yet the destructive ontology of oil palm is only part of the story. Over the course of the last decade, Marind have received information from agribusiness companies, NGOs, and state representatives about oil palm's origins, growth, biotic affordances, industrial cultivation, and global trade circuits. I myself participated in this process by imparting, at the behest of my friends, information about oil palm's natural history, commercial uses, and transnational supply chain. In the last few years, some community members have visited monocrop oil palm plantations in Java as part of company-sponsored comparative-study fieldtrips known as studi banding. Many young Marind have also attended plantation management and agronomic science trainings facilitated by corporations as part of their promotion and recruitment campaigns. In the absence of other reliable sources of income in rural Merauke, some have started to work as pesticide sprayers and harvesters within plantations, where they interact daily with oil palm and participate in its transformation from plant to commodity.[14]

The knowledges that Marind acquire in these settings are transforming their understanding of oil palm's reality, exemplifying what David Gegeo and Karen Watson-Gegeo call the "indigenization of knowledge." The indigenization of knowledge, Gegeo and Watson-Gegeo explain, refers to the reconstruction and resignification of outsider knowledge to fit within Indigenous and local practices, experiences, and culture (2002, 381). On the one hand, this process of indigenization constitutes a form of epistemic self-determination, whereby Marind actively refashion and repurpose introduced knowledge in a distinctively Marind way. But indigenizing oil palm also complicates the plant's otherwise seemingly straightforward contrapuntal relationship to the sago palm.

Kosmas, a young Marind man who had just graduated from high school and taken up employment as a plantation laborer, explained to me how his understanding of oil palm had changed over time: "Oil palm is greedy and selfish, for sure. But the more time I spend with oil palm, the more I have come to realize that this is not the whole story. Oil palm is also a victim. Oil palm comes from a faraway place, but oil palm is also our kin. Oil palm, too, is a tree of many lives." For Kosmas and others, oil palm is a destructive assailant—*but not only* (cf. de la Cadena 2014). At the same time as Marind deeply resent the plant for its devastating effects and self-interested proliferation, many *also* express pity and compassion toward it in light of its exploitation at the hands of humans, institutions, and machines.

For instance, when I explained the palm oil production process to Marind friends, many of them deplored the fact that every stage of the plant's growth is determined, or controlled, by human agents. Christofus, a young man from Mirav, lamented the fact that the temporality of oil palm's existence is subjected to countless biological and technological manipulations that regulate its development, form, and uses from seedling to commodity. The plant, he had learned from corporate representatives, is artificially bred through controlled pollination and its fertility and yield are boosted with a range of chemical fertilizers to achieve corporate targets of production.

Meanwhile, villagers who had visited the palm oil processing mill described how the cash crop's own lucrative wetness, palm oil, is forced out of its body through high-heat, high-pressure processes of mechanical extraction—sterilization, threshing, steaming, mulching, boiling, cracking, filtering, stripping, winnowing, crushing, diluting, purifying, clarifying, fractionating, churning, pressing, and more. Care enacted toward oil palm, as Marind plantation workers like Kosmas often reminded me, is selective and conditional on the profit the cash crop can generate. Seedlings that fail to achieve yield targets are culled en masse. Unproductive or diseased mature stands undergo a radical dismembering. Felled, deboled, defronded, chipped, and stacked by the voracious mandibles of trunk shredders, these faulty palms are pulverized in under four minutes.[15]

The fate of oil palm seeds and fruit, too, matter to this story. For instance, Beata, a young mother of two from Mirav, described to me how oil palm seeds are uprooted and transported to Merauke from distant places in trucks and boats. During agronomic trainings she had attended in Merauke City, Beata had learned that these seeds grow in sanitized plastic bags and industrial nurseries rather than in their native ecologies. Unlike sago suckers that remain organically connected to their families, these seeds, Beata explained, are displaced by the millions from their home environments and separated from their kin. It is important to note that the life of seed-propagated oil palm is not perpetuated through its offspring, as is the case in sago's vegetative reproduction, but rather terminated when its productive years have elapsed and the palm is felled. Meanwhile, Beata's cousin, Nataliya, who had also participated in the agronomic trainings, recounted with sorrow how oil palm fruit are ripped from their parents every fifteen days during harvesting, and then unceremoniously flung into overloaded trucks "like dead bodies" (macam jenazah). What happens to these fruit in the mill where they are transported is abu-abu. People see only dark puffs of smoke, trucks driving back and forth at all hours of the night and day, and finally, the deep-red crude palm oil extracted from the fruit, which Nataliya and others called "oil palm blood" (darah sawit).

My companions also criticized the monocrop model for forcing oil palm to grow alone rather than in reciprocal relation with other plants or animals. One who felt this way was Miriam, a middle-aged schoolteacher and the designated spokesperson for Marind women during meetings with state and corporate representatives in Bayau. As we walked along the boundaries of the plantation encircling her home village, Miriam stopped in her tracks and looked at the rows of palms stretching out into the horizon. She mused, "Oil palm, too, has a family and a home, somewhere far away. Oil palm, too, has a story. But where is its family, its land? Does it not miss them? Has it ever known what it is to be wild? How can it be happy, growing alone in the plantation? How can it grow well, cut off from the forest?" Miriam's reflection suggests that the loneliness of the plantation does not only do violence to the diverse forest lifeforms excluded from its midst. Monocrops also heighten the vulnerability of commodity cultivars to pest and disease epidemics by concentrating the opportunities for rodents, weeds, and insects to thrive and eroding the ecosystems that play host to the natural enemies of these pests.

Much as they do in the forest, plantation workers like Kosmas come to know and respond to oil palm's parasites by attuning in bodily and sensory ways to their presences and traces within the landscape. As we walked through the plantation, for instance, my companions pointed out the deep dents in the crown of an oil palm, where rats had nested and fed on the plant's developing fruit bunches, male flowers, and fallen fruit. Together with cuscus and porcupines that devour the terminal bud of young palms, Geronimo explained, rats reproduce extremely rapidly and can bring down hundreds of palms within a matter of weeks. More widespread and damaging in their effects than mammals are insects that repurpose oil palms as incubating hosts and sources of sustenance. Culprits identified in the plantation by Geronimo and other laborers include the chartreuse-green coconut nettle caterpillar that devours the leaves of young and mature palm leaves and wounds the hands that seek to remove it with the venom of its urticating spines. The larvae of furry *Mahasena corbetti* bagworms gestate on the undersurface of oil palm fronds, scraping away at the epidermis of leaflets and eventually causing palm necrosis.

The rhinoceros beetle is the most destructive of oil palm's insect parasites, according to Pius, a young pesticide-sprayer from Bayau. Under the cover of the night, the armored creature burrows through the petioles of young palms and developing spears of the crown into the soft tissues of unopened leaves, mining labyrinthine galleries through the terminal bud, thriving in the rotting stems of felled palms, and laying its eggs in empty fresh fruit bunches. Fungal parasites of the palm, too, can trigger epidemics of dramatic proportions. Stem, root, bud, spear, and leaf diseases caused by fungal epidemics include Fusarium wilt, white stripe disease, brown germ, and Curvularia leafspot and blast. In the plantation, Pius pointed to me the patchy discolorations and sooty molds that had developed along the fronds of trees afflicted by these diseases. Their leaves had begun to fracture and desiccate. Eventually, the palms would suffer generalized chlorosis and die as fungi appropriate the water, minerals, and light necessary for their growth. Most resistant to chemical and biological controls among oil palm's fungal "enemies" is *Ganoderma*, the fungus that had decimated a third of the palms in the plantation near Pius's home village and for which no long-term cure has yet been identified. *Ganoderma*, Pius explained, spreads through the root systems of host and neighboring palms, where it can remain dormant for several years before bursting into action. Fructifications in the form of hard, orange mushrooms then appear along the oil palm trunk. The root cortex begins to rot while fungal sporophores multiply without bounds. The base of the palm stem blackens and exudes a viscous gum. Eventually, the crown of the tree falls off and the trunk collapses.

As Miriam's astute comment implies, the severance of palms from their native home, the modular simplifications that undergird the monocrop model, and the feral ecologies these simplifications enable all can have potentially lethal consequences for cash crops themselves.[16] The entanglements of palms and parasites in particular suggest to some villagers that oil palm, while indubitably harmful in its effects on forest lifeforms, is not itself immune to more-than-human violence. Rather, different necrobiopolitical assemblages alternately enable or undermine the cash crop's proliferation.

The compassion that villagers express toward oil palm stems from the plant's totalizing subjugation to human domination—a subjugation that violates the principle of restrained care, which animates the relations of Marind to forest organisms. Echoing the plight of Ruben the "plastic" cassowary, the plantation embodies in extreme form the loss of freedom of wild beings turned domesticates. Coerced into global circuits of mass production, the vitality of its resident palm is harnessed and engineered to serve human ends, as a replaceable and homogeneous form of abstract labor. Similarly, the plant's growth and death are dictated by the demands of capitalist production, in a parasitic exploitation of the unique temporality of vegetal lifeforms. Grown and manipulated only to be ultimately killed and consumed, oil palm trees are robbed of their capacity to exist as autonomous and consequential historical subjects. Their mode of survival under the plantation regime is at once infernal and artificial.

From this perspective, oil palm bears an uncanny resemblance to the zombies of post-Apartheid South Africa described by Jean Comaroff and John Comaroff (2002), who incarnate persons reduced to alienated ghost labor, abducted from their homes to feed the fortunes of depraved strangers as their privatized means of production. Like the zombie, oil palm–turned–palm oil is an "undead thing" (Haraway 1997, 133)—a captive commodity whose life, death, and after-life as plant, part, and product is exploited and impoverished within global circuits of consumption (see also Collard 2014, 153–54). The zombie figure as "undead thing" disrupts the Life-Nonlife distinction that Elizabeth Povinelli (2016) identifies as a key trait of late neoliberal capitalism.[17] It also captures powerfully the layered and abu-abu ontology of oil palm as exploited and pitied victim, destructive and haunting killer, and lively and lethal capital. As the forest burns to make way for a plant that takes everything but gives nothing back in return, palm oil itself is burned as calories and combustibles by bodies and vehicles in places and by people far removed from its native roots. And while real fires ravage the incinerated landscapes of the Upper Bian, oil palm also fires the imagination of Marind in a complex combination of pity, animosity, and curiosity.

Indeed, many of my companions expressed great interest in the life of this introduced and proliferating plant-being, whose origins, modes of growth, needs, and movements—in other words, whose story—remains largely unknown to them. The mystery surrounding oil palm's story stands in marked contrast to the fleshy stories of sago palms and sago people described previously. Sago stories form a textured sagoscape of connected sites, events, and actors. They embody and communicate the vegetal-animal meshworks of life through which palms become living repositories of relations past and present. In contrast, making stories with oil palm is almost impossible because concessions remain off-limits to nonpersonnel. Under the regimented surveillance of the monocrop landscape, physical encounters with oil palm are rare and dangerous. Consequently, oil palm is devoid of the affectively charged fleshiness that enables organisms in the forest to respect and respond to each other in transformational interspecies encounters. Instead, the plant exists through the meaningful voids and obstructions it creates within the landscape—the vast swaths of razed vegetation, the oily waters of the Bian, the poisoned fish strewn across the riverbanks, and the choking haze enshrouding the villages. The cash crop itself has yet to be planted in many concessions, but its presence is already violently tangible in the haunting rubble and ruination of the forest.

Like Miriam, many among my companions would ponder at length over where oil palm's home is, who its kin are, and what relations it sustains with humans and other beings in its native soils. Speculation abounded over oil palm's multisited flows as plant, part, and product along dispersed channels of provision, distribution, and accumulation. My friends frequently asked me how the plant—an introduced species in Merauke—traveled from its native Africa to West Papua. Some were fascinated by how oil palm had managed to spread so successfully across almost half the globe. Community members wondered how oil palm grows in the other places where it is cultivated industrially—for instance, Columbia, Nigeria, Ecuador, India, and Peru—and more important, whether oil palm is as destructive in these distant countries as it is in Merauke. Others were curious as to how plantations in the Upper Bian connected them to palm oil–consuming communities across the Indonesian archipelago, and beyond vast oceans to the biofuel-hungry nations of Europe and the United States.[18]

The association of oil palm with global connections among Marind resonates with the characterization of coffee among Gimi of Highland Papua New

Guinea (West 2012, 101–29). Like coffee for Gimi, oil palm is devoid of "value" for Marind in that it does not generate gendered human and more-than-human socialities in the way sago does. On the other hand, however, oil palm enables Marind to form relations *outside* their own society with global markets, traders, and consumers that Marind associate with modernity. As with sweet potatoes and coffee among Gimi, sago palm is key to Marind being in the world, but oil palm "makes the world bigger" (West 2012, 29) for them.

At the same time, the opacity of oil palm's story is heightened by the fact that few community members know where oil palm seeds originate, what trajectories they take to reach Merauke, and how they travel beyond West Papua. Throughout its serpentine journey from seed to frying pan, palm oil as a product flows through the palms of multiple actors across various sites (see Chao 2022b). From biotechnology labs to nurseries, plantations, mills, and refineries, the oil travels to manufacturing plants across the globe, mixing with oils and ingredients from other sources, only to disperse and form new liquid mélanges as it is loaded onto ships, planes, and trucks for transport to warehouses and stores. Yet for Marind on the ground, the processes, labors, actors, institutions, and relations that coalesce and congeal in palm oil remain shrouded in mystery. Similarly, the users of palm oil produced on Marind soils live in distant places and their identities are unknown. The plant's ambiguity is thus exacerbated by the invisibility of the human actors driving and deriving from its expansion as producers and consumers. Once again, we are in the world of abu-abu.

The abu-abu being of oil palm is further inflected across scale and substance, giving rise in turn to different affective responses among Marind. When encountering palms as *individual* stands, for instance, curiosity tends to predominate. In the same way that Marind read the pasts and relations of forest organisms through tactile engagements with their individual morphologies, people often stop when traveling along or inside plantations to scrutinize and touch oil palm stands at length—their bark, fronds, and fruit—in an attempt to "know its skin and story." Large-scale monocrop concessions, as oil palm *collectives*, on the other hand, produce fear and anger among those who behold or enter them. The seemingly relentless expansion of these massive biocapitalist formations conjures for many the systematic occupation of the West Papuan territory by Indonesian military forces and corporations. Gendered and generational differences within Marind communities also shape the prevalence of certain depictions of oil palm over others. Oil palm seeds and fresh fruit bunches, for instance, were the object of widespread pity among Marind women, who often described them as oil palm "children" (anak sawit) or "orphans" (anak yatim). In contrast, the characterization of oil palm as a foreign and invasive plant-being was particularly prevalent among male village elders, who had lived through the forceful incorporation of West Papua into Indonesia, the beginnings of large-scale resource extraction, and the influx of settlers into the region.

Finally, clear-cut symbolic and moral counterpoints between oil palm and sago palm are troubled by uncanny *similarities* that Marind identify between these two vegetal beings. Many Marind community members, for instance, noted morphological resemblances between the two palm species and described them as cousins or siblings of the same family. Others compared the fertile wetness of sago pith to the lucrative wetness of oil palm as an oil-producing plant. Both sago and oil palm, my friends recognized, serve an array of purposes as sources of food and other derivative products. Just as sago connects Marind to other New Guinean sago-eating communities, the diverse processed foods containing palm oil that Marind now purchase and eat also incorporate them within *transnational* communities of palm oil consumers—even as their constituents across the globe remain remote and opaque. Perhaps most intriguing to many of my companions was the scientific name of oil palm itself—*Elaeis guineensis*, or the "olive of Guinea." This name, my friends speculated, pointed to the possibility of shared origins and kinships between themselves as Melanesians and their home island of Guinea and the African peoples of Guinea and Guinea-Bissau, where oil palm grows endemically.[19]

Characterizations of oil palm as a kin of sorts were particularly prevalent among villagers like Kosmas who now work within plantations as fresh fruit harvesters and pesticide sprayers. These individuals have chosen to work in the oil palm sector in order to improve their material well-being and provide for their children and families. They interact directly with oil palm and support its healthy maturation through their various practical labors. The material-semiotic relations they have come to cultivate with the plant thus differ from those of Marind who do not work in the plantation sector and who therefore know oil palm only through its destructive effects.[20] As Abraham, another young plantation worker explained, "My job is to care for oil palm and support its growth. I share wetness with oil palm every day when I sweat and toil in the plantation. I have come to know oil palm's wetness and skin, and oil palm has come to know mine. Oil palm has become a kind of kin to me." Kosmas and Abraham's observations thus complicate the otherwise radically different attributes that distinguish sago palm (and sago people) from oil palm and the colonial-capitalist forces driving its expansion. Both "near kin and alien colonist"

(Haraway 2011), oil palm is a problematic other because it embodies at once too much, and yet too little, alterity.

IN THE PREVIOUS CHAPTER, I described how plant-human assemblages in the Upper Bian arise from and produce the storied lifeways of sago palms and sago people. Stories and storying, as Māori scholar Linda Tuhiwai Smith (2012) notes in the context of Indigenous activism, are at once a way of knowing and a way of being—a means of reclaiming the past, testifying to injustice, and forging possible futures (see also Archibald, Morgan, and Santolo, 2019; Kovach 2009). In this regard, stories constitute powerful strategies of Indigenous resistance and self-determination. Stories also actively participate in the worlds that they describe. Fleshier, livelier stories make us care more—or if not more, then perhaps differently (van Dooren 2014, 10; see also van Dooren and Rose 2016, 89).

The desire of many Marind to know the fleshiness of oil palm's skin and story, too, can be conceived as a form of care—a sentiment that, as María Puig de la Bellacasa reminds us, may stem as much from grief and sorrow as from interest and anxiety (2012, 212). Care is thus intimately tied to curiosity, a disposition Michel Foucault describes as "the care one takes of what exists and what might exist" (1997, 325). In this regard, Marinds' curiosity and compassion for oil palm offer a powerful, if troubling, counterpoint to the prevailing characterization of plantations as radically incompatible with the capacity to love and care (Haraway, cited in Mitman 2019, 6). For Marind *do* care for oil palm—albeit in deeply conflicting ways and in spite of oil palm's ravaging effects and unwillingness to care in return. Part of this care entails a refusal on the part of Marind to remain indifferent to the lifeworld of oil palm *beyond* what they themselves experience in its destructive presence. Indeed, Marinds' animosity toward this introduced and harmful plant-being goes hand in hand with a recognition that oil palm, too, has a storied existence—with other beings, in other places, and at other times. Imagining what that storied existence might or could be in itself constitutes a form of care—one that is at once radical and speculative.[21] This kind of care thrives in shadowy spaces of absence, distance, and loss. It takes root in an agroindustrial nexus that, by its very infrastructure and logic, severs plants and persons from their ability to make fleshly stories together. Within this same nexus, care for oil palm offers a tentative architecture of hope—a hope for different more-than-human stories, relations, and futures.

As the story of oil palm remains abu-abu to most Marind, the possibility of establishing mutually beneficial relations with it—the "what might exist," to use

Foucault's language—remains uncertain. Yet many among my friends recognize that the plant is here to stay. It is a vegetal being that demands imaginative accounting *for* and practical commoning *with*.[22] Amid the troubled terrains of emergent capitalist landscapes, Marind must find ways to relate and survive with the significant otherness of oil palm—be it through outright acts of defiance and survivance or tentative forms of complicity and compromise.[23] The story of oil palm, then, becomes something *worth knowing* (cf. Meyer 2001, 125). It is fundamental to the forging of a new politics of interspecies habitability.[24] And it introduces the possibility of a more responsible sharing of suffering with oil palm in the making of difficult, yet necessarily joint, futures across species lines (Haraway 2003, 7).

The heterogeneous perspectives and imaginative speculations that coalesce around oil palm and sago palm among Marind draw attention to the consequential, power-laden, and situated nature of human-plant entanglements, or what I call *dispersed ontologies* (Chao 2018a). Dispersed ontologies and their constitutive interspecies engagements are pregnant with the possibility of both fusion and fission. On the one hand, the ontology of the sago grove is dispersed across Marind and the manifold agentive other-than-human beings with whom they entertain loving relations of care and reciprocity. No single entity, human or other, constitutes the center of this dispersed ontology. Rather, all beings participate equally, if differently, in enlivening the forest and each other's existences through the exchange of fluids within a generalized cosmological skin. Just as the ontology of sago disperses across the many species with which it coexists, so too the reality of oil palm multiplies as it manifests and transforms from plant to part and then to product, in an ongoing process of self-différance. Both plants, for better or for worse, constitute plural and disseminated beings.

From driver of ecological destruction to victim of human exploitation, oil palm accrues new and different meaning across the range of affectively charged discourses that surround it.[25] In this regard, the plant constitutes a potent and diverse object of wonder—a disposition that, as philosopher Mary-Jane Rubenstein reminds us, may encompass awe, amazement, marvel, dread, astonishment, shock, and horror (2008, 9; see also Scott 2016). At the same time, the ontology of oil palm disperses beyond Merauke as a vegetal nexus of disparate corporate, governmental, financial, and socioenvironmental interests and agendas—bane to some, bounty to others. Oil palm thus constitutes a heterogeneous boundary object (Star and Griesemer 1989) that is *enacted* into meaningful being through the practices and perspectives of situated individuals, communities, and institutions across space and time.

This multiplicity also characterizes the ontology of oil palm *beyond* its life-course, or *bios*, as a commodity of unmatched ubiquity and versatility. Camouflaged under some two hundred different names and forms, palm oil, too, is a skin-changer. It shape-shifts in myriad guises and uses across the globe. Capable of becoming any and everything, palm oil is chameleonic—it is a "metamorphic sublime" (Taussig 2018, 189). Indeed, despite their aversion to this destructive plant, Marind are aware that they themselves routinely consume palm oil in the form of instant noodles, cookies, and ketchup. As Maria, my fellow schoolteacher, reflected during our trip to the village kiosk to purchase snacks for the students, Marind are entangled daily (albeit often unwittingly) with this oil through countless foods, cosmetics, and toiletries. They absorb the plant's oil-as-wetness within their bodies. In doing so, Marind participate in the plant's dispersed reality as lively capital. They, too, partake in what Simryn Gill and Michael Taussig (2017) call a global "becoming-palm."

At the same time, the concept of dispersed ontologies highlights the *limits* to which interspecies relationings are possible and mutual in the capitalist ecologies of monocrop plantations. In the Upper Bian, oil palm refuses symbiotic relations with Marind and their forest kin. Instead, it ruptures the shared skin of forest landscapes and lifeforms and depletes their bodily wetness. The plant's self-interested and relentless proliferation destroys morally valued interspecies relations of *interessement*, sociologist Michel Callon's (1984) term for the processual locking of roles to different entities that are implicated in and sustain each other's identities. But oil palm itself is difficult to encounter directly as it grows in heavily guarded and privatized concessions or may not even have been planted yet. Bodily relations, such as those that animate the sago grove, are practically impossible to achieve with this elusive being, whom Marind experience primarily through its destructive effects rather than its tangible presence.

The dispersed ontology of oil palm in turn invites attention to the violence enacted both *on and by*, other-than-human lifeforms as they become incorporated and mass-corporatized within global market and trade regimes, or what Kaushik Sunder Rajan and other contributors (Rajan 2012) call *lively capital*. Attending to commodified lifeforms such as oil palm as lively but also *lethal* capital allows us to explore their damaging effects on places and people as unloving rather than—or as well as—"unloved" others. It highlights the violence enacted *toward* living capital as the unpaid labor force of the capitalist system, which is manipulated and exploited by humans in the production of Cheap Natures (cf. Moore 2015). It also foregrounds the diverse ways in which species become enlisted and transformed by capitalist-nationalist projects while also themselves shaping these human-driven projects through their own, sometimes unexpected (and unexpectedly destructive) agencies and effects.[26]

Oil palm in Merauke embodies the three coexisting aspects of lethal capital outlined earlier. On the one hand, the plant benefits from the forces of global capitalism driving its proliferation. Humans and machines have become increasingly dependent on the food and fuel it provides. As controversial as oil palm monocropping might be, the plant remains the most optimal of all vegetable-oil crops in terms of land-to-yield ratio. Some people might argue that the biological affordances and mass-domestication of oil palm is, in fact, a clever, if unconscious, evolutionary strategy on the part of the plant itself to promote its expansion—a form of reverse or mutual domestication.

At the same time, the machinations of human-driven capitalism that enable oil palm to proliferate en masse at the expense of the forest and its lifeforms *also* dictate the particular ways in which this expansion takes place and at what cost to a plant that is cultivated only to be ultimately pulverized into profit. Oil palm unleashes unrestrained violence on the landscapes it invades and colonizes. But its own life is also subjected to regimes of *violent care*, a concept deployed by environmental humanities scholar Thom van Dooren (2014) to describe the imbrication of care with coercion in interspecies relationships. Indeed, many Marind themselves recognize that the harmful effects of agribusiness arise from oil palm's entrapment within particular modes of political-capitalist exploitation, rather than, or alongside, the destructive attributes of the plant itself.[27]

In line with Marinds' own heterogeneous conceptions of oil palm, the concept of lethal capital helps highlight the contextual and transformational ontology of species and capital as loving or unloving and lethal or lively in the emergent ecologies of the present era. Indeed, Marind themselves rarely stick to any one particular account when describing oil palm's being. Rather, they shift constantly and contextually *between* different facets of the plant's dispersed ontology. Recent events or the acquisition of new knowledge related to oil palm, in particular, often result in an emphasis on one dimension of oil palm over another. For instance, the notion of oil palm as a "victim" became widely prevalent when a fungal epidemic destroyed some three thousand oil palm trees in a nearby plantation. Conversely, when some fifty hectares of land were razed overnight to make way for oil palm without the local communities' knowledge or consent, oil palm became once again the "killer of the forest."[28]

Oil palm in Merauke, then, is a concomitantly known, unknown, and imagined object of wonder, whose relationship to Marind is mediated by complex and contextual material-semiotic processes. The radically simplified ecology in which the plant grows produces a contrastingly heterogenous ecology of affects among Marind, replete with fear, pity, resentment, and wonder. Such intensified expressions of wonder speak to a radical reconfiguration of more-than-human realities and relations in the colliding worlds of monocrop plantation and multispecies forest. They also suggest that oil palm matters to Marind not just as a counterpoint to sago palm, but also as an abu-abu pressure point—one that destroys the liveliness of sagoscapes and everything they represent, yet one that also enables connections to an expanded world through its global circuits of production and consumption.

Much like the transspecies valence of wild/domesticated and native/alien distinctions among villagers and village animals described earlier, Marind discourses about oil palm also highlight the limits of the *species* as a category of analysis and being. The morality of oil palm differs across its various manifestations as plant, part, product, individual stand, or collective monocrop. In refusing to flatten oil palm to a singular and bounded ontology, Marind refuse to limit the possibilities of their own moral dispositions toward it. This is, to borrow philosopher Alexis Shotwell's (2016) terms, a stance *against* purity and *for* multiplicity—one of perpetual ontic differentiation and deferral, which in turn constitutes a form of epistemic resistance to the reductionist logic of the plantation model and its colonial-capitalist underpinnings.[29]

In contrast, polyvalent discourses of wonder surrounding oil palm and their diverse affective dimensions create what Marisol de la Cadena (2017, 2) calls "ontological openings" by making space for compassion and curiosity, alongside resentment and dread, toward this uninvited vegetal being of abu-abu skin and wetness. Recent events and new knowledge acquired by Marind from corporations, NGOs, and anthropologists, play an ongoing and transformative role in fleshing out the shifting ontoepistemic contours of oil palm's contested reality. The plant's ontology, in other words, is not an "either or" between different states of being, but rather a series of apposite yet accretive "ands"—"ands" that trouble the contrapuntal relation of oil palm to sago palm. At once lively *and* lethal, assailant *and* victim, plant *and* person, foreigner *and* kin, oil palm, too, inhabits plural realities across space and time. Oil palm, too, to return to Kosmas's words, is "a tree of many lives."

The Empty Sago Grove—The Dream of Agustinus Gebze

I was walking through the forest with my wife, my three children, my two brothers, and my uncle. We had been walking for a very long time. The children were complaining that they were tired. My feet were sore, and my mouth was dry. Finally, we came upon a sago grove. In the grove, it was cool and fresh. The wind was singing. The sago was swaying—*ssshhh, ssshhh, ssshhh.* We were happy. My children ran around with excitement, little feet pattering—*tuk, tuk, tuk.* We could forget about oil palm here.

My wife slumped down beside a cluster of sago suckers, panting heavily. She rubbed her stomach, where our fourth child was sleeping. We looked for sago palms that were nearly about to flower, because those are the ones filled with the most flour. We found one close to a rivulet and my wife said it was a good tree to fell. My brother and I started chopping the bole, but we were surprised because the trunk sounded hollow. It was like a <u>kanda</u> drum with a faulty snakeskin drumhead. We were confused. I touched the bark of the tree with the palm of my hand and the tree collapsed—*tsshhuuu.* It cracked open and we saw that it was empty on the inside. Just a shell of thin bark, but no pith at all.

We felled another tree, but that one, too, was hollow. We tried another, and another, and another, but all the sago trees were empty of starch, like the skins shed by snakes in the forest. The children began to cry again. Then we realized there was no food to be had here because all the sago trees were sick and had nothing to feed us with. We had to walk on, find another grove. And so, we set off again, heading toward the sinking sun. Walking, walking, walking, but without direction we walked. I walked so slowly, as though something was pulling me back, like a heavy stone in my heart. I knew what that darkness was. I had seen it in the grove. I could see it in myself. I knew that in truth, the sago was no different to us. It, too, was lost and landless. It, too, was empty. It, too, had no wetness. It, too, was without hope—hollow. It, too, had been eaten by oil palm.

7. Time Has Come to a Stop

On December 30, 2015, I accompanied my friends Antonius, Gerardus, and Yosef from their home village of Bayau to Hasanap Sai field in Merauke City to watch President Jokowi deliver a speech titled, "The Dreams of the Indonesian Nation." A local agribusiness company, PT PAL, had facilitated my companions' transportation to the event in the interest, the letter read, of "community advancement" (kemajuan rakyat) and so that Marind would not be "left behind the times" (ketinggalan waktu). We arrived late at the ceremony and the field was already packed with people. More than 1,500 soldiers and policemen had been deployed to supervise the event. Standing on our tiptoes, we could just glimpse the stage.

Two elementary-school students of mixed Papuan and Javanese descent stood on the podium beside the president, who wore a traditional Marind headdress bestowed on him by then-regent Romanus Mbaraka. A large metal cylinder sat at the front of the stage. Brought by Jokowi's delegation, it was referred to in the president's speech and in ensuing news coverage as a "dream capsule" (kapsul mimpi) or "vision capsule" (kapsul impian) containing the hopes of the

Figure 7.1. A modern totem. The Libra 969 monument, Merauke City. Photo by Sophie Chao.

Indonesian nation for the next seventy years. As the centerpiece of the Time Capsule Expedition, the capsule had toured 84 Indonesian provinces and 187 cities over the preceding three months. It had started its journey at kilometer zero in the westernmost area of Sabang (Aceh) and traveled 24,089 kilometers to Merauke City at the eastern tip of the archipelago. Here, the capsule would be buried and then unearthed in 2085 to mark the second seventieth anniversary of Indonesia's independence.

A press release read by the Jokowi delegation declared, "Today, through Indonesia's seventieth Independence National Movement, the nation's dreams are concluded in the time capsule. We are unifying the dreams of all citizens from each province" (Reza 2015). In his speech, the president affirmed that Papua was the appropriate place for the time capsule to rest as it was a land rich with the natural resources needed to fulfill the dreams of the nation. The governor of Papua province, Lukas Enembe, too, shared his hope that "before 2085, all areas across Papua, including West Papua, [would be] connected by highways" (Janur 2015). The two students standing on the stage then read out the dreams that had been handwritten by Jokowi and that would be placed inside the buried capsule. Indonesia, the statement read, would uphold pluralism, culture, religion, and ethics. It would become a global center of education, technology, and civilization. Indonesia would strive to be free of corruption and invest in equal infrastructure development throughout the country. Eventually, Indonesia would become the most influential country in the Asia-Pacific region and a barometer for global economic growth.

When the ceremony was over, my friends and I retreated to the shade of a mango tree across the road. We squatted at its foot, shrouded by the sweet-pungent smell of rotting fruit and the billowing clouds of dust raised by passing vehicles. Yosef kicked off the shoes he had purchased for the occasion and massaged his sore feet. Hordes of children flooded out of the field, comparing (or, more enthusiastically, taking apart) the plastic ballpoint pens distributed by the event organizers. After a long silence, my companions began to talk about the events of the day.

Gerardus muttered, "Time capsule . . . how strange . . . I've never heard of that before. It sounds no different to a time bomb. Tik, tik, tik." Wiping the sweat off his forehead, Yosef replied, "Tua (elder), you know it's all about politics. Just getting people together, giving out free food and drinks, and pens that work just long enough to sign a land deal with the oil palm bosses. The Time Capsule Expedition is about connecting Indonesia through time and through the Indonesian homeland (tanah air). A symbol of the nation." Gerardus scoffed and responded, "Ha! There are no dreams to be had if time has come to a stop. Hope cannot exist if time has stopped. It's funny that the time capsule should end up buried in our land. Oil palm has eaten up time. Because of oil palm, time has come to a stop. Papua, a land rich with natural and cultural resources, says Jokowi. A land rich with oil palm, he means! A land where time doesn't move. What do they expect by 2085 when 2085 may never come!"

"Your words are true, tua," Yosef said smiling. "But these are the words of little people (orang kecil). No one will listen to us curly-haired and black-skinned ones." Sitting beside me, Antonius was tracing palms in the dust with his forefinger. He looked up and added, "What is Lukas thinking—that highways will make time come back? We have seen enough of roads. The road brings only oil palm and settlers." Yosef turned to me and said, "Namuk (child), did you know the government calls oil palm the tree of hope (pohon harapan)? I heard it from the Head of Spatial Planning. And yet the forest is gone. The forest is our past. If the forest disappears, what future is there for Marind and our amai? Time is in the forest, not in capsules. How can you hope when there is no future, and when the past is lost?" My companions sighed and shook their heads. We stood up to leave. Yosef slapped the dust off his shoes and begrudgingly put them back on. Then he muttered, "A time capsule . . . a time bomb, you mean. . . . Tick, tick, tick, tick, tick . . ." Antonius put his hand on my shoulder and said, "This is our history, namuk. The history of Marind people. First, they came for our birds. Then, they came for our customs. Then, they came for our land. Now, they come for our life, and time has come to a stop."[1]

IN THIS CHAPTER, I examine how the arrival of oil palm reconfigures Marinds' sense of time. In particular, I attend to a prevalent statement among my companions—that since oil palm arrived, time has "come to a stop." This halting of time, as I will demonstrate, differs in several respects from millenarian notions of the end of time identified in other parts of Melanesia. For instance, communities in the Upper Bian do not envisage radical or sudden transformations to come, the return of a glorious past era, or the arrival of material wealth from distant places and persons, as cargo cults frequently entail. Nor do they associate the end of time with the coming of Jesus Christ or the impending judgment of God.[2] Rather, the future itself cannot be anticipated because time itself no longer exists. As Elly, Yosef's twenty-year-old daughter, put it, "If time has come to a stop, how can we even think about the future?"

Alongside the contrapuntal dynamics and abu-abu grayness of the Marind lifeworld, time has been a recurring motif throughout this story. Time manifested in the contested ontology of maps as representations of salvaged pasts or imagined futures. It surfaced in the stunted reproductive time of forest species losing their habitats to oil palm or making new homes in the village. Time also accompanied us when we encountered the multispecies pasts and presents inscribed within the sago grove and the function of sago itself in connecting Marind generations across space and time as "sago people." In this chapter, I focus on the time-stopping effects of oil palm, as these are experienced and interpreted by Marind through their own morally and politically imbued sense of historicity.[3] This historicity is anchored in a broader distinction Marind make between mythical time (zaman mitos) on the one hand and historical time (zaman sejarah) on the other. These temporalities are understood to be coexistent, but they are infused with contrasting moral significance—in other words, they operate as temporal counterpoints.

Felip, an elder from Bayau, once explained the difference between mythical and historical time to me as follows: "The time of myth is a time of giving. It is everywhere. It is in the colors of the forest and the cry of the khaw. It is in the taste of sago, the ripples of the water, and the notches of bamboo trees. The time of myth is like a noken (fiber bag) woven out of sago. Its threads flow through each other. The story of anim and amai are woven together. Like the noken, the threads of the time of myth are all different, but they are strong. They support each other."

Felip's comparison of interspecies relations in mythical time to the interweaving threads of a sago fiber bag bring to mind Donna Haraway's description of the "silk-strong thread of ongoing alertness to otherness-in-relation" (2003, 50) that enables ethical, more-than-human relatings. In the Upper Bian, these relatings encompass the fashioning of plants, animals, and humans by the ancestral creator spirits and their enskillment in the arts of hunting, sago processing, fishing, swimming, flying, and sharing skin and wetness. During mythical time, living organisms, human and other, made their first journeys across the landscape, imprinting it with shared stories remembered to this day in the form of orally transmitted sagas and songs.[4] The time of myth, as Felip describes, is a time of mutual "giving" across species lines. It is associated with creation, growth, and learning. And it is alive in the landscapes and lifeforms of the Marind world—the ancestral spirits of the forest (dema), the plants and animals (amai) of the grove, the humans (anim) roaming the land, the nourishing soil, and the ebb and flow of the Bian River.

The time of history, or zaman sejarah, in contrast, ruptures the life-sustaining threads of mythical time. Four consecutive events characterize the time of history as one of loss, destruction, and plunder—or a "time of taking," as Felip once put it. Like my fellow traveler to the city, Antonius, many Marind described the erosive temporality of history as follows: "First, they came for our birds. Then, they came for our customs. Then, they came for our land. Now, they come for our life, and time has come to a stop." This prevalent statement reveals the different regimes of historicity at play in Marind experiences and assessments of the past and its relation to the lived present.[5] In particular, it situates the destruction provoked by oil palm today within a layered history of Indigenous subjection to parasitic "others"—colonial bird-hunters, deadly viruses and bacteria, the Indonesian State, settlers, and corporations—who have systematically occupied, exploited, or contaminated Marind bodies and environments, to the detriment of their mutualistic relations with each other and with forest organisms.[6] Let us explore these regimes in more depth.

First, they came for our birds. The first temporal rupture in the time of history was triggered by the European plume trade. Raging market demand for bird of paradise feathers from 1908 to the 1920s led to intensive hunting in Merauke and other parts of Dutch New Guinea.[7] Many inland Marind acted as expedition guides in the early years of the plume trade. Their knowledge of the forest ecology and of the customary rituals and restrictions entailed in hunting allowed bird populations to stay relatively stable and relations between Marind and their avian amai to remain harmonious. But soon, the lucrative plume trade attracted European, Australian, Japanese, and Chinese hunters to the shores of Merauke, along with Indonesians from Ambon, Kei, and Timor. Marind lacked the firearms and motorboats that these foreigners were equipped with and consequently became marginalized from bird-hunting expeditions. Bird hunters invaded sacred sites and groves without regard for local customs. To them, the forest was, in the words of one foreign explorer, "a no-man's-land where anyone could shoot where he visited" (A. K. Nielsen, cited in Swadling 1996, 183). Relations between hunters and Marind turned acrimonious, resulting in several attacks and killings. By the time the Dutch authorities banned hunting in Merauke in 1922, bird of paradise populations had fallen dramatically. Dema became angry with Marind and inflicted illness and hunger on them. "The birds went mute," Antonius explained, "and time began to lose its voice. Time began to slow down."

Then, they came for our customs. The Dutch occupation of New Guinea constitutes the second disruption in historical time. Foreign rule and missionization led to the abolition of Marind cultural practices that had once sustained kinship bonds and cross-clan alliances. Sedenterization initiatives transformed formerly scattered and temporary settlements into larger permanent villages, which often cut across customary land boundaries. The ban on headhunting raids, fertility cults, and great feasts, along with the promotion of horticulture (including of introduced crops such as rice), further restricted the movements of Marind across the forest and wetlands. Dema were recast as primitive fetishes to be abandoned in the age of Christianity and civilization. The very survival of Marind was threatened during this time by rampant epidemics of influenza and granuloma venereum, that allegedly depleted over 20 percent of the population of Merauke in the early years of Dutch rule.[8] Many Marind affirmed these diseases were punishments inflicted upon them by dema, because they had failed to protect the forest and sustain local customs.[9] "Dema," Antonius continued, "became mute. The soil trembled from their rising anger. Time was stuck in the mud, unable to move."

Then, they came for our land. The colonization of West Papua by Indonesia represents the third rupture in the time of history. The controversial Act of Free Choice of July–August 1969 led to a massive influx of Indonesian officials, settlers, and military forces into the region, who were quick to exploit the region's natural resources. Legal opacity over the jurisdiction of subdistricts and regencies, the absence of formal titles to customary land, and the restricted scope of the administrative village left vast areas available for allocation by the state to corporations for the development of rice paddies, oil palm plantations, and mining and logging concessions. Migrant populations increased dramatically as a result of transmigration policies and occupied what little arable land was available for cultivation. During this period, Antonius explained, "The land became broken and dry and we forgot how to speak Marind." Again, the wrath of dema intensified, and Marind suffered disease and hunger. "There was so much anger," according to Antonius, "that time was blinded by dust. Time became confused."

Now, they come for our life. For many Upper Bian Marind, the arrival of oil palm constitutes the most recent and dramatic disruption in historical time. My companions themselves experience firsthand the terrifying atemporality of the monocrop landscape when they travel through plantations. "In the world of oil palm," Gerardus explained, "one loses all sense of time because everything looks the same, no matter how far or in which direction one walks. It is a truly frightening thing—to be lost in oil palm. To be lost in time."

As plantations expand, a plant-being of singular growth and form replaces the diversely rhythmic and intersecting life cycles that enliven the forest. For instance, the calls of birds marking the break of dawn, the fall of night, and seasonal transitions become rare. The times of fruit maturation, bird and animal migration, and fish and amphibian reproduction, too, halt indefinitely. As it takes over grove and forest, oil palm severs the shared temporalities of growth, reproduction, and senescence of Marind and their vegetal companion, the sago palm. The obliteration of space and species produced by agribusiness expansion thus constitutes a form of *aenocide*, a term developed by philosopher James Hatley (2000, 30) to refer to the murder of time and human generations, and repurposed by Deborah Bird Rose (2012, 137–39) to describe the mass destruction and disordering of other-than-human organisms' temporalities of life (see also Rose 2012). Once bound like the tight weave of a sago bag, the threads of interweaving human and other-than-human lifeways become disjointed as time itself unravels.

Much like Porgerans in Highland Papua New Guinea, who claim that the land is ending as a result of gold mining and its adverse effects on the more-than-human landscape (Jacka 2015, 237–40), so, too, the ending of time is fraught with ominous significance for Marind, and also for the multiple companion species and elements that enliven the rapidly disappearing forest. It speaks to an abu-abu world where lifeforms and landscapes, and the temporalities they embody, are rendered literally out of joint. As it devours the land and imposes its mono-time of homogeneous growth on the lively temporalities of the forest, oil palm erases the spatially experienced past of human and other-than-human organisms that are embodied in the landscape. In doing so, it forestalls the possibility of a meaningful present and thwarts the shared futures of the forest's dwindling communities of life.[10]

The time-stopping effects of oil palm are further amplified by the invisibility of the human actors driving its expansion. Unlike the bird hunters of the early twentieth century, the colonial agents and missionaries of the Dutch era, and the officials, settlers, and soldiers of the Republic of Indonesia, the people and institutions behind agribusiness projects are largely unknown. Of the policies emanating from Jakarta, the purported benefits of the MIFEE mega-project, the multinational conglomerates active in Merauke, and the global communities of palm oil consumers, my friends say they "see only oil palm." And yet the temporality of oil palm itself remains mysterious. For instance, this cash crop has not yet been planted in many areas or is still at the seedling stage. No one knows how long its maturation lasts or understands the stages of its development. The duration of the concession permits allocated to plantation operators and what happens once they expire are also uncertain.

Compounded with oil palm's ambiguous temporality of growth is its uncanny tendency to appear in unexpected places and at unexpected times—a quality that I characterized earlier as rhizomatic. For instance, new concession boundaries are erected without the knowledge of customary landowners. Sago stands that once flourished abundantly to the polytemporal rhythm of the lifeways of their diverse symbiotes are razed overnight. Even areas marked for conservation by agribusiness companies end up reduced to rubble and ashes in a matter of hours to make way for monocrops. At the same time, the temporality of oil palm itself is determined by the humans who decide when it is planted, harvested, felled, processed, and culled. Its growth and reproduction, too, are dictated by humans in the form of biogenetic experimentation, chemical fertilization, and artificial pollination. The plant kills the time of others. But it has no control over its own temporality.

As we walked home from Hasanap Sai Stadium, my companions insisted we pass by Tugu Lingkaran Brawijaya 969. Known among the locals as Libra 969, this towering edifice was erected in 2014 and constitutes a landmark monument of Merauke City. Meeting this "modern totem," Yosefus emphasized, was important. It would help me understand Marind history. A good half hour and two bus trips later, we reached the foot of Libra 969. No entity, my companions told me, embodies the four episodic ruptures that characterize historical time better than Libra 969. Four large clocks sit at the apex of the building, facing the cardinal points. Garishly painted bird-of-paradise statues rest atop four concrete Marind drums surrounding the monument, representing Marind of the forest, the coast, the swamp, and the island of Yos Sudarso. (*First, they came for our birds.*) The date 1902 adorning the four sides of the pillar commemorates the establishment of Merauke City as a military post under the Dutch. (*Then, they came for our customs.*)

A huge cast-iron figure with the numbers 969 runs down the front of the Libra monument. While a government pamphlet explains that this number refers to the age of the venerable biblical figure Methuselah, my friends affirm it alludes to 1969—the year of the fateful Act of Free Choice. The imposing structure is painted black in the bottom third and white in the upper two thirds, representing the journey of growth of Merauke from backward isolation to successful incorporation into the Indonesian State. The elevated round platform on which it is erected bears the colors of the Indonesian flag, symbolizing, according to the government pamphlet, an "eternally red and white Merauke." Elements of Marind culture depicted in the monument—ritual <u>kanda</u> drums and forest animals and plants—literally concretize the incorporation of West Papua into the Indonesian nation. Finally, the slogan <u>izakod bekai izakod kai</u>, meaning "one heart, one goal" in Marind, runs around a rusty globe adorning the top of the stele, incarnating the unity of Indonesia's citizens. (*Then, they came for our land.*)

Libra 969, my friends informed me, was inaugurated in 2014 at the height of the oil palm boom. Numerous companies had received their permits at this time and begun to clear the forest to establish plantations. In fact, the Libra monument was built primarily with funds donated by oil palm conglomerates, in appreciation of the business contracts they had successfully concluded with Mbaraka and his administration. Inaugurated by the state, financed by corporations, and firmly cemented into the ground, Libra 969 symbolizes the industrial plantations promoted by the government in the name of development and progress. "Just look at the pillar itself," Yosefus exclaimed. "Does it not resemble oil palm, standing so straight and erect and imposing? Is it not also a symbol of hope? Is it not a modern totem?"

My companions saved till last in their explanations what they deemed the most striking feature of the modern totem—the four giant clocks adorning its peak, whose rusty needles have never once turned since the monument was built. The defunct mechanisms of these modern technologies of time incarnate vividly the temporal stagnation that oil palm, as a lethal time-stopper, imposes on Marind and their forest kin. "These clocks," Gerardus explained, "they have always been broken. Four broken clocks. Four periods of destruction. I've heard Romanus has been trying to get those clocks fixed by Javanese settlers for months now. It must be embarrassing for him to see them broken like that—especially after all the tax money he spent on them! But for some reason they've never managed to make the clocks move. Just like time has stopped, the clocks are forever stuck. Now, everywhere, there are plastic flowers. Plastic totems. Plastic cassowaries. Even plastic anim." (*Now, they come for our life, and time has come to a stop.*)

Antonius, Gerardus, and Yosef's criticism of the dreams of Indonesia as incarnated in the time capsule speak to the destructive force of nationalistic and modernist futures inflicted on them and their forest kin by the state's developmentalist agendas. The time capsule, as Gerardus put it, is nothing other than a time bomb. All it promises is more destruction in the now, although always in the name of a better, redemptive place and time to come. It is an object, in Gastón Gordillo's terms, propelled in and toward the future from a wounded present in an attempt to conceal that present's ruptures (2014, 253). The time capsule's visions for the future obscure the devastation wrought on land and life by the promise of progress. "Little people" like Marind have little say over what the dreams contained within it encompass or achieve. Instead, the time capsule imposes a shared Indonesian vision across the nation and its diverse ethnic groups, who are symbolically unified by the capsule's journey from one end of the archipelago to the other. In this regard, the time capsule is akin to the figure of sacrificial love that Elizabeth Povinelli identifies as central to late liberalism's governance of the cultural and political "other." Like sacrificial love, the capsule constructs the present as a mode of pastness by perpetually positioning itself against the redemptive horizon of a putatively perfected future (Povinelli 2009, 82)—a future that conversely preempts the possibility of more complex modes of multispecies dwelling in the fractured present.

In the context of Indigenous historicities, Marinds' affirmation that time has come to a stop brings to mind the temporal effect of Western colonization described by the last Crow chief, Plenty Coups. After the buffalo slaughter and the establishment of reservations, Plenty Coups stated, "nothing happened." This enigmatic statement, philosopher Jonathan Lear suggests, testifies to the destruction of the broader framework of meaning within which specific events gained significance for the Crow people. In other words, the stopping of time conveyed by Plenty Coups referred not to the loss of "happenings" per se but to "the breakdown of that *in terms of which* happenings occur" (Lear 2006, 38, emphasis added). Similarly, the arrival of oil palm erodes the fundamental spatiotemporal frameworks of Marind cosmology and its constituent interspecies relations. The disappearance of the forest is tantamount to the annihilation of the embodied past, which in turn renders the present precarious and the future unforeseeable. Everyday life might continue in the villages of the Upper Bian, but the structures of meaning underlying or framing Marind ways of being are increasingly under siege. In a world of growing grayness, Marind, like the Crow, are "[running] out of *whens*" (Lear 2006, 41, emphasis in original).

DURING OUR CONVERSATION AND over the course of the following weeks, Antonius, Gerardus, and Yosef joked frequently about the time capsule event of December 2015. For instance, they laughed at the fact that a box containing dreams for the future had been buried in a place where time had come to a stop. They found it ironic that a "tree of hope" had been planted in a land without a future. My friends predicted that the expensive highways that the governor of Papua envisioned for the region would quickly become clogged with mud. Eventually, no one would use them. Antonius was adamant that no matter how hard the government tried its efforts would be pointless. The year 2085 would never come. Chiming in, Yosef affirmed that no matter how much money the regent of Merauke spent on laborers and engineers, the needles of Tugu Libra's clocks would never budge. Gerardus, meanwhile, reveled in playing with the double meaning of sawit makan waktu, which can mean both "oil palm eats time" and "oil palm is a waste of time." The pun communicated aptly the futility and delusions of grandeur that characterized agribusiness projects and their destructive dreams of progress. Once an object of frustration, the time capsule and everything it represented for Antonius, Gerardus, and Yosef, gradually transformed into an object of mockery.

This discursive shift points to the twofold, or contrapuntal, meaning of the expression "time coming to a stop." On the one hand, the statement positions Marind as victims deprived of their self-determined past, presents, and futures by developmentalist agendas that were engineered and imposed by state-corporate technocapitalist assemblages. But the statement is more than just an expression of resignation on the part of Marind to the dictates of capitalist modernity and Indonesian nation-building. Rather, the disavowal of the existence and passage of time is, in itself, a form of temporal resistance through which Marind *repudiate* the punitive futures conjured by dream capsules and modern totems.[11] In other words, Marind give *up* on the future to avoid giving *in* to the particular kinds of futures inflicted upon them by the state through its top-down developmental agendas. This politics of refusal can be understood as a weapon of the weak, in James Scott's (1985) terms, or a form of provocative impotence, in Jean-Paul Sartre's (1987, 174) terms. In this covertly self-empowering stance lies the possibility of emancipation from the dystopic regimes of capitalist modernity, whose only certain promise, as Yosef put it, is that of time bombs tick, tick, ticking away.

To refuse, as Carole McGranahan notes, is not just to say no (2016, 319). Among Upper Bian Marind, refusing the passage of time means refusing to find meaning in the present *from the vantage point of an imagined future.* This future-oriented drive is, in many ways, characteristic of capitalist modernity itself—a progress-driven temporal disposition, spurred by a linear imperative in which the future acquires a presumed superiority over the past.[12] Yet paradoxically, modernization projects like oil palm plantations are also defined by their destructively myopic temporal horizons. Based on short-term calculations of economic cost and gain, these projects disregard the enduring and adverse impacts of large-scale ecological transformations, marginalize the temporalities of extant organisms, and jeopardize the longue durée of planetary survival. Capitalism, then, appears to suffer from two distinct yet related pathologies. It is schizophrenic, as Gilles Deleuze and Félix Guattari (2005) suggest, in that it is propelled by conjured visions of the future that are often literally out of touch with the realities and aspirations of marginalized communities, like Marind, who are subjected to capitalism's violent effects. At the same time, capitalism is bipolar. It celebrates and idolizes an ever-unattainable future, while simultaneously neglecting and preempting it by destroying ecologies and relations in the present that make the future itself foreseeable and livable.[13]

In this light, Marinds' affirmation that time has stopped represents their rebuff of the pathological contradictions of modernity as an eternally deceptive hype of exaggerated yet self-obliterating futures. By refusing the passage of time, Marind refuse to situate their pasts and presents in relation to the anticipatory *horizons of expectation* that historian Reinhart Koselleck (2002) considers central to the production of historical time itself.[14] This version of the *end of history* is a far cry from political theorist Francis Fukuyama's celebration of liberal capitalism as the apogee of human civilization. By choosing to remain "stuck in history" (1989, 17), Upper Bian Marind engage in a form of creative sabotage of time itself. They actively embrace, rather than oppose, the denial of coevalness that positions them outside of and prior to the dynamics of *settler time*.[15] In doing so, Marind torque the arrest of time into a provocative form of temporal self-empowerment. Their stance of "no future" constitutes a form of affirmative negativity that challenges the normative temporality of capitalist and colonialist modernity—or what gender and sexuality studies scholar Elizabeth Freeman (2010), in a queer register, might term *chrononormativity*.[16] In adopting this position, Marind *work time* in order to evade being worked *on* by futurities that have all too rarely been theirs to determine and more often theirs to endure (Ahmann 2018, 144, 151).

In this regard, Marind conceptions of time differ from what Arjun Appadurai calls the "politics of waiting" (2013, 126–29) in that nothing is awaited and no one is awaiting.[17] Rather, communities in the Upper Bian practice what Kathleen Millar calls a "politics of detachment" (2014, 48–49), wherein repudiating the toxic temporality of modernity allows one to continue to live in, and despite, the precarious present. Much as philosopher Walter Benjamin (2015) describes revolutions as a cessation of happening that challenges the fixity of the past and transforms the present and future of the oppressed, the arrest of time in Merauke is nothing less than a political move—an assertion of *temporal sovereignty*, to borrow gender and literary studies scholar Mark Rifkin's (2017) term. It is a rejection of colonialist-capitalist time *itself* as a pharmakonic, temporal pressure point that promises to heal and advance but instead provokes rupture and loss among those on whom it is imposed. The arrest of time embodies Marinds' disavowal of a future thrust time and again on them by capitalist modernity—a future, in the words of Gerardus, of highways and oil palm, of plastic flowers and plastic beings.

LIKE INDIGENOUS PEOPLES THE world over, West Papuans continue to be perceived by the Indonesian State and its corporate allies as peoples forgotten by time or stuck in a Stone Age temporality (cf. Kirsch 2010; Rutherford 1996; Slama and Munro 2015). As exemplified by the letter sent to Gerardus and his friends by PT PAL, government and corporate representatives in Merauke routinely label Marind "backward tribes" who are woefully "behind the times." These racial stereotypes are then invoked by the state to justify the very kinds of agribusiness projects that, for many Marind, are causing time itself to grind to a halt. At the same time, the affirmation among many Marind that time has ended gives rise to frustration among NGOs and church organizations striving to help them secure their land and livelihoods in the face of the oil palm incursion. Many members of these organizations attribute the dismissal of the possibility of a future by Marind to laziness or short-sightedness. Others are concerned that it will further entrench the perception of West Papuans as primitive and backward, giving even freer rein to corporations and the state to determine on behalf of Marind what form their future should take. Indigenous activists in the Upper Bian are profoundly aware that their statement may be misinterpreted by NGOs and church organizations, or instrumentally exploited by the state and corporations to justify further developmental projects. However, many stick to their claim that time has come to a stop, and that there is no future for them or for their forests and forest kin. Marina, a close friend and activist from Khalaoyam, once explained her stance as follows: "The future is the same as hope. It is something that is yet to happen, that is far away. For too long, we have been forced to hope. By the government. By oil palm companies. By NGOs. We had no choice. Now we want to be free from hope. Fighting for our land is our right. But refusing to hope is also our right."

Marina's rebuff of the future invites reflection on the ambivalent significance and effects of hope as an inherently future-oriented disposition. Hope may be an enabling drive in the making and imagining of desired futures. As a generative force, hope can set things into motion and keep them in a perpetual state of change within the scope of real possibilities. As a way of being and knowing, hope calls for curiosity and creative imagination in rethinking our position within, and engagements with, the polytemporal worlds we inhabit. The form and possibility of *multispecies* hopes and futures have become evermore critical questions in the broader context of unprecedented and intensifying ecological change—a change that Paul Crutzen, the atmospheric chemist and coiner of the term *Anthropocene*, describes in temporal terms as the *Great Acceleration*. For instance, Eben Kirksey (2015, 6) counters apocalyptic and messianical thinking by drawing attention to hopeful opportunities for new multispecies becomings in the emergent ecologies of late capitalism. Anna Tsing (2015, 22) encourages us to look around rather than ahead for the possibility of livable collaborations in the rubble of technocapitalist empires. Rejecting abstract futurism and its related affects of sublime despair and sublime indifference, Donna Haraway (2016, 4) calls instead for creative kin-making practices across species lines in the ongoing thickness of the present.

Not everyone, however, is hopeful about hope. Becoming caught up in hope can preempt the kinds of mobilization needed to fulfill hope's objectives. It can roil social action in the present and obscure the fraught compromises that the pursuit of freedom invariably entails. Hope, in other words, can be toxic. The capacity to hope itself is always unevenly distributed across the wealthy and the poor, the colonized and colonizer, or what Ghassan Hage calls the *differential of hope* (2016, 466). Suspending hope, in contrast, may enable us to take seriously the possibility of the future, in Sara Ahmed's words, as "something we have *already lost*" (2010, 163, emphasis added). The ambivalence of hope is perhaps best captured by theologian and philosopher Henri Desroches, who writes that nothing happens without hope, but everything that does happen falls short of what was hoped (1980, 40). Hope, as such, can alternately function as a poison or remedy to social suffering. Hope, too, can be a pharmakonic pressure point.

On the one hand, the stance of affirmative hopelessness adopted by Marina and others in the Upper Bian can be conceived as an expression of grieving for beings rendered ungrievable under the hegemonic infrastructures of productivist technocapitalism (see Chao, forthcoming). While Marind persevere with their land rights struggle in the face of agribusiness expansion, they witness and experience the destruction of life and landscapes daily. For the many beings already or imminently displaced or made to disappear by oil palm's arrival, there is little hope to be had and much to be mourned. Each organismic death marks the end of a fleshy story and its constitutive relations, past and present, which in turn preempts the possibility of storied futures across species lines. These deaths are not one-off, singular happenings but rather deeply consequential multispecies events (cf. van Dooren 2014; Rose and van Dooren 2017). Hopelessness among Marind can thus be conceived as a stance of respect toward the dead—a mourning from the irreversible loss of life in an abu-abu world and its consequences for the relations and interdependencies that enable beings, human and other, to exist meaningfully in the first place.

At the same time, Marina's repudiation of hope epitomizes a resistance to the particular *kinds* of ecologies, productivities, and temporalities promised by time capsules and modern totems.[18] Here, hopelessness becomes a covert expression of temporal self-determination through which Marind repudiate what Lauren Berlant (2011) calls the *cruel optimism of modernity*—when what one desires is an obstacle to one's flourishing. By refusing the hopes incarnated in oil palm projects, time capsules, and nationalistic monuments, Marind resist the abduction of the precarious present-as-anticipation by barren futures-as-speculation. *Not* hoping, Marina suggests, is nothing less than a right. This hopelessness does not necessarily imply that things can *still* get better (i.e., there is still hope). Nor does it suggest that things have *already* become better (i.e., there is no longer any need for hope). Rather, Marind creatively deploy hopelessness to oppose the unconditional disappointability and indeterminacy that characterize their experiences of capitalist modernity. To torque the words of Haraway cited earlier, then, the sublime indifference of many Marind to the passing of time is nothing less than a mode of resistance to the sublime despair produced by capitalism as a pervasive yet deceptive mode of abstract futurism.

The transforming temporal textures of the Marind lifeworld explored in this chapter call for greater attention to the more-than-human detemporalizations provoked by capitalist incursions.[19] Such an approach would examine the frictions produced by perspectival divergences across situated groups and individuals with regard to time and its passage, forms, and effects, or what sociologist George Wallis (1970) calls *chronopolitics*. It would explore how variably situated imaginations of the future collide across and within collectives of inequitably distributed power, such as Indigenous communities, the state, and corporations. It would investigate the abu-abu force of hope as an affective and ethical disposition—one that, like the disposition of geographic pressure points, can alternately haunt and heal those on whom it is imposed. It would recognize the possibilities that *not* having a future might hold in an abu-abu world—for humans and others. Amid attritive histories of ontological occupation and multispecies violence, attending to the dynamics of detemporalization reveals how disempowered communities creatively harness hopelessness to reclaim the very terms of their existence.

8. Eaten by Oil Palm

A group of Bayau villagers and I were bivouacking in the forest following a day of hunting and fishing. A heavy haze draped the landscape, hot with fresh ashes from land clearing in the nearby oil palm plantations. The children and dogs lay beside the fire in a warm tangle of bodies and breath. Hushed conversations mingled with the crackling palpitations of dying embers, until eventually the melancholic hoots of a solitary owl lulled us to sleep. In the middle of the night, I was awakened by a panicked, weeping-hiccupping sound and small, sweaty hands digging into my arm. Fourteen-year-old Rosalina buried her head into my chest and sobbed: "I had a dream, sister." Her face glistened with sweat. Her body, frail from a recent bout of malaria, was trembling. I drew Rosalina into my arms and shared wetness by rubbing my cheek against hers to comfort her. Eventually, the child calmed down and began to share her dream. Her words trickled out hesitantly, stutteringly, staining the starless night with blotches of a deeper darkness.

Figure 8.1. Eaten by oil palm. Karolina shares her dream with me. Photo by Sophie Chao.

I died in my dream. Me and several other family members, and others that I do not recognize. It was the middle of the night, and we were on the PT PAL plantation. We were all on our knees in a big circle in the middle of a clearing, surrounded by oil palm. Our heads were bent, and our hands were tied behind our backs. We were barefoot. It was very quiet. There were no birds or wind to be heard. I could feel my father beside me, even without looking at him. I do not know who the others were because they had plastic bags over their heads, tied around their necks with rope.

Everyone was silent. It was hot and dry. My lips were cracked like I had not drunk water in days. I had no more wetness in my body, and my skin was broken. At one point, I changed into the skin of a kewekawe (olive-backed oriole), perched up on the highest branch of an oil palm tree, looking down at the circle of people kneeling on the ground in the clearing. When I was kewekawe, I could see the moon clearly, level with my eyes. At other times, I was an anim in the circle. I could see myself as kewekawe up in the tree. I changed from kewekawe to anim and back, over and over again. My head was spinning. Nothing happened for a long time. It was like time had stopped.

Then, I saw that some of the trees surrounding us were not oil palm. I had to squint because they were black and green and tall and strong in the shadows, and if you didn't pay attention, you wouldn't have seen that some of them were military men. The color of oil palm at night is like that of soldiers' uniforms—black and green. And the spines on the fruit, you have seen them. Sharp like the blades of their bayonets. And the fruit, like bullets. Hard. Round. Black. Sometimes red—like blood.

Looking down as kewekawe, I could see the soldiers in between the oil palm trees, standing around the group. It was so silent, so quiet, out there in the plantation. No one was moving. Not even a little. They were so still I did not know if they were even breathing. Then, I heard a gunshot in the darkness. Pang! My father's body collapsed beside me. Black blood flowed from the back of his head as he lay face down in the soil. Pang! The person beside him fell to the ground too. Again, and again, oil palm shot us and the bodies collapsed. Or maybe it was the soldiers shooting. Still, no one moved.

As kewekawe, I could see the bodies crumble. I could see myself down there—the last to be shot. I did not fly away. I could see my anim skin on the ground below, bleeding. I sat there for a long time, but I never saw the sun rise. It lasted forever—like time had come to a stop.

Rosalina's dream kept me up until the break of dawn. Eventually the other group members began to rouse. The dogs scampered around, licking the children's faces to wake them up. Only Rosalina remained in a deep slumber. I told Evelina, Rosalina's mother, about the nightmare that had disrupted her daughter's sleep. She sighed and said, "She was eaten by oil palm (dorang dimakan sawit). I knew it would happen eventually. Everybody else in the family has been eaten." Evelina's words were interrupted by the earsplitting explosion of a burning bamboo tree. An acrid smell of smoke soon assailed our nostrils. Corporate land clearing had resumed. Rosalina's father, Oscar, dragged himself up on his good leg and packed away his betel nuts in his woven sago bag. He stamped out the fire and said: "It is time to move on. I told you, oil palm never stops eating. It is always hungry for more land. Place has changed, time has stopped. Anim have become plastic and cassowaries eat instant noodles. Oil palm eats land, water, and time. At night when we sleep, it comes to eat us too. Since oil palm arrived, there are new kinds of dreams. Oil palm never stops eating. And so, we must move on."

IN THIS FINAL CHAPTER, I explore Marind dreams of "being eaten by oil palm." During my fieldwork in the Upper Bian, I frequently heard of haunting dreams like the one Rosalina experienced that night in the forest. These dreams are uncanny and dystopic. In them, human bodies become disfigured by oil palm drupes, proliferating subcutaneously like cancerous tumors. Razor-edged palm thorns incubate between sinew and muscle, perforating dreamers' skin like arrows shot out from the viscous entrails of their writhing bodies. Women eaten by oil palm suffer tortured labors, deadly hemorrhages, ripped wombs, and the birthing of monstrously deformed and spinescent fresh fruit bunches, or "oil palm children"—alive and stillborn. Plant and animal kin and deceased relatives of the dreamer appear alongside soldiers, bulldozers, charred sago groves, contaminated rivers, and oil palm. Most dramatically, dreamers witness and experience their own deaths repeatedly from the perspectives of diverse forest beings whose existence, like their own, is jeopardized by agribusiness expansion.[1]

The frequency of dreams of being eaten by oil palm increased dramatically in the last few months of my time in the Upper Bian. Community members traveled up and down the river and across settlements almost daily to hear and share their nocturnal experiences with kin and friends. Behind closed doors, my English classes turned into dream-storying classes. Some children narrated their dreams, while others preferred to draw the black rivers, felled trees, thorny palms, and blotches of blood they had seen during the night. When the rusty school bell rattled out its tired dribble of notes, I erased the dark words jotted across the weary blackboard, picked out here and there from the children's accounts. To avoid raising suspicion among fellow teachers, I replaced them with the anodyne names of colors, animals, and foods.

In the villages, the fall of darkness was shrouded with restless apprehension. People talked for hours in half-whispers about what torment the night might bring, and to whom. At the break of dawn, nebulous dream narratives dispersed across homes and hearths, as those eaten by oil palm revealed the harrowing deeds and deaths conjured in their sleep. Following night-long deliberations about strategic campaigning, when the NGO workers wandered the village with their mobile phones in vain pursuit of signal, villagers gathered and shared their dreams in hushed voices next to dying wood fires, sleeping babies pressed close to women's breasts in the penumbra, the glow of embers shining in the men's coal-black eyes. Some dreams were recounted weeks after they occurred, others the same night. Some were very short—a few fragments of half-remembered places and events—while others formed intricate tapestries of different dreams woven together over the course of several months. My participation in dream-sharing sessions became more frequent after I, too, was eaten by oil palm—a moment that, to many of my companions, proved I had finally become part of their community. In the day, oil palm ate the land as bulldozers razed hectare after hectare of forest. In the night, oil palm consumed the bodies of slumbering men, women, and children haunted by its ghostly visitations. No one, it seemed, was immune to being eaten by oil palm.

If expressing dreams with words often fails to capture what one remembers of them, conveying dreams in writing only further transforms their experiential register. While the final versions of the narratives presented here were cowritten with those who shared them with me, much of their affective texture is lost through being reported in the textual medium. This includes, for instance, the meaningful pauses and tortuously long silences punctuating dream-stories, the repetitions and exclamations peppering my interlocutors' accounts, and the expressive tone, inflection, and pace of their speech. Multiple voices mushroomed in and out of these vivid yet often only vaguely remembered memories. Different individuals participated in molding and remolding the narratives, blurring context and content, and mingling lived and dreamed events in ways that defied clear-cut distinctions between the real and the imagined. Absent from these narratives, too, are the rubbing of skin accompanying the spoken word, the telling quiver of speakers' voices, and the heavy shroud of smells and tears cloaking the crouched bodies of men, women, and children, as they held hands in the dark and shared their story of being eaten by oil palm. Destruction, dread, and death spilled over from reality into these dreams and back, provoked by a being that, as Rosalina's father put it, never stops eating.[2]

Upper Bian Marind describe the dream of being eaten by oil palm as a new kind of dream prompted by oil palm's arrival. Their narratives emphasize the tortuous sensory dimensions of this nocturnal experience, which include feelings of extreme hunger, thirst, heat, and dryness. Almost all dreams take place in the darkness of nights that never end, and whose deafening silence is interrupted only by screams or gunshots. Dreamers report heightened sensations of fear, pain, loneliness, and panic, along with a disturbing loss in motility and spatial cognition. The expressions villagers use to describe these experiences— "becoming strange" (jadi aneh), "stuck" (jadi tertancap), "confused" (jadi bingung), or "not recognizing oneself" (tidak kenal diri lagi)—convey the sense of self-alienation and sensory entrapment entailed in being eaten by oil palm. These harrowing sensations also shape the affectively charged practice of dream sharing, which involves weeping, protracted silences, sighing, shuddering, and the repetition of particularly unsettling sequences of dreamed events.

At the same time as they entail visceral experiences of pain and violence, dreams of being eaten by oil palm also encompass the absolute extinguishment of the senses in the form of death. My companions reported that only in oil palm dreams could one witness or experience one's own demise. Not only this, dreamers live their death in agonizingly repeated loops that are multiplied by the perspectives of the beings whose skin they unwillingly adopt during dreams—for instance, when Rosalina experienced and visualized her death as <u>anim</u> (human) from the perspective of herself as <u>kewekawe</u> (olive-backed oriole). These oneiric deaths are distressing omens of the fate of organisms, human and other, whose survival is threatened by the arrival of oil palm. Ruben, a young man from Mirav who was eaten by oil palm the night after Rosalina, described his repeated demise within the dream as follows: "When you are eaten by oil palm, you see yourself die over and over again. You die as <u>khei</u> (cassowary). You die as <u>basik-basik</u> (pig). You die as <u>anim</u> (human). It is terrifying. You are stuck. You cannot escape. In the real world, you only die once. In dreams, you die many times. That is why since oil palm arrived, there is too much death. <u>Anim, amai</u>, we are all victims of oil palm." While Marind note that dreams of being eaten by oil palm only began in the last decade, they often compare them to customary dreams (mimpi adat), which have existed since time immemorial.[3] Customary dreams are inflicted by <u>kambara</u>, or perpetrators of black magic, and result in the illness or sometimes the death of the afflicted individual. Ritual experts or medicine men (<u>messav</u>) interpret these dreams to identify the responsible <u>kambara</u>, or black magic perpetrator, and the remedies to the dreamer's plight. These may include, for instance, the types of medicinal plants required to heal the dreamer, where these can be found, and how they should be prepared and ingested. In the past, a relative of the <u>kambara</u> victim would sleep beside their corpse and <u>messav</u> would interpret their dream to discover the identity of the culprit. Dreams were purportedly the only certain way of identifying perpetrators of sorcery and imposing legitimate retributive measures upon them. Without the aid of mes-sav, victims would eventually succumb to the illness transmitted to them by kambara through the medium of the dream.

Both customary dreams and dreams of being eaten by oil palm are emotionally harrowing experiences that produce somatic and social dis-ease in the individual's waking life. For instance, people realize they have been eaten by oil palm when they experience a lingering sense of nausea on rousing. Over the ensuing days, they may suffer from a sudden onset of malaria, dengue fever, or cluster headaches. These individuals also become increasingly taciturn and may refuse to participate in social activities such as sharing food, hunting, processing sago, and exchanging wetness. Their bodies begin to dry out and their skin becomes dull. The nightly residues of oil palm dreams thus undermine peoples' reality and relations by depleting their physical energies and jeopardizing their capacity to behave like anim.

Both customary dreams and dreams of being eaten by oil palm also entail the violent subjection of dreamers to the power of malevolent, foreign agents that they cannot control—suanggi and sawit.[4] Unlike kambara, which is a local form of sorcery practiced exclusively by and among Marind, suanggi encompasses forms of sorcery that are considered foreign in origin and superior in potency.[5] These include sorcery introduced in precolonial times by Indonesian traders and sorcery practiced today by non-Papuans who control or exploit Merauke's natural resources (see Chao 2019g). Many of my companions identify uncanny resemblances between suanggi and sawit, or oil palm.[6] Both are ambiguous in their intentions, and destructive in their actions. They have the capacity to be everywhere and nowhere at the same time. They refuse to take responsibility for the consequences of their actions, behave in a selfishly individualistic manner, and create fear, anger, and suffering among their victims. These similarities are also conveyed by the expression "to be eaten," which Marind use to describe both the effects of oil palm dreams and the way kambara kill their victims, by devouring their bodies from the inside out, until their flesh wastes away, leaving behind a tottering husk of skin and bones.

Like Rosalina, who shared her dream with me in the forest, Upper Bian Marind frequently described dreams of being eaten by oil palm as a form of skin-changing because it entails the bodily entrapment of anim by an other-than-human lifeform—oil palm. Just as humans become vulnerable to perspectival capture when they exchange bodily skin and wetness with animals in the forest, so too oil palm dreams are said to be triggered by close encounters with oil palm. These may take place when community members travel to, in, or near plantations, touch or examine the oil palms lining the roads, or consume too many products known to contain palm oil. At the same time, dreams of being eaten by oil palm differ in several respects to traditional skin-changing practices. First, conscious skin-changing can only be achieved by men, whereas skin-changing in dreams also occurs to women and children. Second, while skin-changing is a source of pleasurable freedom when carried out willfully in the waking world, the metamorphoses that take place in oil palm nightmares lie from the outset outside the dreamer's control and are profoundly distressing. Finally, in active skin-changing, individuals exchange their human body and perspective for that of a specific forest creature, such as a cassowary or a pig. In oil palm dreams, on the other hand, individuals wear different skins simultaneously, or in nauseatingly rapid alternation. For instance, the dreamer may be a bird observing his or her human self at the same time as a human observing his or her bird self, such as when Rosalina alternates between kewekawe and anim in body and point of view.

The oneiric perspectivism at play here differs from Amerindian perspectivism in that the latter is characterized by the *incompatibility* of simultaneous points of view (Viveiros de Castro 2015, 178). For Marind, in contrast, dreaming is at once an *altered* and *alternating* state of consciousness. Anim dream *of* and *as* anim, birds, pigs, and other forest creatures. Furthermore, the perspectives dreamers take on are not necessarily nor always human, but rather species-specific. In other words, to dream is to witness one-*selves* (rather than one-*self*) from multiple synchronic points of view, in a form of perspectival tacking across species lines. What human and animal oneiric shapeshifters hold in common is not a universal human soul or perspective on the world, but rather, as Ruben suggests, a shared *vulnerability* in the face of oil palm and its life-threatening expansion—one that is inflected by, yet transcends, bodily and perspectival distinctions.

Oil palm dreams thus extend beyond the human in terms of both their perpetrator (oil palm) and protagonists (the more-than-human lifeforms of the forest). These dreams also transcend the individual dreamers and their nocturnal experiences. On the one hand, Upper Bian Marind recognize that dreams, as these manifest in sleep, are private because they are only experientially accessible to the dreamer. It is also up to the individual dreamer to decide whether to share the dream, to whom, how, under what circumstances, and to what ends. In line with the respect for the opacity of others' minds, the personal nature of the dream discourages people from asking about each other's nocturnal experiences or telling a dream that is not their own. Finally, dreams occur in the state of sleep, which is a time when individuals are temporarily detached from their bodily relations of skin and wetness with other beings. Severed from the social activities that sustain their identity as anim, sleepers find themselves particularly susceptible to the psychic influence of malevolent others.[7]

However, it is the *social* dimensions of dreaming that my companions in the field most frequently emphasized. For starters, the capacity to dream and tell dreams in itself is an important marker in the socialization of individuals into anim. For instance, Marind commemorate the first dream of a child by holding a celebratory feast, or by giving the name of the child to a sago palm that is of the same age. The dreams of children capable of narrating them are taken just as seriously as those of adults. Conversely, children's words may be dismissed if they "do not know how to dream yet" (dorang belum tau bermimpi). Similarly, deaf-mute children are the object of widespread pity among Marind because they are unable to share their dreams or hear those of others and therefore will never become true anim. The ability to dream thus constitutes an important stage in the maturation of humans into fully fledged social beings.

Dreams are also social for Marind in that they are not *caused* by the person who experiences them nor explainable solely in relation to that person's developmental life-course. Rather than projections of the dreamer's individual self or psyche, dreams arise from the lethal oneiric relation *between* vulnerable anim and malignant oil palm. Dreams, in other words, are a liaison dangereuse across species lines. As Felicia, a young woman from Mirav, put it, "One is never alone in a dream (dalam mimpi kitorang tidak pernah sendiri)." The fact that people have little control over their dreams and easily forget the exact events that took place within them or even that they had dreamed in the first place testifies to the loss of agency experienced by individuals afflicted with the oneiric violence of oil palm. When my companions could not recall a dream in part or in full, they said it was because the dream itself did not want to be known or had gone away. The lack of autonomous volition on the part of dreamers is also conveyed grammatically by the passive tense of the expression "being eaten by oil palm," which implies the subjection of individuals to the uncontrollable effects of oil palm. In sum, dreamers do not dream so much as they are *dreamed*. The dream-as-relation evades the self-as-dreamer, even as it is imposed on individuals by the haunting being of oil palm.

Just as the dream is a lethal relation between vegetal possessor and human possessed, so, too, social relations *between* dreamers are necessary for their emancipation from oneiric capture by oil palm. Much as skin-changers may unwittingly become trapped within animal bodies, individuals are often unconscious of the transformations in their behavior and appearance caused by their nocturnal visions. They consequently rely on others to detect the signs of their ensnarement by oil palm. This can be achieved by two means. The first takes place during the state of sleep, when people are particularly vulnerable to oneiric attacks by suanggi and sawit. Community members minimize this threat by sleeping together in small and closely huddled groups rather than on their own.[8] Since dreams of being eaten by oil palm began, villagers in the Upper Bian have also initiated a practice of night watching, whereby individuals take turns overseeing their sleep-mates to detect the signs that indicate oil palm is taking over their sleeping bodies. These include, for instance, rapid eyelid twitching, protracted moans, jerking limbs, back arching, sweating, finger clasping, grimacing, nodding, and irregular breathing. When these symptoms arise, the individuals on watch "save" (menyalamatkan) the sleeper by vigorously waking them and talking to them incessantly. For instance, they may ask the individual about places they visited recently, notable events in their childhood, what they did the previous day, and the names of their family members. Much as enhanced exchanges of skin and wetness resocialize individuals into the community and protect them from perspectival capture by forest beings and foreign people, these questions and interactions help remind dreamers of their human condition and relations, and thereby interrupt their tortuous ensnarement by oil palm.[9]

A second and even more important means of liberating individuals from oil palm is through "dream sharing" (bagi or cerita mimpi), the narration and interpretation of dreams by the dreamer and their social circles. In contrast to Western definitions of dreaming, which are generally limited to the experience occurring involuntarily in the mind of individuals during certain stages of sleep, Marind consider dream narration and interpretation part of dreaming itself. All three elements are encompassed in the expression, "being eaten by oil palm." For instance, over the months following the night spent in the forest, Rosalina's dream lived on as it traveled up and down the Bian River, told and retold by community members who attended in speculative mode to its events and actors. Kin and friends far and wide reflected on the circumstances that may have caused Rosalina's nightly visions. They also identified resonances between her dream and others they had heard of or undergone themselves.[10] Together, community members fashioned the young girl's individual dream experience into a shared dream narrative, in what dream researcher and psychologist Kelly Bulkeley describes as a "back-and-forth dialogue of questioning, answering, and questioning anew; of distinguishing, recombining, and distinguishing again" (1994, 118).

Marind dream-sharing practices resonate with David Gegeo and Karen Watson-Gegeo's characterization of Indigenous epistemological processes as nonlinear articulations, disarticulations, and rearticulations of evidence, which are achieved communally across multiple sets of situated actors (2001, 75–79). The dispersed dream narratives produced by these articulations can be conceived as attempts to overcome what Jeanette Mageo (2004) calls the "mimetic incompleteness" of dream memories. These memories constitute small fragments of a greater whole, whose missing parts must be reconstructed through interpretation. Among Marind, the work of mimetic recovery is a collective endeavor. It seeks to piece together the oneiric fragments of a specific dream and dreamer, but always in relation to the dream memories of other individuals. This process of reordering dream experience into relatively coherent narratives differs from the psychoanalytical work of secondary elaboration in that it is carried out by many dreamers together across time and space, in what might be described as an *inter*psychic, rather than intrapsychic, process.

In this cumulative mode of tertiary elaboration, everyone is at once "therapist" and "client." Dreams take on new life every time they are retold, in an ongoing process of shared interpretation. In this regard, and much like pressure points, maps, and oil palm itself, dream sharing among Marind can be understood in Derridean terms as a collective process of onto-oneiric différance. Meanings accrue to dreams in relation to other dreams and the dreams of others. Their significance is never final or fixed. Multiple layers of interpretation coalesce around dreams, such that their significance is continually and contextually deferred and different. Living on through myriad dispersed and situated narratives, dreams last forever. As Marcus, an elder from Khalaoyam, told me, "You think dreams stop when you wake up. But actually, dreams never stop. They live on through the telling and the sharing. Dreams never really end."

As suggested in this chapter, dream sharing as a collective activity is essential to emancipate the dreamer from oil palm's oneiric hold. However, dream sharing among Marind is not premised on the assumption that dreams have a definitive or authoritative meaning. Indeed, many of my friends were adamant that dream interpretation does not ultimately explain anything and that identifying the true significance of dreams is difficult, if not impossible (cf. Stephen 1995). Furthermore, and unlike customary dreams, dreams of being eaten by oil palm are not indexed against a preexisting hermeneutic lexicon or symbolic charter. They do not serve prophetic functions nor precipitate effects in real life by serving as guides to action (cf. Tuzin 1997). Marind do not conceive dreaming as a medium for the communication or revelation of parts of the self or the world deemed unknowable through ordinary or conscious means (cf. Stewart and Strathern 2003a). Neither does dream interpretation empower dreamers by giving them access to supernatural knowledge or entities in ways that might enable them to transcend the suffering or afflictions they face in everyday life (cf. Lohmann 2003). As Kristina, a middle-aged mother of three from Bayau, explained, "Dreams are not medicine. They cannot cure the pain of life."

For my companions, what dream content *means* matters less than what dream sharing *does*.[11] At the individual level, dream sharing helps victims retrieve their <u>anim</u> self by liberating them from the oneiric grip of oil palm. If those afflicted by oil palm's nightly visitations do not share their dreams, their flesh and fluids remain haunted by the plant indefinitely. At the collective level, dreaming together, to torque historian Benedict Anderson's term, gives rise to "communities of imagination" who are bound by their traumatic experience of being eaten by oil palm. Interactive dream sharing, in turn, creates and strengthens relations between those who participate in transforming individual dreams into collectively fashioned narratives, in what might be described as a form of oneiric solidarity. As Felicia, the young woman from Mirav, explained: "Dreams can kill. But talking about dreams can help. It doesn't heal us. But at least, we know others are also dying in their dreams. We are not alone. We dream together. That is why dreams belong to everyone."

The oneiric solidarity achieved through dream sharing transcends distinctions based on age, gender, and status because no one is immune to being eaten by oil palm. Furthermore, and unlike customary dreams, whose meaning can only be discovered by experienced <u>messav</u> in a two-way relationship between expert-as-interpreter and dreamer-as-interpreted, everyone in the community can participate in speculating about dreams of being eaten by oil palm without the need for a ritual expert. These processes also take place in intimate and familiar settings, such as the hearth, sago grove, and forest, in contrast to the highly ritualized exegesis of customary dreams, which takes place in men's huts or sacred forests. Finally, while customary dreams are inflicted by <u>kambara-anim</u> on fellow community members out of jealousy, vengeance, or otherwise hostile sentiments, dreams of being eaten by oil palm transcend interhuman animosities because they are caused by a being that Marind widely consider to be their common enemy—oil palm. Disputes within and across villages may abound, but in the face of oil palm, as Ruben, the young man from Mirav, put it, "we are all victims."

In some cases, dreaming together, to borrow Perpetua's words, allows people to form relationships otherwise impermissible or difficult to achieve in everyday life. For instance, Mina, my host sister in Mirav, was in love with a young man called Gerardus but was forbidden from marrying him due to a longstanding feud between their families. While Mina described this as the greatest tragedy of her life, she found solace in the fact that the star-crossed lovers could still confide to each other their experiences of being eaten by oil palm. "We cannot make a life together," Mina told me, "or share skin and wetness as husband and wife. But at least we can still share our dreams." Silas, a middle-aged man from Khalaoyam, expressed similar thoughts in relation to the man who had violently speared his younger brother to death some three years earlier. When he recounted the event to me, Silas concluded by saying: "Yes, Pius killed my brother. But Pius and I, we still dream together." The possibility of reconciliation afforded by shared dreaming was also exemplified in situations where disputes between villagers perdured. In such cases, my friends often said they would know if the parties had overcome their differences when they were able to "dream together again."

Dreams of being eaten by oil palm exemplify what Kevin Groark (2017) calls "persecution dreams" in that they constitute acts of aggression that communicate in a covert form the violence perpetrated by oil palm against Marind and their plant and animal kin in the waking world. Rather than communicating interhuman hostilities and desires, however, these dreams are symptomatic of interspecies discord between Marind and forest beings as victims on the one hand, and oil palm as their persecutor on the other. Collectively experiencing and sharing dreams in turn helps people overcome interhuman conflicts that sublimate in the shared trauma of being devoured by oil palm.[12]

And yet the psychosocial function of dream sharing in achieving oneiric solidarity also gives rise to a problematic double bind. On the hand, nobody wishes to undergo the harrowing experience of being tortured and killed by oil palm in the dream-world or plagued by dreaming's disturbing aftereffects in the waking world. Here, dreaming reveals its destructive and haunting facet as oneiric poison and psychosomatic harm. On the other hand, experiencing and sharing dreams is what enables the formation of new kinds of sociality and solidarity among those who see themselves as collective victims of oil palm. Here, dreaming reveals its enabling effects as therapeutic remedy and emotional solace. Conversely, community members who do not share their experiences of being eaten by oil palm or seem immune to such dreams are looked on with suspicion. These individuals may be accused of colluding with the palm oil sector—for instance, by surrendering their customary lands and accepting compensation payments—or of wielding an unknown and foreign form of black magic. If being eaten is deeply traumatic, then *not* being eaten is no better. It undermines the social standing and relations of individuals by casting doubt on their capacity to act and dream like real <u>anim</u>.

ONE LATE AFTERNOON IN October, I sat on the riverbank with Ignatius, an elderly man from Khalaoyam. I was reading out to my friend a printout of the website of KORINDO, the Indonesian Korean conglomerate operating the oil palm concessions near Ignatius' village. Twice, my friend asked me to translate the slogan running across the header of the website: "Planting seeds of far-reaching dreams in Papua." Then, he said: "Dreams, they say. Seeds of dreams. Oil palm dreams. There are so many dreams these days. Time capsule dreams. Company dreams. Government dreams. But dreams can kill. People are dying every night, eaten by oil palm. There have never been so many dreams. There have never been so many deaths."

Ignatius went silent. The sun was sinking in the horizon, the clouds turned incarnadine by its placid descent. A solitary egret swooped across the blushing sky as the intermittent croaking of toads rose from among the thick clusters of reeds lining the riverbank. Ignatius watched the last sliver of gold disappear, smiling sadly. His hand went limp on mine and his shoulders sagged. In the fading light, Ignatius looked smaller, thinner. The hypertrophic scar left by a wild boar attack on his left clavicle was a deep, dark clot on his diminished frame. For a moment, I thought I glimpsed fresh blood upon its mottled surface. I asked Ignatius whether he believed Marind would someday be able to stop the expansion of oil palm and the destruction of the forest. Ignatius replied: "Dreams, dreams—sometimes it is hard to see hope. The government and the companies are powerful. Oil palm is their weapon. You never know who will be eaten next. But as long as Marind continue to share their dreams, then there is hope. By dreaming together, we may find ways of living together. For now, there is no future to be had—only dreams to be told."

Dreams can reveal much about survival amid precarity, the relationship of continuity to loss, and how people struggle, in Eduardo Kohn's words, to "become new kinds of *we*" in the face of radical rupture (2013, 23). For instance, Jeanette Mageo (2003) describes how in Samoa, emotionally perturbing dreams reveal anxieties related to Christian conversion and the Americanization of local culture. Wolfgang Kempf and Elfriede Hermann (2003) explore how, among Ngaing in Papua New Guinea, dreams act as translocal spaces where people imaginatively reflect on and negotiate their relationship with White people and the wider world. Changing sociohistorical, cultural, ecological, and political realities can thus radically transform or heighten the significance of dreams or even make people dream in new ways. I offer the term onto-oneiric imagination to highlight the generative entanglement of the real and the speculative in Marind dreams of being eaten by oil palm.[13]

Combining the Greek root *ont-*, meaning "reality" or "being," and *oneiron*, meaning "dream," the concept of an onto-oneiric imagination invites us to approach dreaming as a culturally shaped and collectively experienced form of creative imagination, triggered by the radical transformations taking place within the Marind lifeworld. In the waking world, dreams of progress promoted by the state and incarnated in time capsules and modern totems, kill off the organisms and relations fundamental to Marind ways of becoming <u>anim</u>. Living landscapes and their inhabitants become haunted by imagined futures that lie beyond their control. Dreams take on equally sinister hues in their nightly form. Here, death haunts the oneiric beings, conjured and controlled by a plant that itself is torqued and possessed by the spirit of capitalism within hegemonic agroindustrial assemblages. The physical propagation of oil palm across tangible terrains is thus doubled up by its ghostly proliferation across the imaginative topography of dreams, prompting what Ignatius described as an oversaturation of death in the sentient *and* sleeping world.

Dreams as an onto-oneiric imagination thus constitute powerful psychic projections of the violent subjugation of Marind at the hands of powerful others, condensed in the polysemic archetype of oil palm as malignant possessor. These anxiogenic and profuse dreamings communicate the effects of rampant deforestation and agribusiness expansion among Marind, whose oneiric deaths embody the collapse of Marind subjectivity as place, persons, and time are disfigured by the relentless proliferation of monocrops. The multiplied deaths of sleepers also forebode in ominous ways the anticipated fate of all beings, human and other, rendered vulnerable to oil palm in the interstitial temporality of dreams as a future anterior—a "*something that will have been*," in Amira Mittermaier's terms (2011, 236, emphasis in original). And just as anxiety differs from fear in that it has no determined object or cause, so too the uncertain effects of dreams speak mimetically to the abu-abu ontology of their perpetrator, oil palm—a vegetal being of alien origins, unknown desires, and mysterious volition. Both oil palm and its onto-oneiric manifestations haunt Marind through their inherent indeterminacy, in a world where ambiguity is the order of the day—and night.

In the realm of the dream as a space of abu-abu, the boundaries between self and other, human and other-than-human, object and subject, present and future, and life and death, dissolve temporarily. Dreamers metamorphose under the influence of a plant that is itself a prolific and ubiquitous shape-shifter, concealed in the processed foods that villagers consume on a daily basis, the soaps and washing powders they purchase, and the oil they cook with. The in-betweenness of dreaming also extends to its interpretations. Opaque and ephemeral in contour, chronology, and content, dream exegesis offers no fixed or final answers. Dreams thus exert their influence on reality not as factual *meanings* but as imaginative *means* through which people map out their changing relationships with their self, the world, and other beings.[14]

The creativity of the onto-oneiric imagination is manifest unconsciously in the content of the dream—its protagonists, events, and sensory affordances. In this regard, the onto-oneiric imagination might be described as *autonomous* in Gilbert Herdt and Michele Stephen's (1989) sense of an imagination that takes place independently of a person's conscious will, and that enables communication to and from deeper levels of self without the person's awareness. At the same time, the creativity of the onto-oneiric imagination is manifest consciously in the dispersed narratives that crystallize around it as a nebulous mesh of relational meanings in flux. Bringing together conscious and unconscious forms of imagination, dreams of being eaten by oil palm are an emergent cultural resource through which Marind reflect upon the growing precarity of their lifeworld in a powerfully veiled and imaginative form of oneiric bricolage. Material and affective geographies of ambiguous relief coalesce in these dreams as amphibious beings whose effects transcend the line between fact and fantasy, consciousness and unconsciousness, and the real and the imaginary.

On the one hand, then, dreams of being eaten by oil palm are idioms of distress through which Marind objectify and reflect upon their increasingly dystopic reality—a reality where time has stopped, people have become plastic, cassowaries eat instant noodles, and oil palm devours humans in the night. The sinister transformations caused by oil palm in the waking world coalesce in the macabre hermeneutics of being eaten by oil palm, in ways that drastically amplify the social stresses faced by Marind and their forest kin on a daily basis. In the dream as a form of oneiric murk, individuals suffer harrowing paralyses, morbid metamorphoses, transformations-turned-deformations, and agonizingly recurring deaths. They experience hunger and thirst caused by a plant that destroys their sources of subsistence and pollutes their rivers. They are overwhelmed by the silence, loneliness, and aridity of a homogeneous landscape in which birds no longer sing and animals face starvation. The dreamers' loss of motility and spatial cognition in dreams also replicates uncannily the disorienting environment of the plantation, which is disturbingly devoid of reference points to situate oneself from in space and time. Dreams of being eaten by oil palm thus constitute what Andrew Lattas calls an "indigenous aesthetics of horror" (1993, 68), referring to harrowing experiences through which people objectify and reflect upon the tensions and traumas of everyday life—even if they do not necessarily provide a release of tension or a resolution of conflict in the waking world.

But dreams are also spaces of possibility in the murk of abu-abu. As Jonathan Lear (2006) suggests, dreams in times of extreme precarity can constitute a form of radical hope in the way they enable people to move forward and imaginatively into the future—even as the form of that future remains unknown or unintelligible. By sharing dreams, Marind attempt to respond creatively to their domination by a proliferating plant of destructive intent, in a world where the state and other actors are simultaneously promising and preempting particular kinds of futures. As Ignatius suggested, interwoven dream narratives open speculative space for new ways of being and hoping together, in the making of as-yet opaque futures. Attending to the onto-oneiric imagination as a psychocultural resource thus generates important insights into the formation of new identities and communities based on collective experiences of suffering. In the absence of hope in the real world, dreaming as an oneiric technology fosters a process of social suturing that enables Marind, as collective victims of oil palm, to sustain social cohesion and solidarity amid and against the haunting force of abu-abu.

Dreams of being eaten by oil palm, however, are not emancipatory or revolutionary freedom dreams in the sense deployed, for instance, by historian Robin Kelley (2002) to describe the catalytic force of radical imaginations of hope and action that spurred Black social movements in the twentieth century. Equally abu-abu in their meaning and effects, oil palm dreams do not prompt or guide particular behaviors or responses on the part of Marind in their everyday life, nor do they dispel the lived anxieties and suffering prompted by oil palm. Rather, dreaming as a form of wonder discourse serves to index the ontological transformations provoked by the expansion of oil palm. These dreams in turn precipitate the emergence of new oneiric communities bound by their shared identity as victims of oil palm. As Ignatius's words conveyed, dwelling collectively in the ambiguity of dreams opens speculative space for new kinds of social relations, imaginaries, and realities. In other words, it is in the freedom *to* dream—the dream not *as* freedom but as space *for* freedom—that hope might once again be conceivable, collective survival possible, and shared futures imaginable.

The fraught yet fertile terrains of Marind dreamscapes suggest that attending to the power of the imagination on its own terms is a necessary counterpoint to approaching reality—or as ontological anthropologists would have it, *realities*—on its own terms. Just as oil palm acquires heightened significance for Marind in light of its opaque origins, ways, and wants, so too the unreal or surreal exerts its influence through the intangible world between worlds of dreams. These dreams communicate to Marind something *about* reality without themselves being real. *Not* knowing what dreams mean is what makes them meaningful. Exploring situated onto-oneiric imaginations thus opens analytical space to examine dreaming as a speculative practice of wonder whose effects stem not from what exists, but what could be. Such an approach would investigate how different societies distinguish the real from the imaginary in the first place, along with their respective and mutual effects. It would explore the interplay of multiple, emergent, and conflicting imaginaries, and their relationship to changing realities. It would make space for not knowing in different ways, knowing differently, and different knowledge. It would allow us to dream—and to be dreamed.

Black Waters of the Bian—The Dream of Elena Basik-Basik

I was sitting in my canoe, paddling, when I saw a dark shadow rise in the horizon. Suddenly, the canoe slowed down. I looked at the dark shadow and saw that it was in the water—black, dark, dirty, thick. As it got closer, I could see that the water was oily, staining the riverbanks with slime as it approached. Looking down at the water around the boat, I saw that the darkness had reached us, spreading like big, black flowers. The boat was retreating, forced backwards by the black water. No matter how hard I paddled I could not make it move forward. I told my daughter to keep her fingers out of the river. Then, the river turned choppy. Its skin was broken by the waves. We saw huge waves heading toward us. That is when I understood that the black water was coming from the oil palm plantation. Every day, the companies throw dirty water and oil into the river. It comes out of a long metal pipe, as wide as a person, from the mill.

Suddenly, something hit the side of the boat. I looked down and saw a body floating just under the oily surface of the river. It was naked and facing up. It was a young man. He seemed asleep but he was dead. His head had bumped into our boat. I saw more bodies flowing toward us with the dark waters, crashing into our canoe with their arms and legs and shoulders. There were men and there were women. All of them were very young. I remember seeing my brother among them, even though he had died in 2000 during the Bleeding of Merauke. He was thrown off the bridge by the military in Merauke City.

All the bodies were bobbing up and down on the water. They looked peaceful, but I knew they had been eaten by oil palm. They were all lying face up, traveling with the dark waters of the Bian. My father used to tell me that down the river and across the banks, there were enemies to watch out for, raids to expect, and wars to fight. If he was alive today, he would know that the enemy across the river is not a headhunter or a warrior or a tribe. It is a strange tree that makes rivers flow upstream and brings with it many kinds of death and darkness. It is oil palm.

Conclusions

In late June 2017, my host sister Mina accompanied me to the airport in Merauke City for my flight back to Sydney. We sat at a small café near the entrance of the building, waiting for the boarding gate to open. Mina was texting her siblings in Jayapura. I was reading an online article in the *Sydney Morning Herald*. Published on February 3, 2017, a few months after the US elections, and written by a fervent Republican, the piece celebrated Donald Trump as the best solution to tackling "free spenders, organizers, race-baiters, intellectuals, and tree huggers." Mina glanced over my shoulder and picked up on the word *tree*, which she had learned during my English classes at the primary school in Mirav. "What's a tree hugger?" she asked. I explained that the expression referred to environmental campaigners and that it derived from activists literally hugging trees to protest again their felling. Mina went silent. The boarding announcement crackled through the megaphone above us. Then, Mina turned to me and said, "Marind are trying to protect the forest from oil palm. I guess that makes us tree-huggers. But what if the trees won't hug back?"

TREES, AND PLANTS IN general, are living embodiments of the principle of relational becoming, which is characteristic of the posthumanist approaches that have inspired this work. For instance, plants demonstrate remarkable phenotypic plasticity as they enter into organic dialogue with their environmental surrounds. Plants' existence is also one of perpetual metamorphosis. Growth and senescence, life and death, are dispersed all at once across their leafy limbs and radiating roots. As multiplicities, plants are always already composed of heterogeneous terms in symbiotic relation. Being one by becoming with many, they have no definable center or point of origin. Neither singular nor plural, plants are suborganismic and supraorganismic ensembles—vegetal dividuals par excellence.

Yet the particular plant that is oil palm teaches us another, less appreciated (but no less significant) dimension of relational living and thinking—its *limits*. In Merauke, oil palm refuses to engage in mutually beneficial relations with the human and other-than-human constituents of the Marind lifeworld. Its alien origins and selfishly rapacious behavior contrast with the attributes of forest organisms, and in particular those of sago—a plant with whom Marind share time, space, and substance. Oil palm disrupts the storied lives, movements, and futures of Marind and their companion species through its inexorable proliferation. It literally and figuratively subverts the flows of flesh and fluid that nourish the soils and rivers, along with their diverse inhabitants. Relentlessly appropriating land and life and returning nothing but death and destruction, oil palm thus embodies the tipping point of interspecies reciprocity. Oil palm cuts the multispecies networks cherished by, and constitutive of, the Marind lifeworld (cf. Strathern 1996). The efforts of Marind to come to grips with oil palm's story are undermined by the fact that the plant itself, in Mina's words, refuses to hug back.

At the same time, oil palm comes to embody the allure and pervasiveness of a plastic world for many Marind. Much like the roads multiplying across the landscape, oil palm is pregnant with the possibility of new connections to faraway places and peoples. Marind who cede lands to agribusiness corporations or seek employment on the plantations struggle to participate in the project of modernity while also sustaining the multispecies communities that make them human in the first place. Longing and belonging thus come into friction as villagers find themselves caught between the rapidly vanishing forest lifeworld and the materialities and connectivities presented by technoscientific capitalism and its alluring visions of progress. When even domesticated animals like Ruben, the juvenile cassowary, choose instant noodles over forest foods, deep-seated dilemmas arise among Marind over the lure of plastic futures and posthuman worlds that are at once promissory and deceptive, resisted and proliferating.

Ambivalent continuities, too, are at play in the Marind lifeworld. For instance, the sinister oneiric skin-changing entailed in dreams of being eaten by oil palm magnifies and distorts the potentially lethal forms of perspectival capture that Marind face in their everyday relations with organisms in the forest. The human-like dispositions of domesticated creatures in the village reverse in uncanny ways the animal attributes adopted by Marind men as they change skin in the forest. Palm oil itself is a ubiquitous skin-changer as it is camouflaged as a versatile ingredient within all manner of foods and goods purchased and consumed by people around the world—including Marind themselves. The arrival of this nonnative vegetal being in Merauke may be recent, but it sits within a long and repeated history of violent confrontations between Marind and their forest kin, on the one hand, and various foreign and unloving others—overseas plume-hunters, Dutch colonizers, and Indonesian settlers— on the other. And despite all the differences in their attributes and effects, sago palm and oil palm, as my friends recognize, exist ultimately as contrapuntal, kindred palms, each imbued with a particular form of wetness—the one life-giving, the other lucrative.

The line between the world before and after oil palm—and all the contrapuntal relations of places, practices, and persons these worlds encompass—is thus far from clear-cut. Rather, this world is one of opaque yet generative impurity, or abu-abu, that predates and is exacerbated by, the arrival of an alien and invasive plant. Epistemological fumblings, ontological conjectures, and speculative imaginations, coalesce as Marind work through the murk of living with oil palm in the unexpected country of capitalist monocultures. Wandering (and wondering) through the skin of this wounded land, Marind encounter plastic cassowaries, hungry palms, modern totems, and deathly dreams. These beings and experiences blur the boundaries between the realms of the real and the imagined, the native and the foreign, the wild and the domesticated, and the mourned and the monetized. They reveal more-than-human becomings to be hazardous business and being-in-relation a potentially dangerous and diminishing endeavor. They are, in one word, uncanny—a term derived from the German word *Unheimlich*, or "unhomely," which powerfully captures the paradoxical mix of the familiar and the strange in the growing grayness of the Marind lifeworld.

Abu-abu in the Upper Bian thus takes the form of indeterminate objects, shady beings, uncertain signs, and a lack of clarity about the current and future order of things. It is an affect and atmosphere that speak to the ambiguity of multispecies living, surviving, and dying in a place where people are trying to make sense of their predicament amidst confusion, alienation, and anxiety. It is an onto-epistemic murk that both subverts and constitutes the Marind lifeworld, generating different ways of acting, resisting, accommodating, and imagining at the edge of chaos. Abu-abu brings to light the contrapuntal relations and categories that matter by the very act of transgressing and torquing them. In doing so, abu-abu itself becomes the counterpoint *to* the counterpoint.

But let us return for a moment to the trees who don't hug back. Mina's pithy comment provides important insights on the more-than-human dynamics of the present-day Upper Bian. First, not all lifeforms can (or want to) participate in reciprocal relations across species lines. In the Upper Bian, the proliferation of oil palm violates the principle of restrained care that enables loving relations between Marind and their forest kin. Second, the incapacity or unwillingness of lifeforms to establish relations with each other is not solely or necessarily a human prerogative. From the perspective of many Marind, oil palm constitutes a vegetal form of significant otherness that refuses to participate in preexisting lifeworlds and instead appropriates the resources necessary to their survival only to sustain its own. Like settlers, the state, and the military, the plant does not belong in Merauke because it fails to engage in the mutual exchanges through which places and persons become physically and ideologically consubstantial. Growing alone in monocrop plantations, oil palm chooses a life of immunity over community in violently exclusive and exclusionary ways—even as its own existence is ultimately negated by the same technocapitalist forces spurring its expansion.

Oil palm's refusal to hug back, in Mina's words, thus bring into question the prevalent framing of multispecies relations in the plant turn and related posthumanist currents within a complex of love, care, and respect that celebrates and encourages interspecies kinship, relatedness, and entanglement. It points to the difficult yet life-sustaining work that goes into making and maintaining boundaries between bodies and species amid ecologies of skin and wetness both nourishing and contaminating. It highlights the importance of distinguishing wanted from unwanted relations across species lines and attending to interspecies relations that are imagined, imposed, or indeed impossible. Plants may be good to think with in eschewing assumptions of human exceptionalism and appreciating our mutual dependencies with often disregarded other-than-human organisms. But as the experiences of Marind with oil palm poignantly convey, not all plants are necessarily good to *live* with.

Rather, becoming-with oil palm in the simultaneously dehumanizing and more-than-human zones of agroindustrial expansion constitutes what Franklin

Ginn et al. call an "awkward flourishing." Awkward flourishings work with and against situated multispecies communities—some of which prosper and some of which flail. Eschewing claims to innocence or universality, awkward flourishings invite us instead to ask, "Who lives well and who dies well under current arrangements, and how they might be better arranged" (Ginn et al. 2014, 115). In Merauke, the flourishing of oil palm involves violence, loss, and suffering, which are inequitably and incommensurably distributed across particular plants, animals, humans, and ecologies. In these tangled knots of life and death, nonhumans are not always the victims and humans not always the culprits.

In this light, assuming that *humans* are primarily capable of and responsible for, violence against other-than-human lifeforms runs counter to many Marinds' conceptualization of oil palm as a being of exceptionally detrimental effects, and of themselves as oil palm's victims. To neglect this reality limits our capacity to think about violence in multispecies terms, in what might be perceived as yet another instance of human exceptionalism. Appreciating the role of plants as agents in our world thus demands that we remain open to the possibility of plants as destructive actors who also participate in biocapitalist regimes of making live and making die. At the same time, it demands that we attend to how racially, historically, and politically imbued differences *within* the category of the human itself inflect situated humans' enactments and experiences of social and ecological violence. In the human-plant entanglements of the Upper Bian, some humans bear the palm in profit, while others must bear with. Failing to consider *equally* the politics within—and not just the poetics of—more-than-human relations, categories, and hierarchies, poses a moral hazard. It risks rendering some human lives "yet *more* unseen" (Galvin 2018, 244, emphasis added).

And yet there is also something meaningfully "unseen" about oil palm in this story. Drawing attention to the fleshy and material entanglements of humans and other organisms, Donna Haraway (2008, 25–26) suggests that the relation is the smallest possible unit of analysis in the vulnerable, fleshly work of encountering and making companion species. In Merauke, however, the ruptures wrought by oil palm on the lifeworld of the Upper Bian are compounded by difficulties that Marind face in encountering and knowing oil palm itself. Community members can only speculate over what oil palm's story might be, and whether the plant can accommodate in less violent ways the storied lifeways of Marind and their forest kin. At the same time, the incapacity—or unwillingness—of oil palm to enlist in, and sustain, these multispecies networks is precisely what makes it so troublingly meaningful to many Marind. Oil palm matters to Marind in light of its tangible destructive effects but also through the generative not-knowings and not-relatings that characterize its abu-abu ontology.

The ontology of oil palm as a plant of ambivalent or nonexistent encounter value for the vast majority of Upper Bian Marind suggests that equal attention needs to be paid to how the *absence* and *impossibility* of fleshy encounters can produce equally meaningful, if only speculative, relations across species lines. Dialogical relations may be heightened in the literal absence of the other who remains nonetheless present as a violent *effect*. To *not* be able to touch or be touched can also be the source of creative imagination—a form of speculative haptics (the imagination *of* what the touch of a possible encounter might entail) rather than haptic speculation (the imagination triggered *by* the touch of an actual encounter). Distance, loss, and imagined or absent encounters matter just as much—albeit intangibly—in the entanglements of species.

By way of pushing these arguments further, one might return to the etymology of "entanglement" to think about the form and limits of relations across species lines. The term entanglement, which derives from *thangul* in Proto-Germanic, meaning seaweed. *Thangul* conjures images of algae mingling with oars, nets, fish, other marine organisms, and its own leafy limbs. Some entanglements, such as those between Marind and their other-than-human forest kin, are mutual and enable their respective constituents to thrive. However, these relations require careful practices of restrained care, and are never devoid of the risk of perspectival capture. Other entanglements are imposed and may be impossible to extirpate oneself from. For instance, many Marind actively resist the expansion of oil palm. Yet they suffer its ubiquitous presence across the landscape and in the products they consume on a daily basis. Entanglements may suffocate some organisms to support the propagation of others—much as the lifeways of Marind and forest species are compromised by the relentless proliferation of oil palm.

Like so many other facets of the Marind lifeworld, then, the entanglements of plants and humans in the Upper Bian are nothing less than pharmakonic. They encompass, in Gilles Deleuze and Félix Guattari's words, "all the attractions and repulsions, sympathies and antipathies, alterations, amalgamations, penetrations, and expansions that affect bodies of all kinds in their relations to one another" (2005, 90). Marind–oil palm entanglements in Merauke reveal that not everyone is relation-*able*—both in the sense of capable of being related *to* and in the sense of being capable of relating *back*. More than this, the entities setting the terms and limits to mutual relations (and relationabilities) are not always, or necessarily, human. Alongside care and love, violence, too, can be a multispecies act.

This book has analyzed human-plant entanglements in lively conversation with emergent theories in multispecies studies. At the same time, its most fundamental criticisms are directed precisely *against* some of the major trends developing within this current. These include multispecies studies' lack of engagement with Indigenous ontologies and epistemologies, its limited consideration of the "human" category in the context of racializing assemblages, its uncritical celebration of interspecies entanglements, its insufficient attention to unloving (rather than unloved) organisms, and its failure to approach violence itself as a multispecies act. In these and other respects, plant-human entanglements in Merauke offer an important counterpoint to the political and ethical drive of much multispecies scholarship to date: namely, how should we love in a time of extinction (Chrulew 2011, 139)?

In the Upper Bian, Marind struggle to sustain their skinships with forest creatures whose freedom and survival—like their own—are threatened by agribusiness. At the same time, the fraught experiences of Marind with oil palm suggest that to stay with the trouble of thriving yet destructive other-than-human entities is perhaps a thornier task. Advocating alliances with other species indiscriminately runs the risk of neglecting the contrasting worldly effects of creatures of empire, on the one hand, and potential nonhuman allies, on the other (Tsing 2018, 75; see also Tsing 2014b, 108). We need to consider critically who benefits from interspecies relations rather than simply celebrate the fact of more-than-human mingling (Star 1991, 43). Indeed, focusing solely on positive or mutualist forms of interspecies relations, or dressing relations in the warming aura of emergence or generativity, can excise the emotional complexity of more-than-human entanglements. In doing so, we risk foreclosing or excluding possibilities for alternative ways of relating, *outside* a reciprocal, kinship-based complex of love, care, and respect.

As I have attempted to show in this story, interspecies relations in Merauke are imbued with different affects and materialities, antagonisms and companionships, histories and futurities—some established, others opaque. Multispecies skinship reveals itself a lively and lethal affair that can be alternately mutual and life-affirming or violent and diminishing. The arrival of oil palm in Merauke also foregrounds how those who become entangled in agroindustrial contact zones—cash crops, chemicals, rice, and settlers—are not always expected, invited, or wanted. Just as dispersion has no permanent center or boundary, so too its parameters and possibilities are contingent. As the colliding ecologies of sago groves and monocrop plantations demonstrate, mixing badly can threaten, compromise, or destroy the relations through which species as storied selves come into existence. The co-opting of strangers in interspecies relationships can fail or turn parasitic and destructive. Dispersed or nondual worlds that come into being through more-than-human relations in perpetual flux may not always be able to fit all exigencies, or to contract or expand their scales to absorb and accommodate all kinds of others—particularly when these others, like oil palm, refuse to reciprocate in lively dispersion. Dispersed ontologies are thus less akin to choreographies than to improvisations in that they are not always successfully or evenly achieved across species lines. Such improvisations prove even more difficult to achieve when the storied lifeways of their constituents, as with oil palm in Merauke, remain largely elusive to those subjected to its effects.

This story has demonstrated that taking love seriously in plant-human relations also requires taking seriously its affective counterparts—hate, pity, ambivalence, and grief. Not all plants are nonappropriative, selfless beings that offer themselves with unconditional generosity to their environment and its inhabitants. Sago commits acts of love, while oil palm commits acts of violence. If, as adherents of the plant turn argue, plants merit greater moral consideration as lifegiving and sentient organisms, attending to the plantiness of vegetal lifeforms—such as oil palm in Merauke—who exclude other organisms (human and other) from their lifeworlds, while themselves being subject to the violence of technocapitalist machinations, becomes both necessary and challenging. It offers a provocative counterpoint to the question of how we should love in a time of extinction: namely, how, and *who*, should we love, in a time of destructive, other-than-human *proliferation*?

The more-than-human and dehumanizing dynamics of agroindustrial expansion in Merauke invite us to reflect more broadly on what an affirmative biopolitics might look like in a world where achieving immunity from oil palm seems impossible and creating community with it equally uncertain. These dynamics also problematize the meaning and form of care and respect as an interspecies practice. How does one respond in epistemic, affective, and ethical terms to organisms like oil palm that wreak havoc on forest ecosystems and human communities, yet that are themselves exploited by technocapitalist and human-driven agendas and visions? Whose lives and deaths matter in capitalist ecologies where the flourishing of some species drives others to the brink of extinction? How does one exercise care in grappling with species (human and other) as biological entities, cultural categories, and power-laden hierarchies?

These questions call for interdisciplinary investigations into the (anti)social worlds of capitalism's more-than-human protagonists and to the ontology of violence as a multispecies act. They demand critical attention to the experiences of Indigenous peoples who have historically recognized diverse nonhuman entities as persons and relations, yet whose own humanity is undermined by entrenched regimes of color and capital. They foreground capitalist natures as lethal assemblages that strip places and persons of their life-sustaining attributes and affordances, with intergenerational impacts on both human and other-than-human beings. At the same time, they reveal how capitalist natures replace multispecies networks with differently, yet equally "lively" meanings, possibilities, and imaginings. Cutting more-than-human networks, in other words, is a violent but often generative process—one that enables new and different forms of liveliness, both epistemic and ontic.

Marinds' imaginative characterizations of oil palm, for instance, complicate the prevailing depiction of monocrops as impoverished realms devoid of semiotic diversity and reveals them as animated instead by manifold, heterogenous meanings in transition and friction. At once material and symbolic in form, attributes, and effects, oil palm travels and transforms across the forest and plantation worlds it alternately undermines, reconfigures, or enables. It is hypercognized in morally and politically imbued ways by different individuals, whose understanding of the plant's reality transforms upon the acquisition of external knowledge that is then internally interpreted, refashioned, and contested.

By recasting plantations as agentive assemblages animated by multiple more-than-human actors, Marind imaginatively subvert the notion of "nature" as a passive and singular object of human control. In doing so, Marind epistemologies constitute a form of resistance to the simplifying logic of the plantation model and its colonial-capitalist undergirdings. After all, oil palm, too, is a "tree of many lives" that self-differs across its diverse material and moral manifestations and its equally diverse human perceivers, in ways that belie the treatment of life solely in terms of "species" categories. Just as there is no singular or static Marind "world of vision" (Viveiros de Castro 2011, 133), so too, there is no such thing as oil palm "in general" (Ramos 2012, 483).

The lively yet lethal ecology of plantations thus invites us to attend to the *value* of different *forms* of life, and not just life in the abstract. In the colliding worlds of multispecies forest and monocrop plantation, choices must be made about who to love and grieve, because beings—human and other—do pass out of existence, and life alone is not a good (enough) criterion of selection in deciding which worlds get to matter and which ones don't. Such an approach calls for complex tales of more-than-human care and violence that deploy a rigorous imagination and reflexive care to grapple with the situated intricacies of agential life. It invites us to think, act, and respond in abu-abu spaces—those spaces lying somewhere between the indiscriminate celebration of interspecies entanglements and the paralyzing politics of despair that arises when it seems nothing can be done. It invites the crafting of better, bitter stories that acknowledge the insights produced by states of seemingly unsurmountable abjection, the more-than-human practices of survivance that people forge within them, and their limits.

At the same time, the relations of Marind to forest beings and oil palm, respectively, highlight the need for a more nuanced and situated conception of posthumanity itself. Rather than taking as their ontological starting point inextricable entanglements of nature, technology, and culture, Marind strive to *retain* what makes them human in relation to plants and animals in the face of technocapitalist nature-cultures and historically ingrained racializing assemblages that deny them full personhood—and indeed, humanity—before the law. Marind do so in a number of creative ways. They contest the putatively transcendent yet lifeless gaze of mechanical drones. They reject the modernist dreams and capitalist temporalities incarnated in concrete monuments and time capsules. They condemn rice-human assemblages and the actors, practices, and powers associated with them. They refuse the companionship of all-too-human animals in the village. Instead, Marind produce living maps that sing and sway and sputter. They find solidarity and hope in the dystopic world of dreams. They celebrate sago-human assemblages that speak to their distinctively Papuan pasts, relations, and sense of belonging. They exhort plastic animals to return to the forest and retrieve their wild and native selves. In the ruin and rubble of the plantationscape, Marind even find ways to extend care, curiosity, and compassion toward oil palm itself.

Through these everyday practices and discourses, Marind resist and oppose the technoscientific transformations that some scholars deem inherent to the posthuman condition. Just as the category of the human is neither singular nor static, so too posthuman ontologies, I argue, must be defined and deployed in ways that account for the particular *kinds* of relations, forces, and actors that they entail—and whether they are embraced or eschewed. Making posthumans in the plural, in other words, is a necessary counterpoint to making natures in the plural.

The story I have shared is not one of feral flourishings in the ruins of capitalism or of traditional ecological solutions to natural resource depletion. I have not indulged in a primitivist narrative of culture death or offered an empowering account of Indigenous environmental activism. Nor do I ask that you necessarily believe everything in this story—that oil palm is an agent, that humans can turn into cassowaries, or that time has come to a stop. Adopting an ontological stance, as I have done, does not imply treating alternative theories of reality as unequivocally real, but as if they *could* be. Rather, I have focused on how an out-of-the-way Indigenous community "make do" within the context of a slow and more-than-human omnicide. I have attempted to bring to life the story of a people, place, and plant with whom we all are inextricably, if unknowingly, entangled through our daily acts of consumption, and whose out-of-the-way existences sit within a broader process of planetary undoing. I have explained Marind worlds not to explain them away but rather to immerse the reader in the murk of figuring out what is at stake—lives, futures, relations, and bodies—in abu-abu worlds at once and never just local or global. For if there is one thing the planetary environmental crisis has taught us, it is that we are all subjects implicated in one another's lives, albeit unevenly, differently, and often elusively so.

Some may wonder whether I give too much agency to oil palm in this story. Is there something ethically questionable, you might ask, about portraying plants as immoral when, by and large, plants continue to be relegated to the status of sessile organisms (think, vegetative) or decorative backdrops to the putatively more interesting activities of the animal kingdom? Surely, Marind perspectives on oil palm speak more to Marxian commodity fetishism than to ontological difference? And in "real-world" terms, might framing violence as a multispecies act not end up furthering the absolution and evasion of humans and human institutions from their responsibilities of care toward each other, the environment, and the planet?

What I can offer is not a steadfast answer to these questions but rather an invitation to reconsider their premises. To question the degree of agency of other-than-human beings, for instance, may speak less to the ethnographic material and analysis presented in this story, than to assumptions made about agency itself—namely, that it is primarily and necessarily a human attribute. In this story, taking seriously the possibility of plants as agents has *also* meant taking seriously the worldview of a people themselves systemically denigrated and treated as disposable under the racialized logic of colonial-capitalism. At the same time, the agency that Marind attribute to oil palm does not preclude their awareness of the human forces driving the particular way in which oil palm grows and affects them. To ignore this would be to reinforce precisely the kind of primitivism that Marind have and continue to face in their fraught interactions with the state and corporations. Instead, I have aimed to show that the agentivity and morality of oil palm are neither singular nor bounded. Pity and compassion sit alongside resentment and fear in Marinds' affective responses to this plant that is at once a driver and a victim to the forces of technocapitalist modernity.

As for the political stakes of framing plants as agents, Marind themselves are acutely conscious of the politics of ontological disclosure and performance as they struggle for recognition in the face of multinational companies and government agencies. Their pursuit of social and ecological justice entails astute negotiations and collaborations not only with oil palm, but also with oil palm's human and institutional allies. In these complex entanglements of plants, humans, and capital, how Indigenous activists engage in strategic self-representation before powerful and predatory audiences—human and other—can profoundly determine the shape of reality itself. Taking seriously the agency of plants and people thus means working through agency's uneven, dialogical, and performative distribution, across *and* within species lines.

What, then, might the critical plant turn I call for, entail? It would attend to plant-human entanglements' necessary situatedness—not as a weakness or limitation, but simply what is. It would move away from the wholesale embrace of relations (multispecies or other) and attend to whose lives and deaths matter, and for whom, within the necrobiopolitics of capitalist natures. It would recognize that interspecies relations are never just good or bad—that there is always more to the story. It would reflect on the potential and limits of the "species" as a unitary category of analysis and being in the context of trans-species indexicalities and individual specificities. Just as species are always more-than-one, so too the moral, affective, and political responses evoked toward, with, and against them are often plural and positional. Whether and what multispecies entanglements are desired "depends on the how and the what and the when" (Roberts 2017, 596). It is important that a critical plant turn attend to the difficult, daily labors—material and imaginative—that go into making and sustaining plant-human entanglements that are desired and enhancing and avoiding or making do with those that are imposed and diminishing. Addressing these questions will make for an anthropology beyond the human that is not only more critical, but also more capacious.

But what, one might ask, of oil palm's reality *outside* of its entanglements with humans—Marind and other? How, in other words, do we take seriously the multiple and often contested realities of humans about other-than-humans *and* take seriously the realities of other-than-humans beyond their human context? Given we will never be able to access oil palm's world of vision, do our attempts to understand it always and ultimately, take us back to the human? Addressing these questions, of course, requires asking *which* humans' worlds of vision are at play, rather than assuming a singular and homogeneous "human" perspective in the first place. It requires acknowledging that human worlds of vision themselves are internally complex, conflictual, and transitional. Among Upper Bian Marind, such worlds of vision imbue entities like oil palm with multiple, contemporaneous vegetal identities—as immoral organisms, uncanny kin, and exploited victims. These ontoepistemologies—plant and other—are always provisional, or in medias res. They transform in light of new information, encounters, and events, in a process of ontogenesis shaped as much by what is known as what is as-of-yet unknown, already imagined, and ongoingly contested.

Ontoepistemic dispersion lies at the heart of cosmopolitics, a concept deployed by Isabelle Stengers (2010) to challenge the givenness of a common world and the scope of its legitimate participants. Closely aligned with Stengers, Marisol de la Cadena (2015) calls for a pluriversal politics in which the voices of other-than-human entities, or "earth-beings," would count. Such a move, de la Cadena suggests, is challenging, but not impossible, and necessary. Together, cosmopolitics and pluriversal politics encourage us to recognize the coexistence of multiple divergent worlds and their potential articulations across and within species lines. They call for us to rethink ontology in political terms and politics in beyond-the-human terms. They invite us to approach the cosmos as a multiply authored and contested composition, rather than a singular and axiomatic reality.

As I have shown in this account, Upper Bian Marind conceive the form and effects of oil palm expansion in more-than-human terms within a cosmos in which the human itself emerges through its relations to other organisms. Oil palm jeopardizes this multispecies cosmos through its destructive effects on places, persons, temporalities, and bodies. And yet despite the destruction it wrecks upon their world, Marind persist in the difficult labor of imagining new kinds of relatedness and ethics with oil palm—if not toward interspecies love, then at least toward a more equitably distributed sharing of suffering across species lines. Inclusion and exclusion, then, are not mutually incompatible stances. The proliferation of oil palm is incompatible with the continuity of the forestworld it threatens and all the values, textures, affects, and bodies this forestworld represents. But oil palm is also imaginatively included in Marind discourse and speculation as a polyvalent object of wonder. In these interspecies entanglements, oil palm's story matters. It, too, has a part to play in the making of less violent common worlds.

What form, then, might a pluriversal politics take in the Papuan plantationscapes that we have journeyed through together? In many ways, the answer to this question remains as abu-abu as the worlds and beings such a politics might involve. Recall here the musings of my friend Ignatius as we sat together on the banks of the Bian that dusky afternoon in late 2016. Marind, Ignatius tells us, are themselves are still looking for the right path. In this book, we have traveled someway down this tentative path in the company of Ignatius, of fellow Marind villagers, of many other significant others. Furtive, shadowy clues have surfaced along the way, guiding us toward the possibility of a Papuan pluriversal politics. Forging a path toward such a politics in this abu-abu world would have to account for storied palms and gnawing bulldozers, whistling bulbuls and weeping cassowaries, plastic birds and rusty clocks. It would have to make room for rivers that flow upstream and totems made of concrete, for maps that won't sit still and plants who won't hug back. It would require thinking with skinships and scarships, locks of hair, and hungry roads. It would involve grappling with and against the animacy of flesh and fluids in the making of more-than-human worlds. It would need to accommodate Marinds' plural theories of reality, while also carving out political space for the power of dreaming over and upon changing realities.

Pluriversal politics in the Papuan plantation nexus demands attention to how politics itself emerges in the rubbing against each other of multiple ontologies and histories whose pluralities and separations offer an important antidote to the hegemonic single-mindedness of plantation logic but whose overlaps and concrescences also play a major part in shaping Indigenous lifeworlds. Within these overlapping worldmaking projects, certain earth-beings emerge as exceptionally harmful toward humans whose very own humanity remains under the siege of racial colonialism and technoscientific capitalism. In the Upper Bian, oil palm refuses to share space and time with other beings. It literally gets under the skin of organisms whose flows of substance and sentience it diverts, disrupts, and destroys, sowing fear and violence on the troubled terrains of a politically volatile region, whose peoples continue to be denied their right to self-determination. At the same time, oil palm invites attention to the multiple and conflicting realities of commodified lifeforms across time and space. At once driver and victim of destruction, the crop's ambivalent ontology calls for critical reflection on who gets to speak for, with, and against, lethal capital. Certainly, navigating the politics of situated knowledges and other-than-human animacies at the cosmopolitical roundtable will be challenging. Yet only by working transversally across porous ontoepistemic divides will ways of commoning with oil palm be able to flourish—otherwise, and wisely.

ON MY LAST DAY in the field, I sat under the old mango tree in Khalaoyam village with Elena, the old woman who had once tended to Ruben the plastic cassowary. Bodies had begun to sag in the settling torpor of the midday sun. A patchwork of pale-yellow leaves and plastic chewing-gum wrappers spread around the gnarled buttress roots of the mango tree. Candy and chlorophyll, each host to its own congregation of festive flies. Elena ground her arthritic frame down to a seat and carefully rolled herself a clove cigarette. Puffing clouds of sweet-scented smoke, she said to me: "You have learned many things about oil palm in your time here. You have heard many stories. As many as the threads running through my noken bag here. The truth is, Marind do not know the story of oil palm. But maybe one day oil palm will know us, and we will know oil palm. Then, there will be new skins to be shared. Then, there will be new stories. Then, time may become alive again."

Elena was one of many villagers who read in the disappearance of Ruben the cassowary a glimmer of hope for Marinds' own future emancipation from Indonesian rule. She was also one of a few who confided in me that although time has stopped, its flow might once again resume. As Elena suggests, the world-shattering advent of monocrops may still give rise to unexpected collaborations that enable more livable shared futures. Budding organismic assemblages may work together to remake the living landscape of the Upper Bian (at least a little) differently. Ecological time bombs may be defused in time to enable explosive forms of hope more equitably distributed across lifeforms and environments. The rusty needles of the clocks of Libra 969, too, may someday begin to turn. In the face of hegemonic state and other institutions of power, refusal in the now may constitute a creative and strategic way of making other kinds of hope possible in the future. In the meantime, multispecies patches of forest remain precious refuges of practice, growth, and memory for Marind like Evelina and Gerfacius, with whom we visited the grove, and who insist they will continue to bring their children to the forest so long as the forest is still standing. But finding hope-in-hopelessness in the blasted landscapes of capitalist expansion will also require that Marind find ways to relate to, and better understand, the needs, growth, and stories of oil palm—a plant whose lifeway currently undermines the very possibility of multispecies hope. Whether, and what, that future can be, will depend as much on Marinds' attempts to know oil palm, as on the plant's own willingness to look back in mutual response and respect. Then, and only then, time will tell.

Endings—The Author's Dream

I was walking through an oil palm concession in the company of a plantation field manager, a short, bespectacled Indonesian man in his mid-forties. The man pointed out a couple of oil palm trees afflicted by basal rot. Their trunks were caved in and frittered to the touch. At their base a proliferation of sporophores burst out like vicious eczema. The tips of their discolored fronds almost reached the ground, defenseless to the ravages caused by the notorious wood-rotting fungus *Ganoderma boninense*—the "cancer" of oil palm, the man described.

Eventually, we came upon a clearing. Here stood a single monumental sago palm in full bloom—a gentle giant in the midst of the plantation. The man removed his glasses and wiped them of perspiration and dust. He asked me what it meant to be alive. I said it was to have a heartbeat, to feel the blood flowing through your veins and limbs. He replied, "If that is so, then what about this?" and axed down a huge chunk from the sago tree's bole with the machete that had appeared in his hands. In the same movement, he raised the machete and brought it back down again, this time hacking off my left arm with a clean cut at the shoulder between the humeral head and greater tubercle bone. I fell to the ground, writhing. He asked me, "Now, is your arm there still alive? Is it still part of you? When does it begin to die? When does it stop living?" The machete rose again, bringing down with it another chunk of sago trunk, before rising again to slice off my right arm. I tried to scream but it was as though the air had petrified into a solid sphere enclosing us, all sound ossified out of existence. He asked me again, "And what about now? Are you still you without those arms? How much of you is left living without them?" Then, he asked where I had learned to speak Indonesian so fluently. Walking around my body, the man carefully took aim and hacked off my right leg, then my left one, wrenching out the blade wedged deep into the soil in between the blows, then repeated the same movement in slicing off two more chunks of wood from either side of the sago's bole, leaving tree and I stumps of ourselves. "You see," he said, "the boundary between what is alive and dead, what is you and not you, is a fine one. Life has no center and no contours. You are made up of many parts, but I can take many away and you will still think you are alive. And so, we will keep cutting the forest, and so we will keep felling the people: for there is much you can take, before life is truly gone from your veins and your roots."

The man walked away, whistling Indonesian pop-singer Cita Citata's lascivious hit, "Sakitnya Tuh Disini" ("The Pain Is Here"). I lay on the floor next to the sago stump, each of us bleeding our own fluids of blood and sap, until they mixed the one with the other, forming a meandering meshwork of leaking arteries that transformed into thick pneumatophores, the winding aerial roots that sago develops when growing in mangrove areas. The pneumatophores rose high up to form a circle above our dismembered bodies, resembling a huge, gnarled ribcage. The air was saturated with the deep, moaning creak of the stretching vegetal limbs. Having reached their apex, the ribs closed into each other like a pair of hands joined in prayer. They sank down slowly, wrapping us both in their bony embrace, pulling us deep into the soft, peaty soil, until we disappeared in a fragrant bed of moss and leaves.

I acknowledge the Gadigal people of the Eora Nation and the Darramuragal people of the Darug Nation, on whose unceded lands this book was crafted. I offer my respects to Gadigal and Darramuragal elders past, present, and emergent, and to their kin—human, vegetal, animal, and elemental. The lands of Gadigal and Darramuragal were taken without consent, treaty, or compensation. They are lands stolen and lands of ongoing Indigenous survivance, resilience, and continuance.

This book has flourished in the nourishing shade of countless precious friends and colleagues. I was first introduced to the complexity of plant-human relations by my postgraduate mentors at the University of Oxford, Laura Rival and Elizabeth Ewart. Six years later, Macquarie University provided me with an intellectually rich environment to explore human-vegetal worlds as a doctoral student. I extend heartfelt gratitude to my supervisors, Eben Kirksey, Jaap Timmer, and Eve Vincent, for their unfailing support throughout my candidature, and for trusting that this project was meaningful and possible. I also thank Payel Ray for her energizing presence and my PhD examiners for their constructive feedback: Nils Bubandt, Michael Scott, and Rupert Stasch.

A three-year postdoctoral research fellowship at the University of Sydney's Department of History allowed me to focus on bringing this book to fruition. I thank in particular my mentor, Warwick Anderson, for his incomparable personal and intellectual support throughout this period, as well as my co-mentors, Paul Griffiths and Sonja van Wichelen, and the generous participants of a first-book peer-review symposium held in November 2019: Roberto Costa, Ben Robin Dean, Margaret Jolly, Jaya Keaney, Paul-David Lutz, Timothy Neale, Hans Pols, Thom van Dooren, and Christine Winter.

Colleagues from a wide range of disciplines provided critical feedback on the arguments and ideas presented in this book at various guest seminars and conferences. These include presentations at the University of Sydney; Macquarie

University; University of New South Wales; Australian National University; Alfred Deakin Institute; University of Melbourne; Asia Research Institute at Yale–NUS College; Institut des Hautes Etudes pour la Science et la Technologie; Institut International pour les Etudes Comparatives; Centre for International Forestry Research; Australian Anthropological Society; Sydney Pacific Studies Network; Multispecies Justice Collective; American Anthropological Association; Society for Social Studies of Science (4S); Association for Asian Studies; Association for Social Anthropology in Oceania; Asian Studies Association; European Association of Social Anthropologists; Asian Studies Association of Australia; Temporal Belongings Network; Australian Food, Society, and Culture Network; Concordia University; Universidade de Lisboa; University of California, Berkeley; London School of Economics; Universitas Udayana; Kunsthistorisches Institut in Florenz—Max-Planck-Institut; Carleton University; University of Wisconsin–Madison; Kunsthistorisches Institut in Florenz; University of Waikato; TBA-21 Academy; and Sydney Festival. I thank the organizers and the participants of these events for their incisive questions and comments.

At the Sydney Southeast Asia Centre, I benefited immensely from the mentorship and advice of Michele Ford as well as fellow early-career researchers and professional staff: Avril Alba, Ariane Defreine, Rosemary Grey, Elizabeth Kramer, Minh Le, Jessica Melvin, Natali Pearson, Josh Stenberg, and Kristy Ward. At the Charles Perkins Centre, I extend my thanks to David Raubenheimer and Stephen Simpson. At the University of Sydney's Department of History, I thank in particular Leah Lui-Chivizhe, Frances Clark, James Dunk, Julia Horne, Roy MacLeod, Michael McDonnell, Mark McKenna, and Penny Russell. At the Sydney Environment Institute, I have deeply enjoyed thinking-with Eloise Fetterplace, Monica Gagliano, Liberty Lawson, Iain McCalman, Astrida Neimanis, Elspeth Probyn, Killian Quigley, Sue Reid, David Schlosberg, Michelle St Anne, Gemma Viney, Dinesh Wadiwel, Sam Widin, and Genevieve Wright. Danielle Celermajer and Christine Winter—through multiple overlapping crises, we have trodden ancient and new paths together. You taught me to be true to myself and to the many beings entrusted to my heart and pen. You blessed me with the tremor of recognition for a life in so many ways completed by your presence.

Other precious colleagues in this journey include Chris Ballard, Karen Barad, Ruth Barcan, Jennifer Biddle, Marisol de la Cadena, Jennifer Deger, Elizabeth Duncan, Peter Dwyer, Ute Eickelkamp, Stefanie Fishel, Kate Fullagar, Anne Galloway, Shaila Seshia Galvin, Anika Gauja, David Gellner, Radhika Govindrajan, Benjamin Hegarty, Annamarie Jagose, Stuart Kirsch, Emma Kowal,

Tessa Laird, Tess Lea, Fiona McCormack, Debra McDougall, Tanya Murray Li, Deborah Lupton, Kristina Lyons, Monica Minnegal, Ursula Münster, Alyssa Paredes, Jess Pasisi, Robbie Peters, Jemma Purdey, Lisa Stefanoff, Alice Te Punga Somerville, Deborah Thomas, Anna Tsing, Adrian Vickers, Lee Wallace, and Megan Warin. Gratitude also goes to Cory-Alice André-Johnson, Dominic Boyer, Chip Colwell, Keridwen Cornelius, James Faubion, Cymene Howe, Amanda Mascarelli, Danilyn Rutherford, Isabel M. Salovaara, Emily Sekine, and Shelmith Wanjiru from the Society for Cultural Anthropology and the Wenner-Gren Foundation for Anthropological Research.

This research was made possible by a number of generous grants and scholarships from the Australian Ministry of Education and Training (Endeavor International Postgraduate Scholarship 2015–2018), the Wenner-Gren Foundation for Anthropological Research (Dissertation Fieldwork Grant 2015–2016, Engagement Grant 2019, and Post-PhD Research Grant 2019–2020), the Australian Research Council (Discovery Project, "The Promise of Justice," 2020–2022), the Janet Dora Hine Postdoctoral Fellowship (2019–2022), the Charles Perkins Centre (Postdoctoral Startup Allowance 2019–2022), the University of Sydney's Faculty of Arts and Social Sciences (Faculty Research Support Scheme 2019–2020), the Sydney Southeast Asia Centre (Conference Fund 2019), the Asian Studies Association of Australia (Biennial Conference Postgraduate Award 2017), and Macquarie University (Fieldwork Research Grant 2015–2018).

Ethics approval for the fieldwork upon which this book is based was received from the Macquarie University Human Research Ethics Committee on March 31, 2015 (Reference Number 5201500051). I thank Duke University Press, the Asian Studies Association of Australia, the Australian Anthropological Society, Macquarie University, and members of their respective judging committees, for recognizing the merits of my research with the award of the inaugural Duke University Press Scholars of Color First Book Award 2021, John Legge PhD Thesis Prize 2020, Best PhD Thesis Prize 2019, and Vice-Chancellor's Commendation 2019, respectively.

A version of chapter 2 appeared in *Anthropology Now* and a version of chapter 4 appeared in *Ethnos*. Modified extracts from articles and essays published in *Art+Australia, Cultural Anthropology, Environmental Humanities, HAU: Journal of Ethnographic Theory, Journal of the Royal Anthropological Institute*, and *The Living Archive: Extinction Stories from Oceania* have been incorporated in specific sections of the book. I thank these journals for their permission to reprint these materials. I also thank Synergetic Press for allowing me to reprint in modified form sections of "Sago: A Storied Species of West Papua," from *The Mind of Plants: Narratives of Vegetal Intelligence* (edited by John C. Ryan, Patricia Vieira, and Monica Gagliano), in chapter 5. At Duke University Press, I extend my heartfelt gratitude to my editor, Ken Wissoker, for his precious mentorship and support throughout our journey together and for helping to bring this book to full fruition. I thank the Duke University Press team for their incredible support, and, particularly, Ryan Kendall for guiding me through the production process. I thank Susan Albury, Brian Ostrander of Westchester Publishing Services, and Nicole Balant for their meticulous copyediting. Immense gratitude goes also to the two anonymous reviewers of this manuscript for their copious and constructive feedback.

Friends who supported me throughout this journey include Justa Hopma, Lena Nguyen Horneber, Brigitte and Kyomi Mimasu, and Jelle Wouters. Karin Bolender and Laura McLauchlan—my "Grasshawg girls"—helped me grapple with the complexities of multispecies ethnography over countless emails and Skype sessions. Just as important to my thinking were the generous inputs of fellow early-career researchers and Higher Degree Research students, including Matt Barlow, Emile Boulot, Sria Chatterjee, Emily Crawford, Jen Dollin, Zsuzsanna Ihar, Anna-Katharina Laboissière, Daniel Ruiz-Serna, Daniel Tranter, Blanche Verlie, Jamie Wang, and Katie Woolaston. I also thank those friends who helped to disseminate my research findings in international media outlets: Deanna Catto, Liane Colwell, Alexander d'Aiola, Kate Evans, Meri Geraldine, Rebecca Gidley, Amy Gunia, Farid Ibrahim, Sarah Jacob, Hanna Jagtenberg Tom Johnson, Elly Kent, Klas Lundström, Shaheryar Mirza, Fidelis Satriastanti, Sanja Savkić Šebek, Rachel Smolker, Hanne Worsoe, and Yifan Wu.

My access to the field was enabled by the sponsorship of the Merauke Secretariat for Justice and Peace and the Jakarta-based NGO, PUSAKA. From these institutions, I thank, in particular, Anselmus Amo, Emil Ola Kleden, Nicodemus Rumbayan, and Franky Y. L. Samperante for their precious friendship and moral guidance. The topic and location of my doctoral research were significantly inspired by my prior work with the UK-based human-rights NGO Forest Peoples Programme. I extend my heartfelt thanks to the members of this organization, including Patrick Anderson, Tom Griffiths, Louise Henson, Justin Kenrick, Tom Lomax, Julia Overton, Sarah Roberts, and particularly Marcus Colchester, whose deeply reflexive approach to Indigenous advocacy has been a constant source of wisdom and inspiration.

My parents, Jacques and Dominique, have been a source of unfailing patience, love, and encouragement throughout my life and particularly during the compilation of this book. I cannot thank you enough for being who you are and holding me as you do. None of this would have been possible without you. My brother, Emmanuel, supported me throughout the trials and tribulations of research with his wit, humor, and unspoken care. Your sparkle was a constant guiding light for me. Thank you for keeping me grounded and alight all at once. As for my partner, Jacob, where do I start? More important, may it never end.

In the field, I extend my deepest thanks to the communities of Bayau, Khalaoyam, and Mirav, who hosted me despite the many risks entailed, and without whose cooperation this research would have been impossible. For reasons of safety, I cannot name you all. Indeed, I cannot name you at all. With pseudonyms applying, I extend particular gratitude to Darius and Theo, who taught me how to walk the forest and listen to birds, and to my host sister, Mina, whose incisive reflections on anthropological practice were central to rethinking my encounters with humans and plants. I also thank the children of the Mirav primary school for entrusting me with their stories, drawings, songs, and dreams. As for Gerardus, Gerfacius, Pius, and Rosa, who passed away during my fieldwork, I will always remember you for your immense kindness and spontaneous laughter. Finally, I thank the rivers, soils, wind, and groves of the Upper Bian and the plant companions whose story I have sought to tell in this book. Each has helped me rethink life, and our place within it, in ways that extend far beyond the scope of scholarly practice. All shortcomings in this book are mine.

INTRODUCTION

1   In this book, *Merauke* refers to the regency of Merauke (kabupaten Merauke)
    and *Merauke City* refers to the regency's capital city and main urban center (kota
    Merauke). The *Upper Bian* (Bian atas) refers to a vast region of forest, swamps, and
    marshlands lying along the northern reaches of the Bian River in Merauke. This area
    covers some 8,593 km², or 18 percent, of Merauke and sits 300 kilometers north of
    Merauke City. I use the term *West Papua* to refer to both the Indonesian provinces
    of Papua (propinsi Papua) and West Papua (propinsi Papua Barat). The eastern half
    of New Guinea island, which encompasses these two regions, was known as Irian
    Barat during the Sukarno era (1963–1971) and Irian Jaya during the Suharto era
    (1973–1999). With the exception of major cities and provinces, pseudonyms have
    been used for all persons and places cited.

2   I underline terms in Upper Bian Marind and leave roman terms in Indonesian,
    or logat Papua, the Papuan creole of Indonesian. The native tongue of Upper Bian
    Marind belongs to the Trans-New Guinea phylum and is spoken from the coastal
    areas of Merauke in the south to the mouth of the Digul River and in the Fly River
    region of Papua New Guinea. While older-generation Marind still speak the Bian
    dialect of Marind, logat Papua has become the lingua franca between different
    ethnic groups in the area, between subethnic groups who speak different Marind
    dialects, and between older and younger generations within the same community.
    Considered to be distinctively Papuan by my interlocutors, logat Papua, or Papuan
    Malay, is a creole of Indonesian that emerged during the first wave of Indonesian
    transmigration into Merauke in the 1970s, when Upper Bian Marind first came into
    contact with non-Papuan settlers, and prior to that, through their interactions with
    coastal and urban Marind. Most of my interactions and interviews in the field took
    place in this idiom, with the exception of certain key Marind terms for which no
    logat Papua translation exists or that my interlocutors deemed important to express
    in their native tongue for cultural and political reasons. In other cases, the sacred
    nature of Marind concepts and practices—skin-changing, stories, and wildness, for
    instance—and their association with male knowledge and male spaces meant that
    I could only access what my companions believed to be their closest equivalent in
    logat Papua. The English terms I offer for these words are therefore based on logat

Papua translations rather than the original source language. Sections of long prose and poetry featured in this book were compiled in the form of fieldnotes and audio recordings and then transcribed and translated from logat Papua into English by the author, with terms originally pronounced in Marind underlined.

3 I borrow the term *out-of-the-way* from Anna Tsing (1993) and Paige West (2006a) to describe places on the periphery of capitalist world systems, where people creatively channel their marginality into creative forms of interpretation, critique, and protest. Out-of-the-way places challenge the assumed stagnancy and homogeneity of the periphery. They reveal the periphery to be animated by plural ways of knowing and being, which are achieved through transcultural dialogue across diverse sets of actors and forces. Out-of-the-way places thus reveal the inherent instability of political meanings and the capacity of marginalized peoples to destabilize seemingly hegemonic forms of authority. On the related notions of margins and frontiers elaborated by Tsing and West, see notes 12 and 68.

4 The unprecedented scale and impacts of plantations have recently brought scholars in the environmental humanities to coin the current era the *Plantationocene*. The term, according to feminist Science and Technology Studies (STS) scholar Donna Haraway (2015, 162n5), denotes a spatiotemporal formation rooted in racialized forms of colonialism, which entails the mass substitution of biodiverse forests with industrial monocrops, to the detriment of the human and nonhuman organisms that forests sustain (see also Chao 2022a). The concept of the Plantationocene is helpful in tempering the human-centric optic of the Anthropocene and foregrounding the pervasiveness of plantation logics and legacies in the past and present. However, I do not deploy it as a conceptual frame in this book. Like the many other taxonomic candidates vying to capture the essence of the current (s)cene—Anthropocene, Capitalocene, Chthulucene, Plasticene, Technocene, and the list continues—the Plantationocene demands a reduction of everything to some "thing." In doing so, it excludes or obscures empirical realities and attendant genealogies of thought and practice that speak directly to the arguments presented in this book. As Janae Davis et al. (2019) note, the Plantationocene discourse privileges multispecies dynamics over racial politics, does not engage with the gendered and bodily effects and affects generated by plantation violence in the Caribbean and United States, and does not account for the wide variety of preexisting critiques of plantation formations in Black and feminist studies (see also Jegathesan 2021). Elided within the multispecies core of the Plantationocene, too, are Black modes of interspecies intimacy, creativity, and resistance toward plants, objects, and other nonhuman lifeforms in colonial plantation zones (see, for instance, Carney 2020; King 2016; Wynter 1971). Also neglected is the vast and distinctive body of literature on South Asian plantation trajectories and ecologies, which foregrounds the centrality of the plantation and its gendered labor dynamics in the formation of postcolonial nationhood, social justice, and agroindustrial sustainability (Besky 2013; Jegathesan 2019; Sen 2017). New World plantations in particular are important to highlight here because of their fundamental impact on modern social and economic arrangements in the Western hemisphere and elsewhere (see Benítez-Rojo 1996; Mbembe 2003; Mintz 1985; Trouillot

1988, 1997). In excluding these literatures and experiences, the Plantationocene ends up replicating what geographer Kathryn Yusoff (2019) and others identify as a critical and consequential flaw of the Anthropocene (another term that I intentionally avoid in this book)—namely, a neglect of the historical construction of Blackness as non-White, therefore nonhuman, and therefore passive, geologic matter (see also Silva 2017; Wynter 2015).

At the same time, transatlantic plantation formations and their legacies followed distinctive historical and geographical trajectories leading to those described in this account. Under European colonial rule, plantations entailed the uprooting of Black people from their native soils to the United States and the Caribbean and their enslavement as undifferentiated flesh (Spillers 1987) and fungible property (Hartman 1997). In the Melanesian context, historiographies of trading networks confirm the status of New Guinea as an exporter of slaves to the Moluccan world in the precolonial period. The regional slave trade, however, was formally abolished by the Dutch administration as part of a broader process of missionization and civilization (see L. Giay 2016; Timmer 2011). Prior to its abolishment, the slave trade centered primarily on the coast—notably, the Raja Ampat Islands (a vassal of the Tidore Sultanate) and the Onin Peninsula—and did not affect Upper Bian Marind, whose lives and pasts I recount in this book. In the contemporary Papuan plantation sector, Indonesian settlers are invariably privileged over Indigenous Papuans in terms of employment opportunities. Marind are thus exempt from the onto-epistemology of "labor" that Black literary studies scholar Shona Jackson (2012, 54) identifies as central to the colonial (and postcolonial) order and the configuration of the modern, disciplined human—albeit on the premise of an equally discriminatory (if differently colored) racial divide (see also Tsing 2009; Wolfe 1999). Finally, while state and gendered violence certainly figure prominently in West Papuans' everyday lives and political landscape, these forms of violence did not begin with the inception of monocrops. Nor, arguably, can they be compared to the spectacular forms of terror documented in European colonial plantations—from the mutilation of slave bodies to the systemic rape of women and the conscription of the unborn to slavery as speculative labor (Caldwell 2007; Glymph 2012; Morgan 2004).

5   For overviews of the polemics surrounding oil palm in Southeast Asia, see Pye and Bhattacharya (2013) and A. Rival and Levang (2014).

6   For instance, Anna Tsing describes plantation science as a "hegemonic, extinction-oriented creed" rooted in the absolute domination of plants by humans (2011, 19). Philosopher Michael Marder decries the capitalist plantation model for violently homogenizing the distinctive modalities of growth and reproduction of plants and for reducing them to food and fuel destined for human consumption (2011, 469–70). Political scientist James Scott condemns agronomic science for radically simplifying nature, and excluding knowledges, practices, and ecologies that lie outside its productionist paradigm (1998, 262–306; see also Shiva 1993). Political ecologist Michael Dove characterizes the plantation as a regime of discipline of plants and people that privileges the crops and technologies of powerful outsiders and deprivileges the crop- and place-specific knowledge of local smallholders in ways that are "inimical

to . . . [the] existence of alternatives" (2019, S310). In a similar vein, Donna Haraway describes plantations as a "system of multispecies forced labor" and a realm of "out-and-out exterminism" in which the "capacity to love and care for place" is negated (cited in Mitman 2019, 10, 6).

7   See, for instance, Besky (2013); Cramb and McCarthy (2016); Dove (2011); Jegathe-san (2019); Li (2014); McCarthy (2010); Sen (2017); Stoler (1985); Tammisto (2018a).

8   The ontological turn is often described as an approach that "takes seriously" the worlds of the peoples we study. In deploying this language and method, I draw from Rita Astuti's useful characterization of what has now become a recurring, yet often glossed-over, expression. Taking worlds and realities seriously, Astuti suggests, means acknowledging the fact that peoples' perspectives are heterogeneous and shifting. It attends to what people themselves have to say and their own interpretations of their words. It recognizes peoples' capacity for critical and creative inquiry into the worlds they inhabit. And it seeks to distil from specific fields ideas of broader relevance and import in understanding the human condition (Astuti 2017, 120; see also Barth 1987; Coburn et al. 2013; Gegeo and Watson-Gegeo 2001).

9   On the possibilities for multispecies justice within plantations as "landscapes of empire," see Beilin and Suryanarayanan (2017); Besky (2013); Chao (2021b); Paredes (2022); Tsing, Mathews, and Bubandt (2019).

10 As Māori scholar Linda Tuhiwai Smith notes, decolonizing research does not mean rejecting all Western theory. Rather, it means putting Indigenous peoples' concerns and worldviews at the center of research and approaching theory and research from Indigenous peoples' perspectives and for their purposes (Tuhiwai Smith 2012, 89; see also Chao and Enari 2021). I-Kiribati and African-American scholar Teresia Teaiwa make a similar point in their call for a broad engagement with theory and theorists of all kinds as an exercise in intellectual agency and a foundation for Indigenous self-determination in the academy. Engaging with White scholarship, Teaiwa writes (2014, 52–53), is a way of recognizing its contribution to Pacific genealogies of thinking while nonetheless retaining sovereignty in the face of "the ancestors we get to choose" (see also Gegeo and Watson-Gegeo 2001; Turner 2006).

11 In this, I follow Aletta Biersack's (2006a) call for a reimagined political ecology that inquires into alterior bodies of practical and theoretical knowledge and, in doing so, decolonizes environmental knowledge from its North-centric monopoly (see also West 2019).

12 Margins, Anna Tsing notes, constitute powerful topographic and conceptual sites from which to rethink the nature and specificity of local and global formations and to question the stability of these categories in the first place. To focus on the margin, Tsing continues, does not mean reducing marginal peoples and places to icons of stability or radical difference nor to symbols of modernity's "dying Other" (1993, x). Rather, margins draw our attention to modes of "creative living at the edge"—both their constitutive differences and internal tensions and their historical positioning within broader regional, national, and global dynamics (Tsing 1994; see also West 2016).

13 For a critique of the "ontology *of*" approach, see Carrithers et al. (2010, 172–79, 194). The approach I describe here has long been a defining feature of anthropological practice, with the possible exception of structural anthropology, However, I intentionally characterize Marind as *ontologists* in order to mark my departure from the prevalent treatment of ontologies as static, apolitical, and bounded meta-concepts, abstracted from the everyday lives of the people said to inhabit them, or what Eduardo Viveiros de Castro (2012, 64) calls "virtual ontologies" (see also Descola 2013; M. Scott 2007). Rather, I approach ontologies—as transforming ways of being in the world and as discourses *about* these transforming ways of being—as they are experienced, produced, contested, and theorized by Marind themselves through their situated actions and reflections (see also Meyer 2001; Erazo and Jarrett 2017; Willerslev 2004).

14 In his ethnography of coral gardens, for instance, Bronislaw Malinowski (1935) described how Trobriand economic life, social relations, and political organization revolved around horticultural crops and their upkeep. James Fox (1977) analyzes the cultural history of Roti and Savu Islanders in southeastern Indonesia as adaptations to the ecology of the lontar palm. Laurentius M. Serpenti (1965) highlights comparable links between root-crop cultivation patterns and social structure among the Kimaam on Frederik-Hendrik (now Yos Sudarso) Island, west of Marind territory (see also Barker 2008; Panoff and Barbira-Freedman 2018; Peluso 1996).

15 See, for instance, Bashkow (2006, 184); Battaglia (1990, 49); Bonnemère (1994); Descola (1986, 166, 175, 197, 215–17); Halvaksz (2013, 149); Kahn (1988, 44); Nimuendajú (1939, 90; 1946, 60); Rival (1998); Tuzin (1972, 234); West (2012, 119).

16 See, for instance, Dundon (2005); Heckler (2004, 243–48); Christine Hugh-Jones (1979, 114–32, 200–217); Pouwer (2010); Stasch (2009).

17 See, for instance, Bonnemère (1996a); Ellen (2006); S. Hugh-Jones (1979, 165–73); Leenhardt (1979); Mondragón (2004); Mosko (2009); van Oosterhout (2001, 31–50); Peluso (1996); Russell and Rahman (2015).

18 See, for instance, Malinowski (1935, 52–55); Gell (1975); Tammisto (2018a, 40–41). Another vast body of literature has explored the transnational trajectories of plants from cash crops to global commodities across space and time. Works in this vein include multisited ethnographies of tea (Besky 2013), maize (Fitting 2011), rice (Ohnuki-Tierney 1993), coffee (West 2012), sugar (Mintz 1985), hoodia (Foster 2017), rooibos (Ives 2017), wheat (Head, Atchison, and Gates 2012), and rubber (Dove 2011).

19 The plant turn is part of a broader interdisciplinary field known as *multispecies studies*, which seeks to displace notions of human exceptionalism by attending to the biological, political, and cultural lifeworlds of animals, plants, fungi, and microbes that humans become-with (Kirksey and Helmreich 2010; van Dooren, Kirksey, and Münster 2016). In light of the radical impacts of human activity on planetary ecosystems, scholars of multispecies studies call for respect, curiosity, and care toward other-than-human organisms and for greater attention to the constitutive entanglements of life—human and other—with the apparatuses of modern science and technology (Despret 2004; Tsing 2014a; van Dooren and Rose 2011). It

is important to note, however, that the recognition and analysis of interspecies dependencies and vitalities is not new to multispecies studies. As Shaila Galvin points out, both earlier environmental anthropology and multispecies studies developed in conversation with concepts and methods derived from the natural sciences. Both are also driven by distinct ethical and political concerns: to displace notions of non-Western primitivism on the one hand and to counter assumptions of human exceptionalism on the other (Galvin 2018, 237).

20 "Sensory ethnobotany," Theresa Miller writes (2019, 5), is an interdisciplinary framework that takes seriously the lived experiences of humans and plants alongside the valuation of these experiences by their human and vegetal protagonists. This approach attends to the forms, values, and meanings of sensory and symbolic relationships with plants for humans, as well as the biotic capacities of plants themselves to respond to human value systems through processes of growth and development. Sensory ethnobotany is also historical in that it attends to transformations and continuities in human-plant relationships over time and across different social, political, economic, and cultural contexts.

21 Natasha Myers coined the term *Planthroposcene* to describe the emergence of new scenes and ways of seeing across human- and plantworlds. Rather than designating a time-bound era, the Planthroposcene, Myers writes, is an "aspirational episteme and way of doing life"—one that demands that we "find better ways to get to know plants intimately and on their terms . . . outside of the rhythms of capitalist extraction" (2017a, 299–300).

22 For examples of science- and conservation-focused studies in the plant turn, see, inter alia, Hartigan (2017); Hustak and Myers (2012); Myers (2015).

23 Critical race, queer, and crip theorists have widely criticized posthumanism and new materialisms for failing to interrogate the dehumanization of "Man's human Others" *alongside* the nonhuman (Haritaworn 2015, 212). Critical race scholar Neel Ahuja, for instance, points out that little has been written on the colonial genealogies of the posthumanist turn in contemporary scholarship and invites us to think about intra- and interspecies entanglements through their histories of colonial warfare and racialization (2016, xiv–v). Jamaican novelist and philosopher Sylvia Wynter reminds us that the "human" often refers to a particular "ethnoclass" (i.e., Western, White bourgeoisie) that "overrepresents itself as if it were the human itself" and hence seeks to secure its own well-being at the expense of other racialized humans, other-than-humans, and the more-than-human collective (2003, 260; 2015, 196). In doing an anthropology "beyond the human," Zakkiyah Imam Jackson notes, we need to ask first what and, crucially, *whose* conception of humanity we are moving beyond (2015, 215). Any "posthumanist" account needs to attend to the structural violence through which humanity itself is gendered and racialized in ways that exclude, inferiorize, objectify, debilitate, and animalize some (sub)human lives over others based on assumed worth, ability, or productivity. Ignoring such historical and contemporary modes of exclusion within the human category risks reinforcing the very same kind of Eurocentric transcendentalism that posthumanism seeks to disrupt (Z. Jackson 2015, 215).

Meanwhile, Indigenous scholars have called for greater engagement on the part of multispecies ethnographers with Indigenous peoples, philosophies, practices, and protocols, which have always recognized and related to other species, ecosystems, and elements as agential and social beings. Among them, many have critiqued posthumanism (and the ontological turn) for forcing or obscuring Indigenous modalities of thought, practice, and agency within Western categories, while ignoring matters of race, history, and sexuality (see Hunt 2013; Sundberg 2013; TallBear 2015; Todd 2015; Watts 2013).

Within the plant turn specifically, Sarah Ives (2014) cautions against celebrations of multispecies belonging and relationality that obscure complicated, contested, and sometimes violent biopolitics. Meanwhile, Ruth Goldstein (2019) voices concern that "plant turn" scholarship—among other theoretical turns— emanating from Euro-American settings might contribute to an ongoing colonial practice of drawing from, but not acknowledging and citing, Indigenous theories and cosmologies, which have long recognized plants as consequential agents and relations. In doing so, the plant turn, Goldstein cautions, may end up replicating and exacerbating the totalizing and historical erasure of Indigenous knowledge and practice under colonial ecologies and ecological science (see also Foster 2017; Galvin 2018; Myers 2017b). For related critiques of posthumanism, see Benjamin (2018, 51); DiNovelli-Lang (2013, 142); Gilroy (2017); Ogden, Hall, and Tanita (2013, 13); Puar (2017, 25–26); Weheliye (2008, 321).

24 This ethos of relationality lies at the heart of Indigenous epistemologies and ontologies across the Global North and South, and constitutes a central element in Indigenous peoples' collective advocacy toward self-determination (see Deloria 1999; Durie 2005; Meyer 2001; Stewart-Harawira 2012, 2018; TallBear 2011; K. Teaiwa 2014; Winter 2019a, 2019b).

25 Donna Haraway deploys the term *cyborg* to describe the breakdown of material and imaginative boundaries separating the human from the animal and the human from the machine in the late twentieth century (2013, 174–79).

26 In exploring Marinds' fraught relationship to modernity and its attendant social and environmental transformations, this book addresses a central theme in both the anthropology of Melanesia and in Melanesian anthropology (see Morauta et al. 1979). For key texts, see Bamford (2007); Bashkow (2006); Errington and Gewertz (2004); Kabutaulaka (2015); Knauft (2002a, 2002b); LiPuma (2000); Narokobi (1980); West (2012).

27 These imposed transformations exemplify what historian Patrick Wolfe calls the settler-colonial "logic of elimination." The logic of elimination is premised on a negative articulation between settler and Indigenous society that legitimates settler expansion through processes of invasion, confrontation, assimilation, and repression (Wolfe 1999, 27–30, 167–69). Rather than an event, then, the logic of elimination operates as a structure—one that remains as the ideological foundation of settler-colonialisms past and present (see also Wolfe 2006, 388).

28 On plantations as racialized ecologies of empire, see Allewaert (2013); Benítez-Rojo (1996); Mbembe (2003); McKittrick (2013); Thomas (2019). Cameroonian philosopher and political theorist Achille Mbembe (2003) identifies the plantation as one

of the earliest instances of systemic and institutionalized necropolitical experimentation (see also Rusert 2019). Expulsed from humanity, slaves within the plantation became subjects of domination, alienation, and social death, whose lives were sustained only insofar as they remained useful as labor and property. Slave life, Mbembe writes, is life lived as if in a state of permanent injury, or "death-in-life" (2003, 21)—one that foregrounds necropolitics as the indissociable counterpart of biopolitics.

29   Scholar of critical race studies Alexander Weheliye deploys the concept of "racializing assemblages" to analyze how colonial logics discipline the category of humanity itself into "full humans, not-quite-humans, and nonhumans" (2014, 4).

30   On racism as structural violence in West Papua, see Butt (2005); Kirksey (2017); Kirsch (2010); Munro (2015b, 2020). The racialization of West Papuans can be traced back to European classifications of the nineteenth century, where the term *Papua* came to define a racially distinctive area encompassing island Southeast Asia and the western Pacific, variously named *Papuanesia* or *Oceanic Negroland* (Ballard 2008). The racial distinction of Papuans from other Pacific and non-Pacific peoples was perpetuated during the Dutch colonial era and became institutionalized in the policies and practices of successive structures of governance in Indonesia, as well as in public discourses and scholarship on West Papua. After Indonesian independence and up until the Act of Free Choice of 1969, racial inferiority served to legitimate the acculturation of Papuans under *Indonesianisasi* ("Indonesianization"), a government-endorsed process designed to incorporate West Papuans into the Indonesian state through formal education, national media, economic development, and transmigration (Gietzelt 1988). In its present manifestations, the logic of racism remains premised on the systemic primordialization, bestialization, and infantilization of West Papuans as peoples in an arrested stage of cognitive and physical development, whose appearance and behaviors are comparable to those of lowly animals, notably monkeys, pigs, and dogs. This logic operates by way of counterpoints that valorize specific and intersecting categories relative to others—Malay versus Melanesian, agricultural-capitalistic versus hunter-gatherer, educated versus uneducated, dark-skinned versus light-skinned, and so forth (Giay and Ballard 2003). Even within the Oceanic ethnic landscape, as Pacific scholars have noted, Melanesians are frequently regarded as racially inferior to lighter-skinned Polynesian peoples (Gegeo 2001, 502; Hau'ofa 2008, 6; Kabutaulaka 2015, 122–26). While certain Papuan political activists have sought to subvert racializing assemblages by celebrating West Papuans' distinct racial identity—manifest through their black skin and curly hair—as the basis for their vision of an independent Papuan nation, racism remains a deeply divisive and contentious issue in the region (see Chao 2021g).

31   On the historic struggle of Indigenous peoples for self-determined decolonization and the ongoing legacies of racialized imperialism in the postcolonial Pacific, see, inter alia, Banivanua-Mar (2016); Durie (2005); Stewart-Harawira (2005); Trask (1999).

32   As Malaitan anthropologist David Gegeo and coauthor Karen Watson-Gegeo note, it is not just the fact of Indigenous knowledge that matters, but also the ways in

which that knowledge is produced, interpreted, and then applied by Indigenous peoples (2002, 403). Indigenous ontologies thus cannot be dissociated from Indigenous epistemologies or from the particular ways in which Indigenous peoples (re)create, (re)theorize, and (re)structure knowledge via cultural discourses and mediums and within situated sociopolitical, economic, and historical contexts (see also Gegeo 1998; Gegeo and Watson-Gegeo 2001).

33 For examples of the largely positive moral framing of plant-human relations in the plant turn, see M. Hall (2011); Lewis-Jones (2016); Miller (2019); Myers (2017a).

34 For examples of the love-care-respect complex underlying such approaches, see Atleo (2012); Plumwood (2002); TallBear (2016); Todd (2017).

35 Here, I take up Eva Giraud's call to centralize and politicize the frictions, foreclosures, and exclusions that (multispecies) entanglements inevitably entail, which are often obscured in uncritical celebrations of relationality and its ethical potential (2019, 2–3).

36 I borrow the term *friction* from Anna Tsing to describe the sticky materialities of practical encounters that give grip to global connections in local contexts. Friction, Tsing writes, foregrounds how situated projects and multiscalar interactions come to define movement, cultural forms, and agency. Friction can slow things down by restricting our capacity to move, both imaginatively and physically. But friction can also make movement easier and more efficient, keeping global power in motion. At once confining and generative, friction inflects historical trajectories through contingent processes of enablement, exclusion, and particularization (Tsing 2005, 6).

37 Critical theorist Mary Louise Pratt deploys the term *contact zones* to describe sites of colonial encounter where "disparate cultures meet, clash, and grapple with each other, often in highly asymmetrical relations of domination and subordination" (2007, 7). Following Donna Haraway (2008) and others, I expand the concept of contact zones to encompass the array of more-than-human actors that together participate in shaping the multispecies dynamics of the Upper Bian.

38 As deployed by Edward Hviding (2003) in his analysis of overlapping conservation, logging, and ecotourism projects in the Solomon Islands, "projects of desire" foreground the conflicting aspirations of different sets of actors across sites and scales (see also Tsing 2000). Projects of desire can take material or immaterial form and can be oriented toward the immediate present or the distant future. Projects coalesce as bundles of ideas and practices that are negotiated and realized in particular times and places. What counts as a project depends on what one desires to know, how one understands the relationship between the local and the global, and where one situates oneself within planet-wide interconnections and their attendant frictions (Tsing 2000, 347; see also West 2006a).

39 Stuart Kirsch (2006) describes this approach as "reverse anthropology," defined as an anthropology that takes as its starting point Indigenous peoples' own theories of socioenvironmental change and its causes (see also R. Wagner 1981).

40 On the impacts of colonization on human-forest relations across Oceania, see, inter alia, Bell, West, and Filer (2015); Hviding and Bayliss-Smith (2018); Jacka (2015); K. Teaiwa (2014).

41 I borrow the term *plantation zone* from literary scholar Monique Allewaert (2013) to describe tropical or subtropical places whose economic and political structures have been fundamentally reshaped by the plantation form and its colonial-capitalist undergirdings.

42 In attending to the affective textures of the Marind lifeworld, I seek to address what Tongan and Fijian anthropologist Epeli Hau'ofa identifies as a critical omission in anthropological representations of Pacific cultures. Doing justice to the richness of these cultures, Hau'ofa notes (2008, 3–11), means writing about everyday expressions of love, kindness, consideration, and altruism—about humor and morality and the good and the bad, along with the diverse forms of bodily and emotive communication that accompany the spoken word and attendant conceptual categories (see also Barker 2007).

43 In her published letter to communities and researchers, Eve Tuck (2009) critiques "damaged-centered research" for creating one-dimensional representations of marginalized communities as depleted, ruined, hopeless, and vanishing. Instead, Tuck invites analytical and ethnographic attention to the complex desires, personhood, and survivance strategies of Indigenous communities as they find, create, and sustain meaningful lives amid institutional and everyday forms of oppression and invisibilization (see also joannemariebarker and Teaiwa 1994).

44 I borrow the notion of "impasse" from cultural theorist Lauren Berlant to describe the historical present as a moment where existing social imaginaries and practices no longer produce the outcomes they once did and no new imaginaries or practices have yet been created to replace them. Echoing Marinds' concept of abu-abu, the historical present as impasse, Berlant writes, is a "middle without boundaries, edges, a shape." It names a "thick moment of ongoingness, a situation that can absorb many genres without having one itself . . . [a] space where the urgencies of livelihood are worked out . . . without assurances of futurity . . ." (2011, 200; see also Stengers 2015). An impasse can take the form of a situation following a dramatic event, such as the arrival of oil palm in the Upper Bian, when one loses the sense of what must be done and yet must find ways to adjust. It dissolves preextant certainties and categories and forces us to engage in the necessary labor of improvising in a world devoid of guarantees.

45 My use of the term *Papuan Way* is inspired by Papua New Guinean philosopher and politician Bernard Mullu Narokobi's influential concept of "the Melanesian Way," a spiritual vision that Narokobi identified as central to the creation of a culturally self-aware and self-determined postcolonial subjectivity among Papua New Guinean peoples. Resonating with the notion of the counterpoint deployed throughout this book, the Melanesian Way, Narokobi writes, is anchored in a recognition that opposites can and must coexist, in the image of human life itself and its "inconsistencies, contradictions, emotions, reason, and intellect" (1980, 4, 8; see also Narokobi 1983, 22–39). Only through a process of collective and dialectical reasoning can these opposites be transformed into a collective Melanesian future that is at once synthetic, ancient, and forward-looking (see also Dobrin and Golub 2020; Kabutaulaka 2015).

46 In his critique of "ethnographic authority," or the portrayal of the cultural "other" as bounded, abstract, and ahistorical, James Clifford draws attention to the contrapuntal relation at play in the intersubjective dialogue between the informant and the ethnographer and in the ethnographer's representation and analysis of the informant's representation and analysis. Ethnography, Clifford argues, must be understood "not as the experience and interpretation of a circumscribed 'other' reality." Rather, ethnography must be approached as a "constructive negotiation involving at least two, and usually more, conscious, politically significant subjects," operating within specific historical relations of dominance and dialogue (Clifford 1983, 133, 119; see also Crapanzano 1992; Wolfe 1999).

47 For critiques of the polarization between "dark anthropologies" and "anthropologies of the good," see Jacka (2019) and Knauft (2019).

48 The acquisition of land for land-banking purposes rather than crop development, or what John McCarthy, Jacqueline Vel, and Suraya Afiff (2012) call "virtual land grabs," has been widely reported across Indonesia's palm oil, rice, Jatropha, and carbon sectors.

49 Inspired by the rampant haze produced by forest fires across Indonesia in the 1998 El Niño drought year, Anna Tsing suggests that crises of visibility—material and symbolic—arise when proper standards of visibility come into conflict or are overlaid. Such periods also generate or reveal contestations over what *should* be visible or invisible, to whom, and to what effect (Tsing 2005, 43–45).

50 The Marind notion of abu-abu brings to mind Michael Taussig's concept of "epistemic murk," the breakdown of knowledge through which colonial modes of production and governance become fused with terror, violence, and chaos and simultaneously come to define and blur the line between reality and representation (1987, 121–22).

51 As Lucas Bessire notes, epistemic murk does not preclude the possibility of becoming-through-negation for peoples who inhabit spaces of loss, death, and destruction. Rather, epistemic murk can be a creative means through which people come to redefine and reconstitute their worlds in the face of seemingly unsurmountable odds (2015, 224–27).

52 On the ecology of Merauke, see Bowe, Stronach, and Bartolo (2007).

53 *Metroxylon sagu* derives from *metra* meaning "pith" and *xylon* meaning "plant tissue."

54 In this book, *oil palm* refers to the oil palm tree and *palm oil* to the oil obtained from the tree's kernel and fruit. The scientific name of oil palm, *Elaeis guineensis*, derives from the Greek *elaia* for "olive" (on account of its oil-rich fruit) and *guineensis* in reference to African Guinea, where the plant grows endemically. The species is also called the African oil palm to distinguish it from the American oil palm (*Elaeis oleifera*) of South and Central America and the maripa palm (*Attalea maripa*) of South America and Trinidad and Tobago. In its native West and Central Africa, palm oil is a source of cooking oil, medicinal ointments, toddy wine, fuel, and housing material. It once featured alongside yams, fish, salt, cloth, and metals in a lively local trade network along the tributaries of the Forçados River in

the Niger Delta. The commodity's first movement beyond Africa was as a foodstuff aboard ships during the slave trade. The Industrial Revolution heightened commercial demand for the product as an ingredient of margarine, candles, glycerin, machine lubricants, soaps, and tin plating (Henderson and Osborne 2000; Jones 1989).

55 The ethnonyms Marind Bian and Marind-deg distinguish Upper Bian Marind from coastal Marind (<u>Marind-duv</u>) of the south, riverine Marind of the eastern Maro River (<u>Marind-kanum</u>), and Marind of the swamps (<u>Marind-bob</u>) in western Kimaam. The self-identification of Marind in relation to the landscape and its ecologies resonates with similar naming practices across New Guinea—for instance, the "mangrove people" of the Sepik Estuary (Lipset 1997), "river people" of East Sepik (Silverman 2018), and "coastal people" of Kamu Yali (J. Wagner 2018).

56 For a compilation of Marind clan affiliations, see van Baal (1966, Annex IV a–d). <u>Amai</u> also include natural elements such as the moon, the sun, and the wind. However, Upper Bian Marind noted that these were of secondary importance compared to plant and animal <u>amai</u>. Throughout my fieldwork, I heard <u>amai</u> species and their respective attributes widely discussed by men and women of different generations. In contrast, my understanding of the role of <u>dema</u>, the ancestral spirits of Marind and their <u>amai</u> kin, was limited by the fact that such knowledge is transmitted primarily by male village elders, exclusively through the sacred medium of origin stories, and usually inside the customary men's house.

57 The forest plants and animals from whom Marind derive their food feature a combination of native species and species introduced into New Guinea both during prehistorical human migrations and by Pacific islanders, Europeans, and Asians during and after the 1870s.

58 For examples of such colonial depictions, see Baxter-Riley (1925); Boelaars (1981); Haddon (1891); van der Kroef (1952). Colonial representations often produced and perpetuated stereotyped depictions of native peoples as alternately romanticized or vilified primitive tribes. These depictions were, in turn, instrumental to the empire's civilizational mission and perdure in contemporary discourses about Indigenous Papuans in Indonesia (Kirksey 2002; Kirsch 2010; Rutherford 2015). As products of militaristic forms of colonialism, they must be treated with the utmost caution.

59 The exonym *Tugeri* was also the name by which Marind were known among the Trans-Fly peoples to the east, against whom Marind directed their headhunting raids (Ernst 1979, 34; Hitchcock 2009, 89n1).

60 Jan van Baal was controlleur of Merauke from 1936 to 1938, then adviser on native affairs to the newly established Government of Netherlands New Guinea (1950–1952), and finally governor of Netherlands New Guinea (1953–1958). His monograph—the first and only comprehensive account of Marind society—incorporates earlier accounts of Marind language and culture produced by the Missionaries of the Sacred Heart, notably Jan Boelaars and Jan Verschueren, as well as the works of Swiss anthropologist and collector Paul Wirz and German anthropologist Hans Nevermann.

61 Father Hoeboer of the Missionaries of the Sacred Heart named the area referred to in this book as Mirav during the 1930s, as part of the Dutch initiative to concentrate scattered and semipermanent settlements into larger permanent villages. The villages of Khalaoyam and Bayau were also established as part of the colonial endeavor to sedentarize Marind into larger administrative units.

62 Prior to Dutch rule and as early as the seventeenth century, the coastal Marind were engaged in the trade of items including copper cannons, choppers, textiles, beads, and earrings with peoples of eastern Indonesia and the Kingdoms of Ternate and Tidore, and later on with the Makassars, Buginese, Arabs, and Chinese (Pouwer 1999, 160).

63 Growing European demand for bird-of-paradise plumes led to an unprecedented influx of hunters into Merauke during this period, including Europeans, Chinese, Japanese, Australians, and Indonesians from Ambon, Kei, and Timor. But by the time Great Britain passed its Importation of Plumage Prohibition Bill in 1921, the trade had fallen into decline. It was banned in Merauke a year later and formally ended across Dutch New Guinea by 1931 (Swadling 1996, 175–204).

64 On the geopolitical history of West Papua, see, inter alia, Brathwaite et al. (2010); Budiardjo and Liem (1988); Chauvel and Bakti (2004); Tebay (2005).

65 Initially implemented under Dutch rule, government-endorsed transmigration into West Papua came into full swing in the 1970s and 1980s. In 1988 and 1989 alone, Merauke was scheduled to take in an estimated 500,000 people (Monbiot 1989, 39; see also Arndt 1986, 161–74).

66 Indigenous and Western scholars and activists have characterized the systemic and naturalized violence perpetrated against West Papuans under Indonesian rule as a "slow-motion genocide" (Banivanua-Mar 2008; Elmslie and Webb-Gannon 2013; McDonnell 2020; Ondawame 2006; Tebay 2005; Wing and King 2005). While some of my Marind interlocutors deployed the language of genocide in describing Indonesian occupation and its impacts, many rejected this idiom because of its anthropocentric focus. Colonization, these individuals argued, obliterates, not only Indigenous peoples, but also the sentient ecologies central to Indigenous peoples' sense of self and continuity. Other Marind, meanwhile, distinguished intentional massacre from territorial expansion as the prime driver of colonization—an argument that Patrick Wolfe makes in his distinction between elimination and genocide. As Wolfe (2006) explains, settler-colonizers are concerned with the destruction of Indigenous societies only to the extent that is required for, and enables, settler possession and exploitation of the land. By the same token, settler-colonialism allows for the recognition of Indigenous peoples' rights only so long as this recognition does not challenge settler-colonizers' territorial and political interests (see also Coulthard 2014; Povinelli 2002; Simpson 2014).

67 On state-corporate power dynamics in postauthoritarian Indonesia, see Hadiz (2010); McCarthy and Moeliono (2012); Robison and Hadiz (2004).

68 West Papua thus constitutes a classic example of what Anna Tsing calls "resource frontiers," which are sites where entrepreneurs, armies, and governments actively reconfigure putatively "discovered" natural resources and landscapes, such as

forests, seas, and mountains, into corporate raw material. Central to this conjuring is the political construction—and subsequent naturalization—of resource frontiers as zones of wilderness in need of exploitation and transformation (cf. Tsing 2003). As ideology and language, the frontier connects diversely situated local and global actors through asymmetric relations of accumulation and dispossession (West 2016, 23, 27). This settler-colonial logic, Fijian historian Tracy Banivanua-Mar notes, further legitimates the displacement and enculturation of racialized Indigenous peoples as putative subjects of civilizational development (2012; see also Banivanua-Mar and Edmonds 2010).

69 The violations of Marinds' rights to consent, land, livelihoods, and food caused by MIFEE, together with the threat to their collective survival as a people posed by population dilution and migrant influx, were highlighted in three submissions from civil society organizations to the UN Committee on the Elimination of Racial Discrimination (CERD) under its Urgent Action and Early Warning Procedures in 2011, 2012, and 2013. They were further reiterated in two formal communiqués to the Special Rapporteur on the Right to Food and the Special Rapporteur on the Rights of Indigenous Peoples in 2011 and 2013. However, the Indonesian government has not responded to the concerns raised on the basis of these submissions by the CERD committee in 2011 and 2013, nor to the Special Rapporteurs' joint statement of 2012 regarding the potentially adverse effects of MIFEE on the food security of some 50,000 people (see Avtonomov 2013; de Schutter and Anaya 2012; Kemal 2011). For the original UN complaints and statements, see Chao (2013), Forest Peoples Programme (2013); Forest People's Programme, Sawit Watch, and Aliansi Masyarakat Adat Nusantara (2012); and MacKay (2011a, 2011b).

70 I convey this reconstruction as a Eurasian female anthropologist striving to reconcile Anglo-European forms of research and interpretation with Indigenous epistemologies and ontologies and operating within a discipline that has historically reproduced (or been used to reproduce) the hegemonic and extractive processes of colonial settlement. While Indigenous epistemologies and ontologies have profoundly informed my thinking, being, and relating, the reconstruction I offer is not an Indigenous account in the sense elaborated by David Gegeo (2001) and others—namely, research undertaken by Indigenous scholars, on Indigenous terms, and both with and for Indigenous communities (see also Coburn et al. 2013; joannemariebarker and Teaiwa 1994). In engaging with Indigenous accounts produced by Pacific and Melanesian scholars throughout this book, I recognize the importance of Indigenous scholarship as a form of resistance against historically entrenched processes of colonization and their extractive forms of "knowledge capitalism" (Stewart-Harawira 2013). I depart from the prevalent positioning of Indigenous peoples within anthropology as research-subjects only, rather than as equal research collaborators and knowledge producers. I also push against the disciplinary framing of West Papua as part of "South East Asian Studies" or "Indonesian Studies," which itself constitutes an artefact of settler-colonialism. Instead, I approach West Papua first and foremost as an Indigenous and Melanesian place—one that endures despite the dispossessory and discriminatory violence of colonial rule.

71 Villages hold ambivalent meanings for many Marind (cf. Barker 1996; Stasch 2013). As I explore in chapter 4, villages are considered spaces for human dwelling (tempat orang tinggal) in that they are inhabited by Papuan and non-Papuan settlers. In this regard, villages stand in contrast to forests and groves, which are populated by Marind and their plant and animal kin. Marind also describe villages as official spaces (tempat resmi) because they are the locus of administrative institutions—for instance, schools, clinics, corporate headquarters, police stations, and military posts. Indeed, much of everyday life in the village tends to revolve around these institutions in the context of children's schooling, patients' visits to the clinic, villagers' meetings with government and company representatives, and more or less official visits from the police and military. The village is also where foods procured from the forest are cooked, shared, and consumed; where villagers gather to chat on the front porches of their homes; where Marind women sell forest vegetables and fruit to settler families; and where basic goods are purchased from settler-owned kiosks. But while the village offers Marind access to public services, it is also widely perceived as a site of control and surveillance from the State and corporations—one that is increasingly inhabited by non-Papuan settlers. The village, I was often told, is a lonely (sepi) place where there is little to do and where people easily get bored (jadi bosan). For reasons I will explore in the chapters to come, the vast majority of Upper Bian Marind prefer to spend their time in the forest. Indeed, many people only return to the village if they need to travel by road to urban areas, such as Merauke City and Jayapura. As a foreign researcher, limiting my time in the villages was also an issue of safety and precaution in that it enabled me to stay "under the radar" and avoid notice by the police, the military, or the corporations.

72 Here I refer to the antiracism protests of July–August 2019 that took place across the Indonesian archipelago. These protests were triggered by racist attacks on Papuan students in Java and prompted an intensified military clampdown in West Papua (see Chao 2019d, 2019f, 2021g). They also coincided with my last visit to the region, after which I was officially blacklisted by the Indonesian government because of my research and advocacy.

73 Anishinabe scholar Gerald Vizenor deploys the term *survivance* to foreground the ways in which systemic cultural genocide generates new spaces of synthesis and renewal that go far beyond the notion of basic survival. Survivance, Vizenor writes, means "a native sense of presence, the motion of sovereignty and the will to resist dominance. Survivance is not just survival but also resistance, not heroic or tragic, but the tease of tradition [that] outwits dominance and victimry" (2000, 93). The notion of "survivance" is closely linked to Kyle Powys Whyte's concept of "collective continuance," or the creative capacity of Indigenous communities to "make adjustments to current or predicted change in ways that contest settler-imposed hardships and other oppressions, establish quality diplomatic relationships, bolster robust living in the face of change, and observe balanced decision-making processes capable of dealing with difficult tradeoffs" (2018b, 69). On Indigenous resilience, creativity, and endurance in the Pacific, see, inter alia, Barker (2008); Durie

(2005); Hau'ofa (2008); Hviding and White (2015); Povinelli (2011); Simpson (2017); Stewart-Harawira (2018).

74 On Marind ritual and myth, see Ernst (1979); Knauft (1993); van Baal (1966).

75 Epeli Hau'ofa has criticized the perduring emphasis on warfare, headhunting, and ritual within anthropological research in the Pacific (2008, 3–10). This emphasis, Hau'ofa notes, perpetuates distorted stereotypes of Indigenous cultures as static and exotic and thereby denies Pacific peoples some of the most fundamental and dynamic aspects of their humanity. Paige West puts forth a similar argument in analyzing how the "representational rhetorics" of nature, culture, savagery, and discovery have dispossessed, and continue to dispossess, Papua New Guineans of their bodily, territorial, and epistemic sovereignty (2016, 63–86). Disarticulated representations of New Guineans produced by anthropologists, West writes, not only essentialize New Guinean peoples but also serve as weapons of dispossession for other institutions and actors (see also Durutalo 1992; Narokobi 1976; T. Teaiwa 2006).

76 As Paige West (2016) and Anna Tsing (2000) remind us, the "global versus local" is, in itself, a false dichotomy. Interconnections, circulations, and flows of people, things, ideas, and practices long predate the era of "globalization." They point instead to the simultaneity of the local and the global in their diversely situated and contested manifestations.

77 See also H. Davis and Todd (2017); Jolly (2018); Todd (2015). Kyle Powys Whyte's (2018a) reflection on climate change as a settler-colonial temporality speaks more broadly to the ways in which Western framings of time and history have furthered the erasure of Indigenous peoples, places, and practices. As Epeli Hau'ofa notes in the Pacific context, the hegemony of mainstream Western historiography fails to reflect Indigenous peoples' cyclical notions of time and consequently undermines their collective capacity to define and construct their pasts, presents, and futures in self-determined ways (2008, 60–79; see also Banivanua-Mar 2012; Lempert 2018; Obeyesekere 1992; Te Punga Somerville 2018; Tuhiwai Smith 2012; Winter 2019b).

78 I returned to the field in July 2019 with the support of an Engagement Grant from the Wenner-Gren Foundation to share my research findings with my host communities and to decide together on the content and structure of the book before you (see Chao 2019b).

79 This 2018 documentary, *Declaration of Land as Our Spiritual Mother*, is available at https://www.youtube.com/watch?v=74Zo-cNY8U8.

80 On the importance of politically engaged and reflexive anthropology as a form of reciprocity, a labor of translation, and a mode of communitarian research, see Biersack (2006a); Gegeo (2001); Greenough and Tsing (2003); Kirsch (2018); Tuhiwai Smith (2012); West (2019).

1. PRESSURE POINTS

1 December 1 marks the date in 1961 when West Papuans first raised their national flag, the Morning Star (Bintang Kejora), after being promised independence by the Dutch government. Eight years later, however, hopes for West Papuan independence were crushed by the Act of Free Choice, when 1,021 voters were appointed

by the Indonesian government to vote in favor of Indonesian rule. Every first of December since, West Papuans have commemorated their stolen freedom by raising the Morning Star—an act classified as treason under Indonesian law and punishable with a jail term of up to fifteen years. Every year, these largely peaceful movements result in violent clashes with state military and police forces.

2   Carlos Mondragón (2009) describes a similar renewal and re-creation of meaning through movement among Torres Islanders in Vanuatu. By moving across land and seascapes, Mondragón writes, Torres Islanders kinesthetically resituate themselves within ongoing flows of persons and things, producing a world animated by both practice and signification. The centrality of movement is also described by Astrid Anderson (2011) in the context of Wogeo cosmologies in northern Papua New Guinea. Here, too, places are rendered meaningful through people's activities on, in, and with the land and its constitutive beings, producing what Anderson calls "landscapes of relations and belonging" (the title of her book; see also West 2006a, 27–28).

3   The intrinsic connection that Marind identify between place and identity resonates with the concept of kula ni fuli among Kwara'ae in Malaita, as described by David Gegeo. Kula ni fuli, or "place situated in source," refers at once to a geographical location; a genealogy within a past, present, and future kin group; the rights and responsibilities afforded by descent and marriage; and the cultural and spiritual knowledge acquired in and from one's native place (Gegeo 2001, 493–94; see also Mondragón 2009; O'Hanlon and Frankland 2003; Rumsey and Weiner 2001; Stewart and Strathern 2003b).

4   Historian Christopher Ballard (2002) deploys a similar approach in analyzing the topographies of terror produced by state and military forces around the Freeport gold mine in northern West Papua. As is the case in Merauke, topographies of terror around the Freeport mine are subject to multiple and differing interpretations on the part of their inhabitants.

5   In using the language of pressure points, I transpose to a topographic context the fundamental principles of acupressure and acupuncture, where the term *pressure point* originates. Activating key points in the body via acupressure or acupuncture serves a curative purpose. Pressure applied with fingers or needles can relieve stress, strengthen the immune system and blood circulation, and reduce physical and emotional pain. The same logic applies to pressure points in martial arts, but with the converse end of inflicting pain. Here, manipulating vital or weak points in the body causes suffocation, injury, or fainting and can allegedly trigger paralysis or death. As with acupuncture, activating these points can affect energy flows and bodily functions far beyond the specific location where pressure is applied and to alternately therapeutic or harmful ends.

6   The pharmakon, philosopher Jacques Derrida writes, is both a drug and a remedy. As the différance of difference, the pharmakon is generated by, and in turn generates, ambiguity through its at once beneficent and maleficent potencies (Derrida 1981, 429; see also Stengers 2010, 28–41).

7   Michael Taussig deploys the concept of a "nervous system" to describe states of permanent emergency where disorder simmers under the surface of apparent

hegemony and control. The nervous system, Taussig writes, is like a giant spring, compressed yet ready to burst at any moment. It is an anatomy animated by tension, fear, and paranoia, in which order and certainty are besieged by perpetual flux and opacity. In this state of doubleness, normalcy and panic sit awkwardly alongside shock and disorientation (Taussig 1992, 10, 18).

8 The promise of the road as a conduit for expanded social relations among Marind brings to mind Paige West's description of "modernity" among Gimi in Highland Papua New Guinea. For Gimi, access to modern things and knowledges depends on, and is indissociable from, the creation of social relationships with an array of outsiders—for instance, conservation ecologists, activists, and mining-company employees (West 2006b, 308–9; see also Kirsch 2006).

9 Similarly, Paige West notes how Gimi who leave the village to work in urban centers are said by elders to "become someone else" as a result of their prolonged coexistence with non-Gimi people. Urban centers enable these individuals to expand and diversify their social networks and knowledges but also undermine their capacity to behave and interact *as* Gimi—both with each other and with the living landscape (West 2016, 133).

10 On the contradictory potential of roads as conduits and obstacles to modernity in Melanesia, see Beer and Church (2019); Eves (1998, 85–92); Hayano (1990); Kirksey and van Bilsen (2002); LiPuma (2000); MacLean (1994); O'Hanlon and Frankland (2003); Silverman (2013).

11 Kopassus hold a notorious track record of egregious human rights violations in Papua, as well as in Aceh and East Timor. This includes their participation in Operation Trikora in West New Guinea in 1961–1962, the East Timor invasion of 1975, and the massacre of alleged communists in 1965. Marind frequently compare Kopassus to suanggi, or perpetrators of foreign forms of black magic, because of the elusive way they carry out their deadly operations. For instance, they leave no trace of their passage other than the abandoned bodies of their victims deep in the forest or along secluded riverbanks and a single bullet lodged in their forehead. It is also believed that they can make themselves invisible and be in two or more places at the same time (see also Kirksey 2012, 114–18).

12 Similar perspectives have been documented among other Indigenous groups whose landscapes are being converted into zones of capitalist production. For instance, in the context of monocrop soy expansion, Indigenous Wichí of northwestern Argentina conceive the bulldozer as a machine of spatial destruction that "crushes objects in order to transform their form and smooth out space" (Gordillo 2014, 269). In Paraguay, Indigenous Ayoreo describe bulldozers used by ranchers to clear the forest as "attackers of the world." These forest-flattening machines pursue Ayoreo relentlessly, consume their land, and constitute a vehicle and sign for the end of time (Bessire 2015, 128–31).

13 On the association of corporate projects with improved futures in Melanesia, see Halvaksz (2015); Jacka (2015); LiPuma (2000); West (2012).

14 The conflicting significance of plantations among Marind resonates with the characterization of soy plantations among Indigenous Toba in Argentina as places of

estrangement, death, and terror on the one hand, and as sources of material wealth, abundance, and power on the other (see Gordillo 2002). It also brings to mind the ambivalence of oil palm plantations among Mengen in West New Britain as places of controlled and alienated labor, sources of desired monetary income, and generative sites where new social relations can be produced outside the village space (see Tammisto 2018b).

15 Jenny Munro (2013, 2018) describes similar experiences among Highland Dani studying in Java, whose efforts to acquire new knowledge, literacy, and employment are routinely impeded by systemic inequalities dividing Indigenous residents from more privileged Indonesians. Prejudicial hiring practices, racial discrimination, and security-sector violence, Munro notes, set Indigenous women and men up for failure and disappointment in the face of modernity and its inflated possibilities.

16 On the exclusionary dynamics of corporate conservation in Merauke, see Chao (2019e, 2021f). On the challenges of reconciling environmental conservation with extractive capitalism in Melanesia, see, inter alia, Hviding and Bayliss-Smith (2018); West (2006a).

17 Military bodies established across Merauke include the National Indonesian Army Military Police Forces (PUSPOM TNI), the Regional Military Command Units (KORAMIL), the Strategic Reserve Military Forces (KOSTRAD), the Military Marine Corps (KorMar), the Special Forces Commandos (Kopassus), and the Mobile Brigade Corps (Brimob).

18 On plantations as state-like formations, see also Tammisto (2016).

19 As political scientist James Scott notes, the simplified terrain of the plantation facilitates the manipulation of the commodity cultivar while also enabling heightened control and surveillance of the landscape and its inhabitants (1998, 18).

20 I borrow the concept "space of death" from Michael Taussig (1984) to describe places created by colonial-capitalist regimes that operate through the conjoined logics of terror and violence, chaos and uncertainty, and power and precarity.

21 Morten Pedersen invites similar attention to uncertainty as an ontic condition in his analysis of shamanism in postsocialist Mongolia. Attending to uncertainty as a condition of being, Pedersen notes, enables us to focus our attention on the generalized feelings of loss and doubt that accompany sustained states of cosmological turmoil and social breakdown (2011, 8, 39).

22 As Rupert Stasch notes in the context of Korowai perceptions of villages in lowland West Papua, the relational makeup of spatial forms may be more consciously articulated when they are alien and new. As with oil palm plantations in Merauke, the poetic density of such spatial forms arises from the multiple relational connections, exogenous forces, and historical processes that they mediate, which make them at once desirable and shunned (Stasch 2013, 556).

23 Michel Foucault deploys the concept of "dense transfer points of power" in the context of sexuality discourses and their effects on individual and social subject-bodies. States, Foucault writes, produce an ideal, normative citizenry through the formation, regulation, and disciplining of sexuality and reproduction. However, subject-bodies as dense transfer points of power are also sites of contravention,

resistance, and subversion, where power relations are remade and negotiated on an ongoing basis (Foucault 1978, 103). The notion of "dense transfer points of power" has also been deployed by Ann Stoler to analyze how the differential management and governance of emotions, intimacies, and private lives on colonial frontiers serve to consolidate the rule of empire (2001, 831; see also Stoler 2018, 209).

2. LIVING MAPS

This chapter is derived in part from an article published in *Anthropology Now* on April 27, 2017, copyright Taylor and Francis, available online: www.tandfonline.com /10.1080/19428200.2017.1291014.

1 For instance, political scientist Benedict Anderson (1983) identifies the map as an institution of power that profoundly shaped the dominion of colonial states in Southeast Asia. Geographer John Harley (1988) describes maps as weapons of imperialism that legitimized colonial rule by reconfiguring space to serve imperial agendas. Historian Thongchai Winichakul (1994) analyzes the instrumentalization of maps as technologies of territoriality in the creation of the nation as a united physical, imaginative, and affective "geo-body."

2 As Katherine Verdery notes, the power of the state often operates less as a hegemonic and centralized force than as a murky play of compromises, uncertainty, and improvisation among local political actors (2002, 8–9, 27).

3 Other terms for "participatory mapping" include subsistence mapping, resource use mapping, community land use mapping, ethno-cartography, counter-mapping, self-demarcation, and community-integrated or public participation GIS mapping, where GIS stands for geographic information system (Chapin, Lamb, and Threlkeld 2005, 622–23).

4 In Indonesia, participatory mapping continues to play a central role in land-titling efforts and in the formalization of customary (adat) land rights (see Dewi 2016). Participatory mapping first took place in 1992 in East Kalimantan with the support of the World Wildlife Fund and in West Kalimantan with the support of the Indonesian NGO Pancur Kasih. Such NGO-led initiatives became more prevalent following the promulgation of Indonesian regulation No. 69 of 1996, concerning public participation in spatial planning and the subsequent formation of the national Network for Participatory Mapping (Jaringan Kerja Pemetaan Partisipasi, or JKPP). Since its establishment in 1999, the Alliance of Indigenous Peoples of the Archipelago (Aliansi Masyarakat Adat Nusantara, or AMAN), which represents some 2,332 Indigenous communities throughout Indonesia, has adopted participatory mapping as a central strategy in bolstering Indigenous peoples' customary claims to lands and forests. In West Papua, participatory mapping initiatives multiplied following the promulgation of the Special Autonomy Law in 2001, which sought to enhance the decision-making power of Papuans in matters of political, economic, and cultural significance. In Merauke specifically, NGOs that have introduced and supported participatory mapping initiatives since the inception of MIFEE include the national organization Yayasan PUSAKA and the international organizations Forest Peoples Programme, World Wildlife Fund, and Climate and Land Use Alliance, as

well as research institutions such as the Center for International Forestry Research and Musamus Merauke University.

5 My companions' description of the khaw's lifeway is powerfully reminiscent of Kalam naturalist Ian Same Majnep's writings in *Birds of My Kalam Country* (Majnep and Bulmer 1977) and the later *Animals the Ancestors Hunted* (Majnep and Bulmer 2007). In both works, Majnep fleshes out the biotic and ecological attributes of particular bird species and also their spiritual, material, and cosmological significance to, and relations with, Kalam people across space and time. Threaded throughout this narrative as well are Majnep's own personal experiences of meaningful interspecies encounters, which among the Marind find expression in the interpersonal, storied, and affective textures of maps and related cartographic practices and discourses.

6 Steven Feld (1996, 2003) deploys the term *acoustemology* to describe a similar practice of knowing space and species through listening among Kaluli in the Southern Highlands of Papua New Guinea.

7 The sonic "bird's-eye view" of Marind maps brings to mind the "fish-eye episteme" employed by Mestiza Colombian multimedia artist Carolina Caycedo in *Yuma: Land of Friends*, an experimental video that highlights the conflictual struggle between local fishing communities and energy corporations concerning the Magdalena River and its conversion into a source of hydroelectric power. Much like the weaving perspective of the moving khaw, Caycedo's film travels continually across shifting points of view, bottom up, top down, transversal, human and fish, underwater and overwater. It is always somewhat murky and opaque, in ways that at once embody and subvert the muckiness of the neoliberal and colonial world (Gómez-Barris 2017, 91–109).

8 For sample footage, see the "Contoh Pembongkaran Hutan Adat" on Youtube at https://www.youtube.com/watch?v=ML1kyggbwIA.

9 A similar conception of space as movement is evident in the cartographic practices of Ongee hunter-gatherers in Little Andaman, whose map coordinates follow the paths of humans, spirits, and animals across the landscape (Pandya 1990; see also Dwyer and Minnegal 2014).

10 Drawing attention to the link between truth and place, sociologist Thomas Gieryn (2018) argues that the credibility and power of claims to knowledge arise from the geographic sites in which these claims are produced or discovered, celebrated or obscured, and reproduced or contested. Places that succeed in making people believe certain claims over others, which Gieryn calls "truth-spots," conceal the place-dependent nature of knowledge by transforming it into a seemingly universal truth.

11 In a similar vein, environmental humanities scholars Katarzyna Beilin and Sainath Suryanarayanan describe how satellite technologies deployed in Argentinian soy monocultures visualize the dreams of a precision and control-based bioeconomy by substituting and making invisible "what can be seen, heard, tasted, touched, and smelled close to the ground, where suffering and destruction accumulate" (2017, 207).

12   Différance, a concept that is central to Jacques Derrida's theory of deconstruction, speaks to the conjoined and repeating processes of deferral and differentiation through which meaning is created, transcended, and recreated. Différance, Derrida writes, reveals meaning in the loops and interstices between speech and writing, making it present without presenting it as such and always concealing the order of truth at any certain point as something mysterious, indeterminable, and perpetually transforming. Resonating with Marinds' multispecies mapping practices, différance constitutes a form of strategic and adventurous play—a wandering through and beyond meanings that arise from their contrast to other meanings, stripped of the presumption and possibility of mastery (Derrida 1982, 1–28).

13   Similar concerns over the social and political consequences of participatory mapping have been documented among Indigenous peoples in Sarawak, East Kalimantan, Papua New Guinea, and Cambodia (see Cooke 2003; Dewi 2016; Fox 2002; Fox et al. 2008; Prom and Ironside 2005; Sirait et al. 1994).

14   On the temporal dimensions of maps as contested performances of imagined pasts, presents, and futures, see Brody (1981) and Sletto (2009).

15   In this regard, Marind mapping practices speak to the power of real and imagined audiences, which Danilyn Rutherford (2012) identifies as central to West Papua's struggle for sovereignty. In this struggle, Rutherford explains, West Papuans actively and contextually inhabit the gaze of their multiple spectators while also acknowledging its limits.

16   Philosopher Nick Sousanis's (2015) call for an "unflattening" of ontoepistemic regimes is closely allied with Macarena Gómez-Barris's (2017, 1–16) invitation to attend to "submerged perspectives," such as those captured in Caycedo's fish-eye film described in note 7. Submerged perspectives, Gómez-Barris contends, creatively counter the extractive violence of the colonial gaze and allow for a multiplication of ways of seeing and being that, in turn, can help decolonize knowledge and the world itself.

17   For comparable critiques of cartographic stasis, see Pandya (1990); Turnbull (2007).

3. SKIN AND WETNESS

1   For examples of the importance of the skin in the affirmation of moral and social personhood in Melanesia, see Bonnemère (1994); Eves (1998); Frankel (1986); Knauft (1989); O'Hanlon (1989); Read (1995); van Oosterhout (2001).

2   The centrality of fluids in Indigenous cosmologies and corporealities has been widely documented across Melanesia. As is the case among Marind, the material and symbolic circulation of these fluids sustains the health and well-being of human and nonhuman entities, while also expressing the nature of the relations that bind them (see Goldman and Ballard 1998; Stewart and Strathern 2001).

3   Both men and women possess and enhance their blood, sweat, tears, and saliva through their activities in the forest and their bodily interactions with others. Other bodily fluids of cultural significance among Marind include semen and vaginal fluids, substances that are said to distinguish men from women while also

affirming their complementarity (see also Strong 2004). According to colonial sources, the mingling of semen and vaginal fluids through rites of serial heterosexual intercourse served life-enhancing purposes, and semen in particular was considered a potent life force and was used in homosexual rituals to "grow" young boys from novices into fully fledged adults (Knauft 1993; van Baal 1966). Although these substances may still play an important part in processes of making <u>anim</u>, my companions were reluctant to talk about them due to their association with sexual rites that were deemed primitive and consequently banned during the Dutch colonial era. I have therefore excluded them from my analysis.

4 On the procuring and sharing of food as a means of creating social relations in Melanesia, see, inter alia, A. Anderson (2011); Bashkow (2006); Battaglia (1990); Eves (1998); Fajans (1983, 1985, 1988, 1997); Kahn (1996); Manderson (1986); Munn (1992); Schieffelin (1976); A. Strathern (1973); von Poser (2011, 2013).

5 Paige West describes similar forms of consubstantiality among Gimi in Highland New Guinea. When a Gimi hunter or gatherer procures or takes in plants, animals, water, wind, or any other entity within the animate world, each party coproduces the other through an act of exchange and consumption (West 2016, 135–36).

6 The Marind expression "to share the same skin" resonates with Roy Wagner's description of individuals bound by social ties among Daribi of Papua New Guinea as being "one epidermis congruent" (1991, 163–64). It also brings to mind the notion of "extended" or "unifying skins" used by Sveinn Eggertsson to describe how particular social categories among Eastern Min of Papua New Guinea—for instance, clans, initiation cohorts, warriors, and sorcerers—come into being through their constituents' common experiences, knowledges, and taboos (2018, 163–64). The importance of the skin in the construction and expression of place-based personhood and social relations has also been documented beyond Melanesia. For instance, among Aboriginal communities in North Australia, sweat, or wenterre, is central to the affirmation of relations between people and place and is said to maintain the country's life and productivity (Povinelli 1993, 32, 153). Among Okanagan in British Columbia, the skin acts as the interface between bodies and the environment and the expression "our one skin" is used to describe people's relationships to each other, with place, and across generations (Armstrong 2006, 463–65).

7 The wetness of the landscape is reflected in the toponymy of settlements in the Upper Bian. For instance, <u>khei ndiki</u>, the Marind name for Khalaoyam, refers to the grease or wetness of a cassowary that once laid its eggs near the settlement. <u>Ngedi</u> (Mirav) denotes the grove of coconut palms whose sweet water once fed <u>dema</u>, or ancestral creator spirits, as they traveled the land toward Kondo. <u>Bukha</u> (Bayau) refers to a chestnut forest rail—a type of bird—that once stopped to drink at a freshwater spring near the village. I documented several other hydrogenic nomenclatures of this kind across the settlements of the Upper Bian. For comparable examples in Oceania, see Mawyer (2018); Mondragón (2018).

8 On the significance of rivers in the construction of kinship in Melanesia, see, inter alia, Goldman and Ballard (1998); Wagner and Jacka (2018).

9   The characterization of wetness as a transspecies, life-generating substance has been noted in several Melanesian societies. For instance, among Upper Asaro in Papua New Guinea, fluids secreted like grease enable growth and fertility for all living things and find particular expression in the oily shine of jabirisi stones. These prized stones, Thomas Strong (2007) writes, constitute collective representations of the beings whose growth they enable and with whom they share the same life-soul (gwondefiya). Jabirisi stones, in turn, are identified with fat in the skins (hokuwe) of both people and pigs, producing what Strong calls a *sympathetic iconicity* between persons, places, bodies, stones, and grease, as well as other gendered fluids such as semen and blood. In a comparable vein, Jerry Jacka (2018) identifies ipane ("grease," or literally, "like water") as a vital life force among Porgerans. As with wetness among Marind, ipane is distributed across humans, plants, animals, and the land, and it is replenished as people interact with the landscape through their movements and subsistence activities.

10 For examples of children as not yet fully human in Melanesia, see Barker (2008, 76); Fajans (1997, 268–69); Little (2011, 151).

11 The labor of making <u>anim</u> also resonates with the concept of klingnan among Mengen of West New Britain, which Tuomas Tammisto translates as "socially productive activity." Klingnan and the physical activities it entails sustain the collective well-being of the community as well as the morally valued relations of humans to their environment (2018a, 26; see also Tammisto 2019).

12 Here, I draw from the notion of *amniotic onto-logics* articulated by cultural theorist Astrida Neimanis (2009, 163–66) to describe common ways of being expressed *through* water and *across* watery difference.

13 Deborah Battaglia documents a similar sense of self among Sabarl in Papua New Guinea, who recognize their self-worth as generative persons through their ability to make and remake new channels of social relationships. The channeled person, Battaglia writes, is therefore one who is materially, psychologically, and affectively open to engaging in the flow of social relations (1990, 57; see also A. Anderson 2011).

14 The morally imbued exchanges of skin and wetness between landscapes and lifeforms in the Upper Bian are antithetical to Enlightenment philosopher John Locke's theories of property, even as both are premised on the logic of mingling sweat, selves, and soil. Locke contends that we can claim property as our own when we mix our sweat with it in the sense of toiling the land. Following this logic, Indigenous peoples' land can be legitimately appropriated and rendered productive because, as nomadic hunter-gatherers, they do not cultivate the land as settlers do and therefore do not mix their sweat with it (Locke [1689] 2016, chap. 5, sect. 25–51). This principle is also at play in the colonial-capitalistic logic of terra nullius, through which the Indonesian government justifies the appropriation of Marind lands and their reallocation to agribusiness corporations. On the logic of terra/corpus nullius and the erasure of Indigenous peoples' territorial rights in the Pacific, see also Davis and Langton (2016); Pascoe (2018); Povinelli (2002); Wolfe (1999).

15 The only exception to the negative assessment of deteriorating skin and wetness concerns the elderly, among whom the depletion of bodily fluids and flesh is the

natural result of a lifetime's worth of bodily substance-sharing with fellow community members and descendants. Conversely, individuals who show no sign of aging and whose skin and wetness remain intact despite old age are terrifying to Marind. These symptoms suggest that the individual has failed to participate in the acts of bodily exchange essential to becoming <u>anim</u> or is a malevolent <u>kambara</u>, a perpetrator of local forms of black magic.

16 The "four-letter disease" (penyakit empat huruf), AIDS, is associated by Marind with the city, with "foreign people," and particularly with oil palm plantations, whose growing labor force has prompted an influx of sex workers into Merauke. Indeed, the prevalence of AIDS in West Papua has been linked to prostitution, which is, in turn, associated with non-Papuan in-migration and the alleged demand for female sex workers in the mining sector and the military (see Butt 2005, 423–24; Kirsch 2002, 60–63).

17 For comparable associations between physical deterioration and cosmological decline in Melanesia, see Goldman and Ballard (1998); Jacka (2007); Knauft (1989); Stewart and Strathern (2001).

18 Examples of bodily deteriorations provoked by encounters with forms of radical alterity abound in the literature on Melanesia. For instance, among the Inanwatan of the south coastal region of Bird's Head, the arrival of Christianization is associated with a premature rotting of the flesh. Peoples' bodies become smaller and skinnier, their blood becomes less fluid, and their skin becomes more flaccid (van Oosterhout 2001, 25). The Bush Kaliai of West New Britain describe their bodies as becoming smaller and weaker and their gardens less fertile as a consequence of the arrival of White people (Lattas 1991, 241). Meanwhile, Jeffrey Clark examines how the colonial encounter among the Wiru of Highland New Guinea manifested in psychosomatic ways as uncontrollable bouts of madness, waterborne contagions, and the physical "shrinking" of Highlander men, who were disempowered and emasculated under colonial rule (2001, 119–43; see also Strong 2007).

19 Karolina's description resonates with the way Porgerans in Highland Papua New Guinea associate the effects of mining and natural disasters with the "drying out" of land and the gradual weakening of their skin (Jacka 2015, 92, 129–31). It also brings to mind the comparisons made by Colombian peasants between the drainage of swamps to make way for oil palm plantations and the drying out of human bodies which, as among Marind, is also associated with supernatural punishment (Taussig 2018, 23, 94–95).

20 A similar situation is at play across the border in south-central New Guinea, where Yonggom communities place responsibility on the Ok Tedi gold mine for the collapse of the environment and the relations embedded within it (Kirsch 2006, 79–131).

21 Other chemicals commonly used in industrial oil palm plantations include the herbicides glufosinate ammonium, triclopyr, and metsulfuron-methyl and the insecticides cypermethrin, deltamethrin, and trichlorfon.

22 Nick Shapiro and Eben Kirksey deploy the term *chemosociality* to refer to long-standing and emergent relationships and social forms that are produced as a result of chemical exposures and dependencies (2017, 483–84).

23 The "chemical sublime," Nicholas Shapiro explains, arises from the accumulation of minute aberrations in human and other-than-human bodies and their surroundings caused by chemical exposure, which are discerned and lived through processes of "becoming sensitive, embodying atmospheres, somatically judging environments, or becoming corporeally aware of nonhumans" (2015, 369).

24 On the unevenness of global chemical geographies, see Agard-Jones (2013); Choy and Zee (2015); Liboiron, Tironi, and Calvillo (2018); Murphy (2015; 2017).

25 Joel Robbins and Alan Rumsey describe this cultural phenomenon as the "opacity of minds" (2008, 407–8; see also Brison 1992; Hau'ofa 1981; Luhrmann 2012; Robbins 2001; Throop 2010).

26 Marind villagers reported that only men are allowed to change skin because their bodies are stronger than those of women and children, and they are therefore better able to return to <u>anim</u> form after shape-shifting. For instance, the legs of men who run the forest with the strength of the cassowary are less likely to break when they revert to their <u>anim</u> skin. The teeth of those who have foraged with the tusks of the wild boar are less likely to cut through their <u>anim</u> jaws. Similarly, the tiny palpitating hearts beating in the bodies of men who transform into birds can expand and return to <u>anim</u> size more quickly than those of women and children.

27 The potentially dangerous loss of the human point of view in skin-changing resonates with the shape-shifting practices of Amazonian peoples and Siberian Yukhagirs described by Eduardo Viveiros de Castro (1998) and Rane Willerslev (2004), respectively. Marind shape-shifters may find themselves incapable or reluctant to return to their <u>anim</u> self because they are no longer conscious of having lost it in the first place. Like Okto, these individuals become, to torque Rane Willerslev's words, a strange fusion of human, not human, and not *not*-human (2004, 642). This is not, however, because these individuals continue to think of or see themselves as human, as Viveiros de Castro's theory of Amerindian perspectivism posits. Rather, skin-changers in the Upper Bian are lured by the bodies and behaviors of forest organisms precisely because the perspectives they offer on the world differ—indeed, often *exceed*—those afforded by a human corporeality. The incapacity of Marind men to return to their original form is further heightened by the inability of all beings—human and other—to ever be fully certain of their *own* minds in the first place.

28 The multiplication and possibly dangerous transformation of perspectives in Marinds' encounters with forest organisms brings to mind Jacques Derrida's description of the interspecies gaze as a moment that reveals the limits of the human while also offering the possibility of crossing over into the realm of the animal. In this moment of vulnerability or nakedness, in Derrida's terms, "everything can happen to me" (2002, 381). In a similar vein, scholar of literary studies Ron Broglio describes how interspecies encounters at the surface foreground the intersubjective spatial and optical phenomenologies that animate the more-than-human world. In an account that resonates with the dangers of perspectival capture in Marind skin-changing practices, Broglio suggests that the human-animal encounter is not without risk or ambivalence. Rather, it "engulfs and implicates all beings

as participants. There is no safe position, no transcendent space from which to stand and judge or observe" (Broglio 2011, 131).

29 Dividuals, Strathern explains, differ from individuals in that they are constructed from the outset as the plural and composite site of the relationships that produced them. Rather than unique or bounded persons, dividuals constitute social microcosms whose plurality as individuals, or "many," is equal to their unity as relations, or "one" (M. Strathern 1988, 12–14). Roy Wagner (1991) formulates a similar concept in describing Melanesian Big Men as "fractal persons," or entities whose external relationships with others are integral to their self.

30 In highlighting the limitations of thinking about dividuals purely in terms of human selves and relations, I join a number of other scholars who have critiqued Strathern's concept for, among other reasons, assuming the radical alterity of Melanesian people, precluding the possibility of agency and resistance, portraying Melanesian personhood as static and ahistorical, and neglecting the importance of gender and of the unconscious in the differential construction of personhood (see LiPuma 2000; Stephen 1995; Wardlow 2006).

4. THE PLASTIC CASSOWARY

This chapter is derived in part from an article published in *Ethnos* on July 26, 2018, copyright Taylor and Francis, available online: www.tandfonline.com/10.1080 /00141844.2018.1502798.

1 Raskin is the short form of beras miskin, or "poor peoples' rice." The term is used locally to refer to subsidized rice for low-income households and is also the official designation from the government programs distributing it.

2 Here, I find inspiration in geographer Rosemary-Claire Collard's (2014) call to acknowledge the relational autonomy of species through forms of active distancing that respect and preserve their wildness.

3 On the cultural significance of gardens in Melanesia, see, inter alia, Barker (2008); Battaglia (2017); Kahn (1994); Malinowski (1935); Mosko (2009); Sillitoe (1983); Young (1971).

4 Historical precedents further testify to the disinclination of Marind toward horticulture and agriculture. For instance, local communities reportedly showed little interest in the large-scale rice project set up by the Dutch in Kimaam in the 1930s and the project never progressed beyond the experimental phase. Later attempts by the Dutch to introduce food crops such as peanuts and maize under nuclear estate and smallholder schemes were likewise met with limited success. A rice project in Merauke started in 1961, for instance, was terminated after producing only 400 tons due to an insufficient labor force (Garnaut and Manning 1974, 18–19; see also Manikmas 2010; Overweel 1992; Pouwer 1999).

5 On the cultural significance of pig rearing in Melanesia, see, inter alia, Dwyer (1990); Goodale (1995); Rappaport (1984); Sillitoe (2003).

6 I borrow the term *reciprocal capture* from philosopher Isabelle Stengers to describe dialectical processes of identity construction in which each agent has an interest in seeing the other maintain its existence (2010, 35–36).

7   As West Papuan theologian and activist Benny Giay and anthropologist Chris Bal-
    lard (2003) note, shared and ongoing experiences of persecution under Indonesian
    rule are a critical denominator in the construction of Papuans' collective identity
    (see also Mote and Rutherford 2001).

8   Juno Parreñas describes similar practices of "tough love" toward apes among
    orangutan conservationists and caretakers in Sarawak, who seek to foster animals'
    independence and capacity to return to the wild while also speaking in powerful
    ways to longer-standing legacies of colonial rule in Malaysia (2018, 33–59).

9   On transmigration in West Papua, see Arndt (1986); Elmslie (2017); Fearnside
    (1997); Manning and Rumbiak (1989); McGibbon (2004); Monbiot (1989).

10 In this regard, Ruben and his fellow village critters are uncannily similar to the
    "outsider" monkeys invading the mountain villages of India's Central Himalayas,
    as described by Radhika Govindrajan (2018, 90–118). "Outsider" monkeys—whom
    local inhabitants distinguish from "mountain" (*Pahari*) monkeys—are profoundly
    disliked by villagers for their destructive behavior and the fact that they cannot
    be gotten rid of. "Outsider" monkeys also come to act as powerful yet ambivalent
    metaphors for larger political and affective anxieties in the villages of the Central
    Himalayas linked to the loss of *Pahari* cultural identity, the urban migration of
    Pahari youth in pursuit of employment, and the privileging of "plains" peoples'
    interests and well-being over that of "mountain" people by the Indian state.
    Whereas villagers claim kinship with "mountain" monkeys based on their common
    experience of displacement, neglect, and exclusion by the state and "plains" people,
    "outsider" monkeys, much like Ruben in the Upper Bian, are resented because
    they are unwelcome intruders. In both cases, village animals' abnormal behavior is
    associated with their loss of animality and their acquisition of dislikable tendencies
    from equally disliked humans—"plains" people in the Himalayas and Indonesian
    settlers in West Papua.

11 The condition of domesticated cassowaries as "matter out of place" brings to
    mind Ralph Bulmer's (1967) analysis of the liminal status of the cassowary in the
    taxonomies of Karam in New Guinea. Bulmer draws attention to the association of
    cassowaries with "wildness," an attribute of organisms that should not be domesti-
    cated and that cassowaries share with other species including the pandanus palm.
    As with Ruben and other village critters in the Upper Bian, the unique status of
    cassowaries among Karam, Bulmer explains, is best understood in light of their
    particular—and in Ruben's case, changing—relationship to both humans and the
    surrounding environment.

12 The uncanny mirroring that Marind identify between their own fates and those
    of domesticated cassowaries brings to mind the prophetic cassowary myth of
    Ilahita Arapesh in the Sepik area of New Guinea. As Donald Tuzin (1997) explains,
    the decline of the local Tambaran cult, the rise of revivalist movements, and the
    consequent inversion of gender power dynamics among Ilahita appear to have been
    prefigured in their creation myth, in which a cassowary woman kills her husband
    (the first Ilahita Man) and populates the world with their children. While not a
    "cassowary's revenge" per se, Ruben's indexicality with his human counterparts

speaks in a similar vein to the demise of a traditional world order and to the doubts and insecurities of Marind in the context of ongoing colonization and rampant environmental degradation.

5. SAGO ENCOUNTERS

Sections of this chapter are modified from "Sago: A Storied Species of West Papua," by Sophie Chao, in *The Mind of Plants: Narratives of Vegetal Intelligence,* edited by John C. Ryan, Patricia Vieira, and Monica Gagliano, 317–26 (Santa Fe: Synergetic Press, 2021).

1   My use of the term *plot* pays tribute to Sylvia Wynter's (1971) contrapuntal characterization of the plot and the plantation. Plots, Wynter writes, constituted spaces where African slaves in Caribbean colonial plantations were able to cultivate their own foods and maintain mutually nourishing relations with the land and its crops. Akin to the forests and groves that we will encounter in this chapter, plots in the Caribbean plantation nexus acted as potent crucibles for the expression of traditional values, folk culture, and environmental care. They also embodied a multispecies locus of resistance to the extractive logic of industrial monocrops and to the commodification of Black bodies as exploitable labor. In so doing, plots rendered possible oppositional modes of Black life and ecological relationalities, despite and against the violence of the plantation (Wynter 1971, 99–100; see also Carney 2020; Davis et al. 2019).

2   The Marind word for sago, <u>dakh</u>, is also the generic term for "food." This is the case in many sago-based Pacific societies. For instance, the term for "sago" is analogous to "food" among Kaluli of the Southeastern Highlands (Schieffelin 1976, 31), hunger translates as "pain for sago" among Korowai in southeastern West Papua (Stasch 2009, 167), and famine or hunger are conceived as the absence of sago among Muyu refugees in Papua New Guinea (Glazebrook 2008, 96–97) and among Naulu in the Moluccan islands of eastern Indonesia (Ellen 2011, 52).

3   Donna Haraway (2016, 58–98) uses the term *sympoiesis* to describe the coming into being of organisms through their situated and dynamic *relations* to other lifeforms, rather than as singular, self-making, and autopoietic entities (see also Haraway 2008, 88).

4   The multispecies pedagogies described in this chapter bring to mind environmental scientist and member of the Citizen Potawatomi Nation Robin Wall Kimmerer's call for forms of learning that attend to the more-than-human world as a "grammar of animacy" (2014, 48–59). Cultivating such pedagogies, Kimmerer writes, demands that we appreciate the gifts and stories of plants, that we spend time in their company, and that we treat them as persons worth listening to and learning from.

5   Sago groves are classified by Marind as <u>dehar</u>, <u>ahakbar</u>, or <u>makhav</u>. Dehar were created by the <u>dema</u> in times immemorial and continue to be protected and managed by ancestral spirits to this day. These groves are considered sacred and only entered if abundant quantities of sago are needed for upcoming rituals, feasts, or ceremonies. <u>Ahakbak</u> refers to groves managed by the deceased relatives of present-day Marind up to three generations prior. These groves tend to be located at or near

former settlements and bear the same name. <u>Makhav</u> denotes groves managed by members of the three most recent Marind generations.

Whereas <u>dehar</u> and their locations are known from intergenerationally transmitted stories and songs, <u>ahakbak</u> and <u>makhav</u> are usually identified from the morphology—and consequent age—of their constituent palms. For instance, <u>ahakbak</u> have wider trunks and larger pneumatophores and fronds than <u>makhav</u>, whereas <u>makhav</u> have lower pith content and shorter root systems. Sago groves located within fifteen kilometers of Khalaoyam, Bayau, and Mirav are primarily <u>makhav</u> and/or <u>ahakbak</u>, whereas those located further away are primarily <u>dehar</u> and <u>ahakbak</u>. Over time, <u>makhav</u> gradually transform into <u>ahakbak</u> as new Marind generations emerge and transplant suckers from one type of grove to the other. Eventually, the distribution of <u>ahakbak</u>, <u>makhav</u> and <u>dehar</u> takes the form of a series of multiple, concentric circles that in turn reflect the successive location of past Marind settlements.

6 Marind refer to sago palm and its flour as <u>dakh</u>. Terms for specific parts of the tree, the stages of sago processing, and the different by-products obtained are also prefixed with the root <u>dakh</u>.

7 The larvae of the sago palm weevil (*Rhynchophorus ferrugineus*) incubate in rotting sago trunks. Creamy in texture and eaten both raw and cooked, they constitute an important source of protein and fat among sago-based societies.

8 I borrow the term "gastro-identities" from Frederik Errington and Deborah Gewertz (2008, 591) to describe processes of collective identity formation that are achieved through the medium of food and that are intrinsically linked to gastro-geographies (who eats what and where) and gastro-politics (who gets what kind of food, from whom, under what circumstances, and with what effects).

9 In this regard, Marind are akin to Gogodala in the Western Province of Papua New Guinea, who associate sago labors with an increase in physical stamina, and Bian-gai in Morobe Province, who correlate the taste and nourishment of sago starch with the hard work involved in its making (see Dundon 2005; Halvaksz 2013).

10 On the "wet" or "dry" classification of foods and their derivative species in Melane-sia, see Bashkow (2006); Meigs (1984); Sexton (1988).

11 The contrapuntal relationship between sago (people) and rice (people) echoes the classificatory logic of contrasts shaping Orokaiva-Whitemen and taro-rice distinc-tions in the Oro Province of Papua New Guinea, as described by Ira Bashkow. Whereas traditional Orokaiva foods such as taro and pork are positively valued for their heavy, hard, and strong qualities and because they are exchanged and culti-vated by Orokaiva on and from their own land, foreign foods such as rice and canned meat are negatively described as light, wet, and weak because they are purchased in stores and associated with money (Bashkow 2006, 146–63). Much like the indexicality of Marind to sago and settlers to rice, the taro-rice binary in turn sits within and manifests a broader ontology of difference between Orokaiva and Whitemen in terms of their lifestyles, forms of consumption, relations to land, and economic systems (see also Halvaksz 2013; Jolly 1991; Pollock 1992; West 2012). For a contrasting example among Dani in Highland West Papua, where the

adoption of imported rice as a central form of value has been accompanied by the waning of the traditional sweet potato, see Nerenberg (2018, 117–20).

12 On the cultural politics of rice and sago in Indonesia, see Rijksen and Persoon (1991); Soemarwoto (1985). Rice did not feature prominently as a topic of conversation among my companions, in part because the crop is not cultivated in the Upper Bian. However, one might speculate on the basis of Marinds' conceptions of oil palm that rice, too, might be conceived as a dangerously nonreciprocal and unloving plant-being. Both plants are introduced species whose proliferation causes the dispossession of Marind of their lands and natural resources. Both are grown in large-scale monocrops and cultivated by non-Papuan settlers. Both crops are also promoted in governmental discourse as key to the economic development of West Papua.

13 Similar associations among sago, motherhood, procreation, and fertility have been documented across Melanesia. Among Kamoro in the southwest coast of New Guinea, for instance, sago is a dominant feature in initiation ceremonies, where it is ritually exchanged, consumed, and displayed (Pouwer 2010, 61–73). Comparable links between conception, bodily substance, and sago have also been documented among Korowai (Stasch 2009, 148) and Inanwatan (van Oosterhout 2001, 31). Marind conceptualizations of sago, however, differ in that sago is not sexualized in the sense of being classified as either male or female (e.g., Juillerat 1996; Tuzin 1992). While women sustain a particularly intimate connection to sago in light of their reproductive functions, both men and women partake equally in its procurement and consumption.

14 Pascale Bonnemère (1994, 1996b) describes similarly gendered associations between red pandanus juice and human blood among Ankave-Anga in the Gulf Province of Papua New Guinea. In the creation myth of Yafar of West Sepik, the blood of the sago palm complements the blood of coconut palms, each of which gives rise to one of the two Yafar moieties (Juillerat 1996, 40–57).

15 Alison Dundon describes a similar phenomenon among Gogodala in the Western Province of Papua New Guinea, who consider the physical marks left by sago work on women's bodies—for instance, broken and blackened fingernails and toenails, calloused palms, and discolored ankles—as central to the construction and affirmation of their womanhood (2005, 25–27; see also von Poser 2013, 183–85).

16 Marind men's experiences bring to mind what political theorist Jane Bennett describes as the state of enchantment—a "fleeting return to childlike excitement about life" (2001, 5) that enables joyful attachments and ethical engagements within and across species lines.

17 Like Upper Bian Marind, Sabarl Islanders and Korowai in West Papua correlate sago growth with human growth and personify the plant as a surrogate child (Battaglia 1990, 49; Stasch 2009, 166–67). Myths recounting the shared genesis of sago and humans have also been documented across lowland Papua, and particularly in the Sepik region (Pouwer 2010, 72–73; von Poser 2013, 91).

18 The association among childbirth, motherhood, and sago palms is particularly evident among Korowai, who use palm parts in several perinatal acts to help the

newborn enter human relations. For instance, women eat a palm heart soon after they give birth. They use the stripped midrib of a sago leaflet to sever the umbilical cord. Once the umbilical cord dries and detaches, the mother inserts it amid the leaf bases of a sago palm in the rosette stage. Newborns are also placed inside a sago leaf base segment when they make their first entrance into the domestic space (Stasch 2009, 166–67).

19 On the transgenerational management of vegetatively propagated sago in the Pacific, see Barton and Denham (2011); Ellen (2006); Verschueren (1970).

20 In a comparable vein, Ilahita Arapesh in the Sepik area characterize sago as a plant that "looks after itself" and thus epitomizes natural abundance and self-sufficiency, as opposed to taro and yams, which require human cultivation (quoted in Tuzin 1992, 105). The minimal manipulation of sago among Marind also brings to mind the marginal modification of plants described by Laura Rival (2002, 14, 88) among Huaorani of the Ecuadorian Amazon and the concept of self-propagating plants deployed by Anna Tsing (2005, 177–78) to describe vegetal lifeforms among Meratus Dayak in Kalimantan. It further resonates with the concept of mulun sebulang among Kelabit in Sarawak, a term that Monica Janowski translates as "wildness," and that refers to forest lifeforms that grow on their own but that may be *assisted* by humans in their development and propagation (2003, 35).

21 Marinds' corporeal engagements with sago palm also bring to mind STS scholar Eva Hayward's notion of "fingeryeyes," which defines the palpability of interspecies encounters as a simultaneous and mutual touching and feeling (2010, 581).

22 Anita von Poser describes similar proscriptions on antisocial behavior during sago processing and consumption among Bosmun of Papua New Guinea (2013, 100).

6. OIL PALM COUNTERPOINT

1 The association of oil palm with hope in Indonesia resonates with the naming of the oil palm variety cultivated in Colombia's plantations as "Hope of America" (Gill and Taussig 2017, 48).

2 Indomie is Indonesia's biggest instant noodle company and a subsidiary of the Salim Group, one of Indonesia's most powerful conglomerates. The Salim Group is also the owner of Indofood Agri Resources, one of the five largest oil palm–plantation companies in the world. The lyrics cited by Agus are part of the company's official commercial, "From Sabang to Merauke, Indomie, My Taste/Desire," an adaptation of the nationalist slogan first pronounced by President Sukarno during Indonesia's Proclamation of Independence in 1950. On the construction of national and global identities through instant noodle consumption in New Guinea, see also Errington, Fujikura, and Gewertz (2013).

3 On the contrapuntal logic of interspecies correspondence, see also Ingold (2017). Von Uexküll's (1982) concept of the counterpoint has also gained traction beyond the realms of environment and ecology. For instance, Deleuze and Guattari (1994) examine the counterpoints of melodic air and polyphonic motif as they enter into the development of one another in the context of music. Alfred Gell (2006) calls for a contrapuntal approach to art that juxtaposes Western and non-Western artworks

toward a synthesis of meaning across artist, depicted object, and viewer. Elizabeth Grosz (2008) deploys the concept to analyze the relational emergence of bodies and earth in Aboriginal songs and the philosophy of art more generally. James Clifford (1983) draws attention to the contrapuntal relation at play in the intersubjective dialogue between the informant and the ethnographer, and between the representation of both parties. Meanwhile, Edward Said (2012) examines how the lived condition of exile inevitably occurs against memories of home and thus enables a plurality and originality of vision across both new and old environments.

4   The moral valences of the sago palm–oil palm counterpoint among Marind bring to mind mid-twentieth-century Cuban scholar Fernando Ortiz Fernández's *Cuban Counterpoint* (1947), a playful allegory that explores the distinct personas and predilections of tobacco and sugar as concomitant subjects and objects in the making of Cuban culture and history.

5   In a comparable vein, Alison Dundon describes how Gogodala women who migrate to urban areas or overseas constantly lament the absence of sago in their new environments and its detrimental effects on their bodies and health and on those of their kin (2005, 24).

6   On the cultural politics of hunger and satiety in Merauke, see Chao (2019c, 2020a, 2021d).

7   The negative associations of "eating money" have been documented elsewhere in the Pacific. Among Porgerans, for instance, "eating money" or "eating gold" refers to forms of symbolic eating that are considered individualistic, antisocial, and short-term (Jacka 2015, 239–40). Similarly, the expression "eating money" (kai kai selen) among Solomon Islanders conveys the notion that money earned by villagers through large-scale logging projects is being wasted on personal consumption, ephemeral consumables, and alcohol (Dyer 2017, 88–89; see also Hayano 1989; Sexton 1988).

8   Corporate development projects in Melanesia and elsewhere are often accompanied by the fragmentation of community ties and the erosion of communal land tenure. For instance, Edward Hviding (2015a) explores how in the Solomon Islands, large-scale logging has ruptured the collectivities of land and people, giving way to conflict and competition over commodified nature. In a similar vein, Aletta Biersack (2006b) describes how mining operations in Porgera have set local inhabitants against each other in the context of uneven compensation payments. For other examples from the region, see Akin (1999); Filer (2007); Kirsch (2001).

9   The gendered implications of the disappearance of sago for Marind women represents an inversion of what David Lipset (2017) calls the "dual alienation" of Murik men from their traditional masculine identity and from postcolonial modernity in Papua New Guinea. Akin to the environmental transformations taking place in Merauke, the alienation of Murik men is further exacerbated by the climate change-induced inundation of former fishing areas, which undermines their subsistence role and source of modern value in the developing economy.

10   Agus's dismissive comment echoes long-standing derogatory stereotypes of sago-based societies promoted by colonial and contemporary governments in Southeast

Asia. For instance, the Scottish colonial administrator John Macmillan Brown, who worked in New Zealand, attributed "native indolence, degeneration, and sterility" to the sago palm and described sago eaters as "hand-to-mouth loafers." Natives were unable to engage in the civilized activity of agriculture because of their laziness and dependence on forest-derived sago. "Where sago grows," Brown concluded, "there is and there can be no progress" (cited in Tan 1979, 27). This derogatory depiction of sago and sago eaters has been perpetuated by the Indonesian government as part of its promotion of rice and other cash crop development projects across the archipelago. For instance, Gerard Persoon notes how state discourses around rice cultivation in Mentawai placed moral evaluation on sago as a "lazy man's food," in contrast to rice which was promoted as a symbol of civilization (1992, 192). In a similar vein, Celia Lowe (2006, 95) points to how government discourse portraying sago consumption as a sign of savagery, poverty, and backwardness, has led Sama in the Togean Islands to conceive of store-bought rice as a more civilized foodstuff.

11 Historian Alfred Crosby (2003, 52) deploys the term *biological ally* to describe plants and animals that thrived as they traveled alongside human colonizers across geographical realms. Colonial movements, Crosby argues, are never solely human, social, or political, but rather are biological and multispecies in nature. On the enlistment of species and ecologies in empire building, see Kosek (2010); McNeill (2012); Mitchell (2002).

12 See also Lyons (2016) on the characterization of coca in Colombian drug eradication campaigns as "the plant that kills" (la planta que mata).

13 The radical differences that Marind identify between the lifeworlds of sago palm and oil palm can be compared to the distinctive dimensions of *Umwelt* emphasized by Gilles Deleuze and Félix Guattari (2005), on the one hand, and Peter Sloterdijk (2011), on the other. Philosophers Judith Wambacq and Sjoerd van Tuinen (2017) suggest that the difference lies in the nature of these two approaches. Sloterdijk, like Deleuze and Guattari, recognizes the *ontological* necessity of movements that open up organismic interiorities, such as heterogenesis, multiplicities, transformation, and porosity. However, his approach seeks to draw attention to the *existential* difficulties associated with these movements. Deleuze and Guattari highlight the generative nature of organisms' unbounded openness to their environment, whereas Sloterdijk draws attention to how organisms sustain their lifeworlds by closing off and selectively *refusing* participation with other lifeforms and exteriors. While both dimensions participate in the making of species' perceptual lifeworlds, sago foregrounds the multiple connections and movements that enable hospitality between organisms and their environments, while oil palm foregrounds the immunity and negation that Sloterdijk suggests are equally central to organismic worldmaking.

14 On the material-semiotic relationship between Marind laborers, plantation parasites and mutualists, and plantation chemo-geographies, see Chao (2021a). On plantations as ecologies of resistance, see also Allewaert (2013) and Beilin and Suryanarayanan (2017).

15　For similar perspectives among oil palm scientists and nursery workers, see Chao (2018b).

16　Haraway describes the logic of plantations as one of "out-and-out exterminism" in that plantations exhaust soil nutrients, proliferate pathologic pathogens, and thus destroy the grounds for their own survival (cited in Mitman 2019, 10; see also Scott 1998, 268–69). Meanwhile, Tsing, Mathews, and Bubandt invite attention to the interplay of "modular simplifications" and "feral proliferations" in the making of anthropogenic landscape structures. Modular simplifications, as exemplified by monocrop plantations, entail the radical simplification of multispecies ecologies and the regimentation of human and nonhuman life in the name of production and profit. Yet modular simplifications are also the root cause of a range of unintended and uncontrollable social and ecological effects, including disease epidemics, toxic contamination, and parasitic invasions, which undermine the productionist logic of technoscientific capitalism. Feral effects, then, inhabit and are inherently entangled with, the very same modular landscape patches that enable their proliferation (Tsing, Mathews, and Bubandt 2019, S189–90).

17　Biopolitics in the late neoliberal age, Povinelli (2016) argues, centers on a Life versus Nonlife divide, which is itself lodged within a larger, yet largely neglected, binary: that of Life (life versus death) (bios) versus Nonlife (geos), or what Povinelli calls "geontologies."

18　On the creation of "commodity connections" through other foreign foods and crops in New Guinea, see Bashkow (2006); Errington and Gewertz (2008); Errington, Fujikura, and Gewertz (2013); Halvaksz (2013); Jolly (1991); Kahn and Sexton (1988); R. Foster (2008).

19　Indeed, the name "New Guinea" was chosen by Spanish explorer Yñigo Ortiz de Retez in 1545 based on a perceived resemblance between the island's inhabitants and the native peoples of African Guinea.

20　Donna Haraway deploys the concept of "material-semiosis" to counter the notion that organisms are passive entities that matter only in terms of their functional uses and meanings for humans (1991, 197–200). Rather, Haraway contends, organisms are endowed with diverse symbolic and bodily attributes that they *themselves* produce in their situated interactions with other actors (1992, 298). In these interactions, humans become just one of many material-semiotic figures engaged in complex conversation with other players who are always both subjects and objects to one another in their ongoing and situated relatings (Haraway 2008, 211, 242).

21　Radical care, according to Native studies and STS scholars Hiʻilei Julia Kawehipuaakahaopulani Hobart and Tamara Kneese, speaks to new and subversive forms of care that offer glimmers of hope in increasingly precarious worlds—even as they remain embedded with systemic inequalities and hegemonic power structures. As a form of strategic audacity, radical care is a refusal *not* to care—a creative and collective imagining of an "otherwise" despite dark histories and potentially darker futures (Hobart and Kneese 2020, 2–3).

22　Commoning, Lauren Berlant notes, is a necessary pedagogy for living within increasingly malfunctioning worlds and their "messed up yet shared and ongoing infrastructures of experience." Commoning, Berlant continues, helps us "view what's broken in sociality, the difficulty of convening a world conjointly, although it is inconvenient and hard, and to offer incitements to imagining a livable provisional life" (2016, 395).

23 I borrow the term *significant otherness* from Donna Haraway (2003) to refer to other-than-human lifeforms whose disparate biographies are intrinsically interwoven with those of humans, albeit often in deeply ambivalent and violent ways.

24 As Stacey Langwick notes in the context of dawa lishe gardens in Tanzania, creating new spaces and forms of interspecies habitability calls for experimentation, curiosity, and care. It involves asking what kinds of flourishings can be cultivated amid conditions of pervasive toxicity; what relations enable bodies and landscapes to grow ampler, denser, and more potent; and whose modes of ongoingness and continuance such spaces can support (Langwick 2018, 436).

25 The broad gamut of emotional and moral responses that oil palm evokes among Marind brings to mind Peter Gose's (2018) analysis of mountains in the Peruvian Andes that exist ontologically and affectively as both intimately cherished kin, unpredictably violent forces, and foreign entities whose personhood is both ambivalent and partial. As with Marinds' understandings of oil palm, the multiple and conflicting attitudes of Andean peoples toward other-than-human beings, Gose argues, cannot be subsumed under any single ontology, animist or other (see also Hugh-Jones 2019).

26 Other examples of lethal capital might include proliferating cash crops such as soy and Jatropha, mass-reared livestock such as cattle and poultry, as well as their attendant parasites and symbiotes, and the humans differentially invested in their commodified life (and death).

27 Marinds' recognition of the human and capitalist forces driving agribusiness expansion resonates with Métis scholar Zoe Todd's reflections on the oil spill that devastated her native North Saskatchewan River and surrounding ecosystems. After feeling intensive anger toward the oil, Todd realizes that it is not the oil per se that is harmful but rather the way in which oil—a substance itself constituted of the remains of myriad organismic bodies—has been weaponized by petrocapitalist modes of extraction and turned into a contaminant and pollutant. As Todd writes, "It is the machinations of human political-ideological entanglements that deem it appropriate to carry this oil through pipelines running along vital waterways, that make this oily progeny a weapon against fish, humans, water and more-than-human worlds" (2017, 107).

28 On the strategic essentialization of oil palm's plural ontology in the context of Marinds' negotiations with state and corporate actors, see Chao (2020b).

29 Being against purity, Alexis Shotwell argues, means refusing the paralyzing effects of framing contemporary ecological transformations against the supposedly retrievable and redeemable purity of the world that preceded it (2016, 4–5, 15). Taking compromise and complicity as starting points for living in an always already impure world offers a necessary alternative to the counterproductive labor of delineating and delimiting the world into something separable, homogenous, and pure (see also Liboiron, Tirono, and Calvillo 2018).

1   Original: Awal, dorang datang ambil kitorang pu burung cendrawasih. Setelah itu, dorang datang ambil kitorang pu adat. Terus, dorang lagi datang ambil kitorang pu tanah. Sekarang, dorang datang ambil kitorang pu nyawa dan waktu berhenti sudah.

2   On millenarianism and cargo cults in Melanesia, see, inter alia, Burridge (1960, 1969); Kirksey (2012); Lindstrom (1993); Oosterwal (1963); Robbins (1998, 2001); Rutherford (2003); van Oosterhout (2000); Worsley (1968).

3   It is useful to remind ourselves of the relationship between history and historicity here. History, as Chris Ballard outlines, refers to the material past and the active study of that past, whereas historicity refers to "the modes of temporal being and awareness specific to particular communities at particular moments in time" (2018b, 103). These modes of temporal being and awareness inform the production of the historical past by peoples themselves as situated agents of historical action, consciousness, and determination in ways that reveal history to be just as cultural as culture is historical (Sahlins 1983, 1985). Historicity also informs the production of situated "futurities," a term that, Ballard notes, "speak[s] not just to the temporal domain of times to come, whether definite or indefinite, but also to their *analysis*, to questions of prospect or possibility, now and in the past, and to the very quality of being futural" (2018a, 280, emphasis added). Finally, attending to historicity draws attention to the interests shaping the construction and contestation of historical narratives, as well as the forms of power at play in making, remembering, or unremembering history (see also Jolly 1999; Koselleck 2002; Trouillot 1997, 2003).

4   On the importance of orality in Pacific historicity and history's material manifestation in places, bodies, and environments, see Dening (1991) and Hau'ofa (2008).

5   I borrow the term *regimes of historicity* from historian François Hartog (2015) to refer to the ways in which the relationship between past, present, and future is understood *through* and *as* moments of crisis and rupture (see also Koselleck 2002).

6   The characterization of oil palm's arrival as the latest in a series of episodic disasters within Marind historicity resonates with Kyle Powys Whyte's framing of the current climate crisis as potentially just one of *many* apocalypses previously experienced by Indigenous peoples worldwide, or what Whyte calls *ancestral dystopias*. For Indigenous peoples, Whyte writes, climate vulnerability and injustice today may be less of a novel threat than a "colonial déja-vu"—in other words, an intensification of a much longer pattern of anthropogenic environmental change prompted by settler-colonial occupation and appropriation (2017, 2018a; see also Davis and Todd 2017; Hviding 2015b; Hviding and Bayliss-Smith 2018; Todd 2015; Yusoff 2019).

7   Commercial bird-of-paradise hunting has a much longer history in New Guinea. However, Upper Bian Marind identify the boom in the hunt and trade in the early 1920s as the period of greatest loss. Species hunted included the Greater Bird of Paradise, the Raggiana Bird of Paradise, the King Bird of Paradise, and the Twelve-Wired Bird of Paradise.

8 Granuloma venereum, also known as donovanosis, is a sexually transmitted disease caused by the bacterium *Klebsiella granulomatis*. It infects the lymphatics and lymph nodes, causing ulcers in the genital areas as well as abscesses, necrosis, and fibrosis, which can be fatal if left untreated. Cases of lymphogranuloma venereum were first diagnosed by Dutch missionaries in 1916. Compounded with the influenza epidemics of 1919 and 1937–1938, the disease reportedly decimated some 25 percent of inland Marind and contributed to a sharp drop in birthrates. Anti-granuloma campaigns launched by the Dutch government in 1922 entailed an almost complete ban on native patterns of life, including interclan feasts, ceremonial dances, sexual initiation and fertility rites, and the closure of men's houses. While the disease was reportedly brought under control within two years, the Marind population continued to decline until 1948—a phenomenon often attributed by Dutch missionaries and officials to Marinds' promiscuous sexual mores (van Baal 1966; Vogel and Richens 1989).

9 Ancestral punishment remains a prevalent explanation for misfortune, disease, and ritual failure among Upper Bian Marind (see Chao 2019g).

10 The correlation that Marind identify between the destruction of multispecies places and of multispecies pasts (and futures) brings to mind Epeli Hau'ofa's now-famous words: "We cannot read our histories without knowing how to read our landscapes" (2008, 73). To destroy these landscapes, Hau'ofa continues, is to destroy the "age-old rhythms of cyclical dramas that lock together familiar time, motion, and space" and in so doing, to undermine the claims and legitimacy of Indigenous peoples' very existence (2008, 75). Paige West describes a similar notion of time among Gimi in Highland New Guinea. For Gimi, the past is not linear in that actions undertaken in the present and future affect the past, along with all the beings who inhabit it. This temporality, West continues, stands in stark contrast to (conservation) science that, much like capitalism and development in Merauke, strives towards an imaginary future-to-be rather than the nurturing of an embodied and living past (2006a, 215–18).

11 The torquing of time has widely been described as a powerful instrument of resistance and empowerment (see Baraitser 2017). For instance, philosopher Jacques Rancière (1999) suggests that interruptions in the regulatory rhythms of everyday life are the ultimate expression of insurgent politics. Cultural critic and historian Walter Benjamin (2015) denounces the empty and homogeneous arrow of time as progress dictated by the ruling class and celebrates revolution as an organized "interruption" of capitalist temporality. Gender and literary studies scholar Mark Rifkin (2017), meanwhile, explores the possibilities for self-determination and duration that arise for Indigenous peoples who actively position themselves as "out of sync" with settler time. In the Melanesian context, Jaap Timmer (2015) examines how West Papuan historiography incorporates larger regional and world events as a means of tempering and asserting control over the erratic temporality of history. Meanwhile, Eben Kirksey (2012) highlights the temporal dimension of the West Papuan resistance movement, whose activists harbored a messianic spirit pinned onto changing figures of hope and anticipated transformations in the near future.

12 Deborah Thomas associates the emergence of modern, abstract time described here with the rise of the plantation and, in particular, its insistence on temporal linearity, teleological progress, and capitalist productivity. This temporal orientation, Thomas notes, reproduces inequalities under the guise of inevitability, while erasing the foundational violence necessitated and perpetuated by the plantation formation (2016, 182–83).

13 In this regard, capitalist modernity constitutes what María Puig de la Bellacasa (2016, 55) calls a restless *futurosophy*—one that entails being perpetually ahead of oneself without the possibility of ever achieving true gratification (see also Bauman 2012; Hodges 2008).

14 Reinhart Koselleck (2002) identifies two meta-categories at play in the production of historical time: the *space of experience* (Erfahrungsraum) and the *horizon of expectation* (Erwartungshorizont). The space of experience is the arrayed past for a given present, whereas the horizon of expectation is the edge of future possibilities for any given present. Each category, Koselleck suggests, shapes peoples' interpretation of past, present, and future events, and each also transforms over time and in relation to the other. It is in the ongoing tension between past experience and future expectation, Koselleck argues, that historical time is iteratively produced (1985, 275).

15 Johannes Fabian (1983) critiques the denial of coevalness at play in the ontological and representational relationship between anthropologists and their subjects. In these representations, anthropologists and their interlocutors are portrayed as inhabiting different spatiotemporalities—the former in the *here and now*, the latter in the *there and then*. This spatiotemporal difference, Fabian argues, replicates and perpetuates broader power asymmetries between the researcher and the researched. More recently, Mark Rifkin (2017) has contested the claim that Native Peoples should be recognized as coeval with Euro-Americans. This stance, Rifkin contends, entrenches "settler time" by alternatively relegating Native Peoples to the past or forcing Native Peoples into the present in ways that normalize nonnative histories, geographies, and expectations. Taking instead Native Peoples' own, multiple framings and experiences of time as an analytical starting point is critical to recognizing their importance in Native peoples' struggles for self-determination, or what Rikfin calls *temporal sovereignty*.

16 Queer futurity scholarship offers pertinent comparative insights on how the rejection of heteronormative futures and the remembrance of affectively dense and embodied pasts can sustain alternative forms of future building. For instance, Lee Edelman (2004) criticizes the normalization and celebration of *reproductive futurism* and calls instead for an affirmative stance of queer negativity in the form of irony, jouissance, and the death drive (see also H. Davis 2015; Muñoz 2015). Heather Love (2009) invites a queer critical practice rooted in a "backward future" that embraces the past as exile, refusal, and even failure, in order to envision a future other than the present. Elizabeth Freeman (2010), meanwhile, examines the pull of the past on a putatively revolutionary present and the potential of the body as a channel and means of understanding this past. On the altertemporality of subaltern subjects, see also Maggio (2007) and Spivak (1999).

17 Waiting, Arjun Appadurai suggests, is a patient yet active strategy through which disaffected communities oppose the politics of catastrophe pervading their everyday lives (2013, 126–29).

18 Marinds' rebuff of hope as a future-oriented disposition resonates with the widespread rejection of utopianism in the late twentieth century—a phenomenon that geographer David Harvey (2000) attributes to a collapse in *specific* forms of utopia envisioned by ideologies such as communism and neoliberalism and one that philosopher Catherine Malabou (2012) associates with the attritive nature of the Anthropocene as a present literally "out of time."

19 The attention I draw here to dynamics of detemporalization complements Robert Foster's (1999) call for analyses of *deterritorialization*, the inherently partial and uneven processes of spatial reconfiguration wrought by globalization and neoliberal capitalism.

8. EATEN BY OIL PALM

1 I compiled over two hundred dream narratives during my time in the Upper Bian. Recurring motifs in these dreams include: becoming lost in an oil palm plantation (100); death (85); hunger (74); illness (52); heat and dryness (87); darkness or night (120); being alone (42); fear (123); a loss of sense of time, place and/or direction (130); kindred plants and animals (amai) (18); the military (89); bulldozers (15); razed or burned sago groves (32); and miscarriages or abnormal births (45).

2 Marind idioms of "being eaten by oil palm" resonate with the widespread characterization of estates, factories, mines, governments, and corporations as hungry forces that rapaciously consume resources in the form of things, peoples, and environments (see Bayart 1993; High 2014, 24–43; Nash 1993; Peña 1997; Stoler 1985, 197–98). In the plantation context, similar idioms are invoked by Toba of the Pilcomayo River who describe the owners of the sugarcane plantations they once worked in as cannibalistic "Big Eaters" that profited from buying and selling the flesh of indentured laborers. The memory of Big Eaters, Gordillo notes, "points to the cannibalistic aspect of capitalist exploitation: the consumption of the bodies of a social group by the hunger for profit of another group" (2002, 42).

3 Upper Bian Marind identify a third category of dreams that they call "ordinary dreams" (mimpi biasa). These are caused by everyday circumstances such as overtiredness or stress. In them, people relive routine activities and recent encounters. Such dreams have little affective impact on dreamers and are deemed to be of minimal interpretive value.

4 The Malukan-Malay word suanggi is used throughout North Sulawesi, the Bird's Head of Papua, and in some eastern Indonesian communities to refer to witches, sorcerers, or other figures of evil (Bubandt 2014, 26–28). Marind use the term to refer both to foreign "sorcerers" and to foreign "sorcery."

5 The superior efficacy of foreign and introduced forms of sorcery has been widely documented across Melanesia, particularly in contexts of radical environmental and social change (see Zelenietz and Lindenbaum 1981). Among Anganen, for instance, Michael Nihill describes the emergence of a new form of sorcery known as

botol that is said to originate from the outside world and from cities. Unlike native sorcery, the way in which botol functions is unknown to Anganen, which in turn enhances its perceived efficacy (Nihill 2001, 104–8). In a similar vein, Andrew Lattas examines how Kaliai are being invaded by new forms of sorcery that are more diffuse (and therefore destructive) than local forms, and that are practiced in plantations and government stations controlled by powerful and mobile White people (1993, 59–60). Austin Hagwood (2017) documents a similar association in Vanimo, Papua New Guinea, between plantations and new modes of assault sorcery that serve to advance the interests of agribusiness operators and corporate loggers.

6   In a similar vein, Maututu villagers in West New Britain and Yonggom in south-central New Guinea associate industrial oil palm plantations and the Ok Tedi gold mine, respectively, with the figure of the sorcerer (Guinness 2018, 38; Kirsch 2006, 107–31).

7   Michele Stephen describes a similar characterization of dreaming as a potentially dangerous experience among Mekeo in Papua New Guinea. In this state, individuals are particularly vulnerable to the influence and control of other forces and persons, which may continue to haunt them in their waking reality (Stephen 1995, 124; see also Stephen 1982).

8   Yasmine Musharbash (2013) describes a similar characterization of sleep as a potentially dangerous moment of isolation and oblivion among Warlpiri in Aboriginal Australia. Much like Upper Bian Marind, Warlpiri achieve safety in this state by sleeping in groups and being protected by watchers, in ways that express both care and trust.

9   An ironic double bind is at play here. Marind deem it essential to wake the sleeper to prevent their possession by oil palm. Yet repeated awakenings themselves can cause sleep and dream deprivation, which results in more frequent and vivid dream experiences while at the same time negatively impacting the sleeper's memory.

10   This brings to mind Barbara Tedlock's (1981) distinction between intratextual, contextual, and intertextual dream analysis. *Intratextual analysis* refers to the analysis of the dream content alone. *Contextual analysis* refers to the events or circumstances of the dreamer's life and his or her reaction to the dream imagery. Finally, *intertextual analysis* combines a particular dream and its immediate context with that of other dreams. All three levels of analysis find expression in Marind practices of dream-sharing.

11   Marianne George describes a similar conceptualization of dreams among Barok in New Ireland. As among Marind, dreams for Barok accrue creative force less as private or personal experiences than as shared narratives that collectively mirror or rehearse both lived and anticipated realities (George 1995, 30–32; see also Tuzin 1997, 153–54).

12   As João Biehl and Amy Moran-Thomas argue in a biomedical context, symptoms are more than just the incidental markers of disease (or dis-ease). Rather, symptoms and peoples' responses to symptoms provide a means through which they can articulate their relationship to each other and to the social and ecological ruptures of the waking world. As with the therapeutic function of dream sharing among

Marind, the point of such articulations is "perhaps less a matter of finding a voice than establishing oneself as part of a matrix in which there is someone else to hear it" (2009, 282).

13   Following Amira Mittermaier (2011), I use the term *imagination* to refer, not to make-believe fantasies, but rather to a broader range of meanings that encompass a variety of spaces, modes of perception, and conceptualizations of the real.

14   Here, I draw from Douglas Hollan's (2004) concept of "selfscape dreams," or dreams that reflect back to the dreamer how their current organization of self in bodily and psychic terms relates to itself and to other objects, bodies, and beings in the world. Selfscape dreams, as Hollan notes, act as culturally shaped maps of the changing bodily and imaginative contours of the self, in ways that may also reflect a subconscious awareness of dis-ease in the psycho-body.

Agard-Jones, Vanessa. 2013. "Bodies in the System." *Small Axe* 17, no. 3(42): 182–92. doi:10.1215/07990537-2378991.

Ahmann, Chloe. 2018. "'It's Exhausting to Create an Event Out of Nothing': Slow Violence and the Manipulation of Time." *Cultural Anthropology* 33, no. 1: 142–71. doi:10.14506/ca33.1.06.

Ahmed, Sara. 2004. *The Cultural Politics of Emotion*. Edinburgh: Edinburgh University Press.

Ahmed, Sara. 2010. *The Promise of Happiness*. Durham, NC: Duke University Press.

Ahuja, Neel. 2016. *Bioinsecurities: Disease Interventions, Empire, and the Government of Species*. Durham, NC: Duke University Press.

Akin, David. 1999. "Compensation and the Melanesian State: Why the Kwaio Keep Claiming." *Contemporary Pacific* 11, no. 1: 35–67. www.jstor.org/stable/23717412.

Al Jazeera English. 2020, June 25. "Selling Out West Papua." www.youtube.com/watch?v=cBbVu1ZOpYY.

Allen, Matthew G., Michael R. Bourke, and Andrew McGregor. 2009. "Cash Income from Agriculture." In *Food and Agriculture in Papua New Guinea*, edited by Michael R. Bourke and Tracy A. Harwood, 283–424. Canberra: Australian National University Press.

Allewaert, Monique. 2013. *Ariel's Ecology: Plantations, Personhood, and Colonialism in the American Tropics*. Minneapolis: University of Minnesota Press.

Amo, Anselmus. 2014. "Proses Pelaksanaan Program Pemberdayaan Masyarakat Melalui Program Pertanian Sayur Mayur Oleh PT Selaras Inti Semesta." Master's diss., Universitas Indonesia.

Ananta, Aris, Dwi Retno Wilujeng Wahyu Utami, and Nur Budi Handayani. 2016. "Statistics on Ethnic Diversity in the Land of Papua, Indonesia." *Asia and the Pacific Policy Studies* 3, no. 3: 458–74.

Anderson, Astrid. 2011. *Landscapes of Relations and Belonging: Body, Place and Politics in Wogeo, Papua New Guinea*. New York: Berghahn.

Anderson, Benedict R. 1983. *Imagined Communities: Reflections on the Origin and Spread of Nationalism*. London: Verso.

Anderson, Tim. 2015. *Land and Livelihoods in Papua New Guinea*. Melbourne: Australian Scholarly Publishing.

Andrianto, Agus, Heru Komarudin, and Pablo Pacheco. 2019. "Expansion of Oil Palm Plantations in Indonesia's Frontier: Problems of Externalities and the Future of Local and Indigenous Communities." *Land* 8, no. 56: 1–17. doi:10.3390 /land8040056.

Anzaldúa, Gloria. 1987. *Borderlands/La Frontera: The New Mestiza.* San Francisco: Aunt Lute.

Appadurai, Arjun. 1981. "Gastro-Politics in Hindu South Asia." *American Ethnologist* 8, no. 3: 494–511. doi:10.1525/ae.1981.8.3.02a00050.

Appadurai, Arjun. 2013. *The Future as Cultural Fact: Essays on the Global Condition.* New York: Verso.

Archibald, Jo-Ann, Jenny Lee-Morgan, and Jason De Santolo, eds. 2019. *Decolonizing Research: Indigenous Storywork as Methodology.* New York: Zed Books.

Armstrong, Jeanette. 2006. "'Sharing One Skin': Okanagan Community." In *Paradigm Wars: Indigenous Peoples' Resistance to Economic Globalization*, edited by Jerry Mander and Victoria Tauli-Corpuz, 460–70. San Francisco: Sierra Club.

Arndt, Heinz W. 1986. "Transmigration in Irian Jaya." In *Between Two Nations: The Indonesia–Papua New Guinea Border and West Papua Nationalism*, edited by Ronald J. May, 161–74. Bathurst, UK: Robert Brown and Associates.

Astuti, Rita. 2017. "Taking People Seriously." *hau: Journal of Ethnographic Theory* 7, no. 1: 105–22. doi:10.14318/hau7.1.012.

Atleo, E. Richard. 2012. *Principles of Tsawalk: An Indigenous Approach to Global Crisis.* Vancouver, BC: University of British Columbia Press.

Avtonomov, Alexei. 2013. "un cerd Formal Communication to the Permanent Mission of Indonesia Regarding the Situation of the Malind and Other Indigenous People of the Merauke District Affected by the mifee Project." August 30. https://www .forestpeoples.org/nl/node/4650.

Awas mifee. 2012. *An Agribusiness Attack in West Papua.* April 2012. https://awasmifee .potager.org/?page_id=25.

Badan Pusat Statistik. 2017. *Statistik Kelapa Sawit Indonesia 2017.* Jakarta. www.bps.go .id/publication/2018/11/13/b73ff9a5dc9f8d694d74635f/statistik-kelapa-sawit-indonesia -2017.html.

Bakhtin, Mikhail. 1984. *Rabelais and His World.* Translated by Helene Iswolsky. Bloomington: Indiana University Press.

Ballard, Chris. 2002. "The Signature of Terror: Violence, Memory, and Landscape at Freeport." In *Inscribed Landscapes: Marking and Making Place*, edited by Bruno David and Meredith Wilson, 13–26. Honolulu: University of Hawai'i Press.

Ballard, Chris. 2008. "'Oceanic Negroes': British Anthropology of Papuans, 1820–1869." In *Foreign Bodies: Oceania and the Science of Race 1750–1940*, edited by Chris Ballard and Bronwen Douglas, 157–201. Canberra: Australian National University Press.

Ballard, Chris. 2018a. "Afterword: Pacific Futurities." In *Pacific Futures: Past and Present*, edited by Warwick Anderson, Miranda Johnson, and Barbara Brookes, 280–93. Honolulu: University of Hawai'i Press.

Ballard, Chris. 2018b. "Oceanic Historicities." *Contemporary Pacific* 26, no. 1: 96–124. doi:10.1353/cp.2014.0009.

Bamford, Sandra C., ed. 2007. *Embodying Modernity and Post-Modernity: Ritual, Praxis, and Social Change in Melanesia*. Durham, NC: Carolina Academic Press.

Banivanua-Mar, Tracey. 2008. "'A Thousand Miles of Cannibal Lands': Imagining Away Genocide in the Re-Colonization of West Papua." *Journal of Genocide Research* 10, no. 4: 583–602. doi:10.1080/14623520802447743.

Banivanua-Mar, Tracey. 2012. "Settler-Colonial Landscapes and Narratives of Possession." *Arena Journal* 37–38: 176–98.

Banivanua-Mar, Tracey. 2016. *Decolonisation and the Pacific: Indigenous Globalisation and the Ends of Empire*. Cambridge: Cambridge University Press.

Banivanua-Mar, Tracey, and Penelope Edmonds. 2010. "Introduction: Making Space in Settler Colonies." In *Making Settler Colonial Space: Perspectives on Race, Place and Identity*, edited by Tracey Banivanua-Mar and Penelope Edmonds, 1–24. London: Palgrave Macmillan.

Barad, Karen. 2007. *Meeting the Universe Halfway: Quantum Physics and the Entanglement of Matter and Meaning*. Durham, NC: Duke University Press.

Baraitser, Lisa. 2017. *Enduring Time*. New York: Bloomsbury.

Barker, John. 1996. "Village Inventions: Historical Variations upon a Regional Theme in Uiaku, Papua New Guinea." *Oceania* 66, no. 3: 211–29. doi:10.1002/j.1834-4461.1996.tb02552.x.

Barker, John, ed. 2007. *The Anthropology of Morality in Melanesia and Beyond*. Burlington: Ashgate.

Barker, John. 2008. *Ancestral Lines: The Maisin of Papua New Guinea and the Fate of the Rainforest*. Toronto: University of Toronto Press.

Barth, Frederik. 1987. *Cosmologies in the Making: A Generative Approach to Cultural Variation in Inner New Guinea*. Cambridge: Cambridge University Press.

Barton, Huw J., and Tim Denham. 2011. "Prehistoric Vegeculture and Social Life in Island Southeast Asia and Melanesia." In *Why Cultivate? Anthropological and Archaeological Approaches to Foraging–Farming Transitions in Southeast Asia*, edited by Graeme Barker and Monica Janowski, 17–26. Cambridge: McDonald Institute for Archaeological Research.

Bashkow, Ira. 2006. *The Meaning of Whitemen: Race and Modernity in the Orokaiva Cultural World*. Chicago: University of Chicago Press.

Basik Basik, Mina. 2017. "Muting Yesterday and Today: Profile of a *Kampung*." In *Writing for Rights: Human Rights Documentation from the Land of Papua*. Series I. Translated by Indra V. A. Krishnamurti and Andrew de Sousa, 33–55. Jakarta: Institute for Policy Research and Advocacy (ELSAM).

Battaglia, Deborah. 1990. *On the Bones of the Serpent: Person, Memory, and Mortality in Sabarl Island Society*. Chicago: University of Chicago Press.

Battaglia, Deborah. 2017. "Aeroponic Gardens and Their Magic: Plants/Persons/Ethics in Suspension." *History and Anthropology* 28, no. 3: 263–92. doi:10.1080/02757206.2017.1289935.

Bauman, Zygmunt. 2012. *Liquid Modernity*. Cambridge, UK: Polity Press.

Baxter-Riley, Edward. 1925. *Among Papuan Headhunters*. Philadelphia: Lippencott.

Bayart, Jean-François. 1993. *The State in Africa: The Politics of the Belly*. Translated by Mary Harper. London: Longman.

Beer, Bettina, and Willem Church. 2019. "Roads to Inequality: Infrastructure and Historically Grown Regional Differences in the Markham Valley, Papua New Guinea." *Oceania* 89, no. 1: 2–19. doi:10.1002/ocea.5210.

Beilin, Katarzyna Olga, and Sainath Suryanarayanan. 2017. "The War between Amaranth and Soy: Interspecies Resistance to Transgenic Soy Agriculture in Argentina." *Environmental Humanities* 9, no. 2: 204–29. doi:10.1215/22011919-4215211.

Bell, Joshua A., Paige West, and Colin Filer, eds. 2015. *Tropical Forests of Oceania: Anthropological Perspectives*. Acton: Australian National University Press.

Benítez-Rojo, Antonio. 1996. *The Repeating Island: The Caribbean and the Postmodern Perspective*. Translated by James E. Maraniss. Durham, NC: Duke University Press.

Benjamin, Ruha. 2018. "Black Afterlives Matter: Cultivating Kinfulness as Reproductive Justice." In *Making Kin Not Population*, edited by Adele E. Clarke and Donna Haraway, 41–66. Chicago: Prickly Paradigm Press.

Benjamin, Walter. 2015. "Theses on the Philosophy of History." In *Walter Benjamin: Illuminations*, edited by Hannah Arendt, translated by Harry Zohn, 245–55. London: Bodley Head.

Bennett, Jane. 2001. *The Enchantment of Modern Life: Attachments, Crossings, and Ethics*. Princeton: Princeton University Press.

Berlant, Lauren. 2011. *Cruel Optimism*. Durham, NC: Duke University Press.

Berlant, Lauren. 2016. "The Commons: Infrastructures for Troubling Times." *Environment and Planning D: Society and Space* 34, no. 3: 393–419. doi:10.1177/0263775816645989.

Besky, Sarah. 2013. *The Darjeeling Distinction: Labor and Justice on Fair-Trade Tea Plantations in India*. Berkeley: University of California Press.

Bessire, Lucas. 2015. *Behold the Black Caiman: A Chronicle of Ayoreo Life*. Chicago: University of Chicago Press.

Biehl, João, and Amy Moran-Thomas. 2009. "Symptom: Subjectivities, Social Ills, Technologies." *Annual Review of Anthropology* 38: 267–88. doi:10.1146/annurev-anthro-091908-164420.

Biersack, Aletta. 2006a. "Red River, Green War: The Politics of Place along the Porgera River." In *Reimagining Political Ecology*, edited by Aletta Biersack and James B. Greenberg, 233–80. Durham, NC: Duke University Press.

Biersack, Aletta. 2006b. "Reimagining Political Ecology: Culture/Power/History/Nature." In *Reimagining Political Ecology*, edited by Aletta Biersack and James B. Greenberg, 1–40. Durham, NC: Duke University Press.

Boelaars, Jan H. M. C. 1981. *Head-Hunters about Themselves: An Ethnographic Report from Irian Jaya, Indonesia*. The Hague: Martinus Nijhoff.

Bonnemaison, Joël. 1994. *The Tree and the Canoe: History and Ethnogeography of Tanna*. Honolulu: University of Hawai'i Press.

Bonnemère, Pascale. 1994. "Le pandanus rouge dans tous ses états. L'univers social et symbolique d'un arbre fruitier chez les Ankave-Anga (Papouasie-Nouvelle-Guinée)." *Annales Fyssen* 9: 21–32.

Bonnemère, Pascale. 1996a. "Aliment de sociabilité, aliment d'échange: Le Pangium edule chez les Ankave-Anga (PNG)." In *Cuisines: Reflets des sociétés*, edited by Marie-Claire Bataille-Benguigui and Françoise Cousin, 423–34. Paris: Éditions Sépia–Musée de l'Homme.

Bonnemère, Pascale. 1996b. *Le pandanus rouge: Corps, différence des sexes et parenté chez les Ankave-Anga (Papouasie-Nouvelle-Guinée)*. Paris: Éditions de la Maison des sciences de l'homme.

Borras, Saturnino M., Jr., and Jennifer C. Franco. 2011. "Global Land Grabbing and Trajectories of Agrarian Change: A Preliminary Analysis." *Journal of Agrarian Change* 12, no. 1: 34–59. doi:10.1111/j.1471-0366.2011.00339.x.

Bowe, Michele, Neil Stronach, and Renee Bartolo. 2007. "Grassland and Savanna Ecosystems of the Trans-Fly, Southern Papua." In *The Ecology of Papua*, edited by Andrew J. Marshall, 1054–63. North Clarendon: Tuttle Publishing.

Bowker, Geoffrey C., and Susan L. Star. 1999. *Sorting Things Out: Classification and Its Consequences*. Cambridge, MA: MIT Press.

Braidotti, Rosi. 2006. *Transpositions: On Nomadic Ethics*. Cambridge, UK: Polity Press.

Brathwaite, John, Valerie Brathwaite, Michael Cookson, and Leah Dunn. 2010. *Anomie and Violence: Non-Truth and Reconciliation in Indonesian Peacebuilding*. Canberra: Australian National University Press.

Brison, Karen. 1992. *Just Talk: Gossip, Meetings, and Power in a Papua New Guinea Village*. Berkeley: University of California Press.

Brody, Hugh. 1981. *Maps and Dreams*. Vancouver, BC: Douglas and McIntyre.

Broglio, Ron. 2011. *Surface Encounters: Thinking with Animals and Art*. Minneapolis: University of Minnesota Press.

Bubandt, Nils. 2014. *The Empty Seashell: Witchcraft and Doubt on an Indonesian Island*. Ithaca: Cornell University Press.

Budiardjo, Carmel, and Soei Liong Liem. 1988. *West Papua: The Obliteration of a People*. Surrey: TAPOL.

Bulkeley, Kelly. 1994. *The Wilderness of Dreams: Exploring the Religious Meaning of Dreams in Modern Western Culture*. Albany: State University of New York Press.

Bulmer, Ralph. 1967. "Why Is the Cassowary Not a Bird? A Problem of Zoological Taxonomy among the Karam of the New Guinea Highlands." *Journal of the Royal Anthropological Institute* 2, no.1: 5–25. doi:10.2307/2798651.

Burridge, Kenelm. 1960. *Mambu: A Melanesian Millennium*. London: Methuen.

Burridge, Kenelm. 1969. *New Heaven, New Earth: A Study of Millenarian Activities*. New York: Schocken.

Butt, Leslie. 2005. "'Lipstick Girls' and 'Fallen Women': AIDS and Conspiratorial Thinking in Papua, Indonesia." *Cultural Anthropology* 20, no. 3: 412–42. doi:10.1525/can.2005.20.3.412.

Caldwell, Kia Lilly. 2007. *Negras in Brazil: Re-envisioning Black Women, Citizenship, and the Politics of Identity*. New Brunswick: Rutgers University Press.

Callon, Michel. 1984. "Some Elements of a Sociology of Translation: Domestication of the Scallops and the Fishermen of St Brieuc Bay." *Sociological Review* 32, no. 1: 196–233. doi:10.1111/j.1467-954X.1984.tb00113.x.

Carney, Judith A. 2020. "Subsistence in the Plantationocene: Dooryard Gardens, Agrobiodiversity, and the Subaltern Economies of Slavery." *Journal of Peasant Studies.* doi:10.1080/03066150.2020.1725488.

Carrithers, Michael, Matei Candea, Karen Sykes, Martin Holbraad, and Soumhya Venkatesan. 2010. "Ontology Is Just Another Word for Culture: Motion Tabled at the 2008 Meeting of the Group for Debates in Anthropological Theory, University of Manchester." *Critique of Anthropology* 30, no. 2: 152–200. doi:10.1177/0308275X09364070.

Carsten, Janet. 2000. "Introduction: Cultures of Relatedness." In *Cultures of Relatedness: New Approaches to the Study of Kinship,* edited by Janet Carsten, 1–36. Cambridge: Cambridge University Press.

Chao, Sophie. 2013. Statement at the Asia Regional Consultation with the UN Special Rapporteur on the Rights of Indigenous Peoples on the Situation of Indigenous Peoples in Asia. March 12, Kuala Lumpur. forestpeoples.org/sites/fpp/files/publication /2013/03/fppsawitwatch-statementmarch2013.pdf.

Chao, Sophie. 2018a. "In the Shadow of the Palm: Dispersed Ontologies among Marind, West Papua." *Cultural Anthropology* 33, no. 4: 621–49. doi:10.14506/ca33.4.08.

Chao, Sophie. 2018b. "Seed Care in the Palm Oil Sector." *Environmental Humanities* 10, no. 2: 421–46. doi:10.1215/22011919-7156816.

Chao, Sophie. 2019a. "Cultivating Consent: Opportunities and Challenges in the West Papuan Oil Palm Sector." *New Mandala,* August 26. https://www.newmandala.org /cultivating-consent/.

Chao, Sophie. 2019b. "Engaged Anthropology Grant: Sophie Chao." *Wenner-Gren Blog.* Wenner-Gren Foundation, August 6. http://blog.wennergren.org/2019/08/eag_schao/.

Chao, Sophie. 2019c. "Hunger and Culture in West Papua." *Inside Indonesia* 137 (July–September). www.insideindonesia.org/hunger-and-culture-in-west-papua.

Chao, Sophie. 2019d. "Race, Rights, and Resistance: The West Papua Protests in Context." *TRT World,* September 4. www.trtworld.com/opinion/race-rights-and-resistance -the-west-papuan-protests-in-context-29515.

Chao, Sophie. 2019e. "The Truth about 'Sustainable' Palm Oil." *SAPIENS,* June 13. www .sapiens.org/culture/palm-oil-sustainable/.

Chao, Sophie. 2019f. "West Papua and Black Lives Matter." *Inside Indonesia* 140 (April–June). https://www.insideindonesia.org/west-papua-and-black-lives-matter.

Chao, Sophie. 2019g. "Wrathful Ancestors, Corporate Sorcerers: Rituals Gone Rogue in Merauke, West Papua." *Oceania* 89, no. 3: 266–83. doi:10.1002/ocea.5229.

Chao, Sophie. 2020a. "'In the Plantations There Is Hunger and Loneliness': The Cultural Dimensions of Food Insecurity in Papua (Commentary)." *Mongabay,* July 14. https:// news.mongabay.com/2020/07/in-the-plantations-there-is-hunger-and-loneliness-the -cultural-dimensions-of-food-insecurity-in-papua-commentary/.

Chao, Sophie. 2020b. "A Tree of Many Lives: Vegetal Teleontologies in West Papua." *HAU: Journal of Ethnographic Theory* 10, no. 2: 514–29. doi:10.1086/709505.

Chao, Sophie. 2021a. "The Beetle or the Bug? Multispecies Politics in a West Papuan Oil Palm Plantation." *American Anthropologist* 123, no. 3: 476–89. doi:10.1111/aman.13592.

Chao, Sophie. 2021b. "Can There Be Justice Here? Indigenous Experiences from the West Papuan Oil Palm Frontier." *Borderlands* 20, no. 1: 11–48. doi:10.21307/borderlands-2021-002.

Chao, Sophie. 2021c. "Children of the Palms: Growing Plants and Growing People in a Papuan Plantationocene." *Journal of the Royal Anthropological Institute* 27, no. 2: 245–64. doi:10.1111/1467-9655.13489.

Chao, Sophie. 2021d. "Eating and Being Eaten: The Meanings of Hunger among Marind." *Medical Anthropology: Cross-Cultural Studies in Health and Illness* 40, no. 7: 682–97. doi:10.1080/01459740.2021.1916013.

Chao, Sophie. 2021e. "Gastrocolonialism: The Intersections of Race, Food, and Development in West Papua." *International Journal of Human Rights*, August 18. doi:10.1080/13642987.2021.1968378.

Chao, Sophie. 2021f. "'They Grow and Die Lonely and Sad.' Theorizing the Contemporary." *Fieldsights*, January 26. https://culanth.org/fieldsights/they-grow-and-die-lonely-and-sad.

Chao, Sophie. 2021g. "We Are (Not) Monkeys: Contested Cosmopolitical Symbols in West Papua." *American Ethnologist* 48, no. 3: 274–287. doi:10.1111/amet.13023.

Chao, Sophie. 2022a. "Plantation." *Environmental Humanities* 14, no. 1.

Chao, Sophie. 2022b. "(Un)Worlding the Plantationocene: Extraction, Extinction, Emergence." *eTropic: electronic journal of studies in the tropics* 21, no. 1. doi:10.25120/etropic.21.1.2022.3838.

Chao, Sophie. Forthcoming. "Multispecies Mourning: Grieving as Resistance on the West Papuan Oil Palm Frontier." *Cultural Studies*.

Chao, Sophie, Karin Bolender, and Eben S. Kirksey. 2022. *The Promise of Multispecies Justice.* Durham, NC: Duke University Press.

Chao, Sophie, and Dion Enari. 2021. "Decolonising Climate Change: A Call for Beyond-Human Imaginaries and Knowledge Generation." *eTropic: electronic journal of studies in the tropics* 20, no. 2: 32–54. doi:10.25120/etropic.20.2.2021.3796.

Chapin, Mac, Zachary Lamb, and Bill Threlkeld. 2005. "Mapping Indigenous Lands." *Annual Review of Anthropology* 34: 619–38. doi:10.1146/annurev.anthro.34.081804.120429.

Chauvel, Richard, and Ikrar N. Bhakti. 2004. *The Papua Conflict: Jakarta's Perceptions and Policies.* Washington, DC: East-West Center.

Chen, Mel Y. 2012. *Animacies: Biopolitics, Racial Mattering, and Queer Affect.* Durham, NC: Duke University Press.

Choy, Timothy, and Jerry Zee. 2015. "Condition Suspension." *Cultural Anthropology* 30, no. 2: 210–23. doi:14506/ca30.2.04.

Chrulew, Matthew. 2011. "Managing Love and Death at the Zoo: The Biopolitics of Endangered Species Preservation." *Australian Humanities Review* 50: 137–57. doi:10.22459/AHR.50.2011.08.

Clark, Jeffrey. 2001. *Steel to Stone: A Chronicle of Colonialism in the Southern Highlands of Papua New Guinea.* Edited by Chris Ballard and Michael Nihill. New York: Oxford University Press.

Clifford, James. 1983. "On Ethnographic Authority." *Representations* 2:118–46. doi:10.2307/2928386.

Clifford, James. 1997. *Routes: Travel and Translation in the Late Twentieth Century.* Cambridge, MA: Harvard University Press.

Coburn, Elaine, Makere Stewart-Harawira, Aileen Moreton-Robinson, and George Sefa Dei. 2013. "Unspeakable Things: Indigenous Research and Social Science. Révolutions, Contestations, Indignations." *Socio* 2: 121–34.

Colchester, Marcus, and Sophie Chao, eds. 2011. *Oil Palm Expansion in Southeast Asia: Trends and Implications for Local Communities and Indigenous Peoples.* Bangkok, Thailand: RECOFTC, Forest Peoples Programme, Sawit Watch, and Samdhana Institute.

Colchester, Marcus, and Sophie Chao, eds. 2013. *Conflict or Consent? The Palm Oil Sector at a Crossroads.* Moreton-in-Marsh, UK: Forest Peoples Programme, Sawit Watch, and TUK-Indonesia.

Collard, Rosemary-Claire. 2014. "Putting Animals Back Together, Taking Commodities Apart." *Annals of the Association of American Geographers* 104, no. 1: 151–165. doi:10.10 80/00045608.2013.847750.

Comaroff, Jean, and John L. Comaroff. 2002. "Alien-Nation: Zombies, Immigrants, and Millennial Capitalism." *South Atlantic Quarterly* 101, no. 4: 779–805. doi:10.1215/00382876-101-4-779.

Cooke, Fadzilah M. 2003. "Maps and Counter-Maps: Globalized Imaginings and Local Realities of Sarawak's Plantation Agriculture." *Journal of Southeast Asian Studies* 34, no. 2: 265–84. doi:10.1017/S0022463403000250.

Corbey, Raymond. 2010. *Headhunters from the Swamps: The Marind Anim of New Guinea as seen by the Missionaries of the Sacred Heart, 1905–1925.* Leiden: Brill.

Coulthard, Glen S. 2014. *Red Skin, White Masks: Rejecting the Colonial Politics of Recognition.* Minneapolis: University of Minnesota Press.

Cramb, Rob, and George N. Curry. 2012. "Oil Palm and Rural Livelihoods in the Asia–Pacific Region: An Overview." *Asia Pacific Viewpoint* 53, no. 3: 223–39. doi:10.1111/j.1467-8373.2012.01495.x.

Cramb, Rob, and John F. McCarthy, eds. 2016. *The Oil Palm Complex: Smallholders, Agribusiness and the State in Indonesia and Malaysia.* Singapore: National University of Singapore Press.

Crapanzano, Vincent. 1992. *Hermes' Dilemma and Hamlet's Desire: On the Epistemology of Interpretation.* Cambridge, MA: Harvard University Press.

Cronon, William. 1996. "The Trouble with Wilderness: Or, Getting Back to the Wrong Nature." *Environmental History* 1, no.1: 7–28. www.jstor.org/stable/3985059.

Crosby, Alfred W. 2003. *The Columbian Exchange: Biological and Cultural Consequences of 1492.* Westport: Greenwood Publishing Group.

Das, Veena. 2007. *Life and Words: Violence and the Descent into the Ordinary.* Berkeley: University of California Press.

Davis, Heather. 2015. "Toxic Progeny: The Plastisphere and Other Queer Futures." *philoSOPHIA* 5, no. 2: 232–50. muse.jhu.edu/article/608469.

Davis, Heather, and Zoe Todd. 2017. "On the Importance of a Date, or, Decolonizing the Anthropocene." *ACME: An International Journal for Critical Geographies* 16, no. 4: 761–80. https://acme-journal.org/index.php/acme/article/view/1539.

Davis, Janae, Alex A. Moulton, Levi Van Sant, and Bryan Williams. 2019. "Anthropocene, Capitalocene, . . . Plantationocene?: A Manifesto for Ecological Justice in an Age of Global Crises." *Geography Compass* 13, no. 5: e12438. doi:10.1111/gec3.12438.

Davis, Megan, and Marcia Langton, eds. 2016. *It's Our Country: Indigenous Arguments for Meaningful Constitutional Recognition and Reform.* Carlton: Melbourne University Press.

de la Cadena, Marisol. 2014. "Runa: Human but *Not Only*." *HAU: Journal of Ethnographic Theory* 4, no. 2: 253–59. doi:10.14318/hau4.2.013.

de la Cadena, Marisol. 2015. *Earth Beings: Ecologies of Practice across Andean Worlds.* Durham, NC: Duke University Press.

de la Cadena, Marisol. 2017. "Matters of Method; Or, Why Method Matters toward a Not Only Colonial Anthropology." *HAU: Journal of Ethnographic Theory* 7, no. 2: 1–10. doi:10.14318/hau7.2.002.

de la Cadena, Marisol. 2019. "An Invitation to Live Together: Making the 'Complex We.'" *Environmental Humanities* 11, no. 2: 477–84. doi:10.1215/22011919-7754589.

de Schutter, Olivier, and James Anaya. 2012. "UN Rights Experts Raise Alarm on Land Development Mega Projects." May 23. http://www.srfood.org/en/south-east-asia-agrofuel-un-rights-experts-raise-alarm-on-land-development-mega-projects.

Deleuze, Gilles, and Pierre-Félix Guattari. 1994. *What Is Philosophy?* Translated by Hugh Tomlinson and Graham Burchell. New York: Columbia University Press.

Deleuze, Gilles, and Pierre-Félix Guattari. 2005. *A Thousand Plateaus: Capitalism and Schizophrenia.* Translated by Brian Massumi. Minneapolis: University of Minnesota Press.

Deloria, Vine, Jr. 1999. "If You Think about It, You Will See That It Is True." In *Spirit and Reason: The Vine Deloria Reader,* 40–60. Boulder: Fulcrum Publishing.

Dening, Greg. 1991. "A Poetic for Histories: Transformations That Present the Past." In *Clio in Oceania: Towards a Historical Anthropology,* edited by Aletta Biersack, 347–80. Washington: Smithsonian Institution Press.

Derksen, Maaike. 2016. "Local Intermediaries? The Missionising and Governing of Colonial Subjects in South Dutch New Guinea, 1920–42." *Journal of Pacific History* 51, no. 2: 111–42. doi:10.1080/00223344.2016.1195075.

Derrida, Jacques. 1981. "Plato's Pharmacy." In *Literary Theory: An Anthology,* edited by Julie Rivkin and Michael Ryan, 429–50. Malden, MA: Blackwell.

Derrida, Jacques. 1982. *Margins of Philosophy.* Translated by Alan Bass. Chicago: Chicago University Press.

Derrida, Jacques. 1996. *The Gift of Death.* Translated by David Wills. Chicago: University of Chicago Press.

Derrida, Jacques. 2002. "The Animal That Therefore I Am (More to Follow)." Translated by David Willis. *Critical Inquiry* 28, no. 2: 369–418. www.jstor.org/stable/1344276.

Descola, Philippe. 1986. *In the Society of Nature: A Native Ecology in Amazonia.* Translated by Nora Scott. Cambridge: Cambridge University Press.

Descola, Philippe. 2013. *Beyond Nature and Culture*. Translated by Janet Lloyd. Chicago: University of Chicago Press.

Despret, Vinciane. 2004. "The Body We Care for: Figures of Anthropo-Zoo-Genesis." *Body & Society* 10, no. 2: 111–34. doi:10.1177/1357034X04042938.

Desroches, Henri. 1980. *The Sociology of Hope*. Translated by Carol Martin-Sperry. London: Routledge.

Dewi, Rosita. 2016. "Gaining Recognition through Participatory Mapping? The Role of Adat Land in the Implementation of the Merauke Integrated Food and Energy Estate in Papua, Indonesia." *Austrian Journal of South-East Asian Studies* 9, no. 1: 87–106. doi:10.14764/10.ASEAS-2016.1-6.

DiNovelli-Lang, Danielle. 2013. "The Return of the Animal: Posthumanism, Indigeneity, and Anthropology." *Environment and Society* 4, no. 1: 137–56. doi:10.3167/ares.2013.040109.

Dobrin, Lise M., and Alex Golub. 2020. "The Legacy of Bernard Narokobi and the Melanesian Way." *Journal of Pacific History* 55, no. 2: 149–64. doi:10.1080/00223344.2020.1759406.

Dove, Michael R. 2011. *The Banana Tree at the Gate: A History of Marginal Peoples and Global Markets in Borneo*. New Haven: Yale University Press.

Dove, Michael R. 2019. "Plants, Politics, and the Imagination over the Past 500 Years in the Indo-Malay Region." *Current Anthropology* 60, no. S20: S309–20. doi:10.1086/702877.

Down to Earth. 2011. "Twenty-Two Years of Top-Down Resource Exploitation in Papua." *DTE* 89–90, November 2011, Special Papua edition. www.downtoearth-indonesia.org/story/twenty-two-years-top-down-resource-exploitation-papua.

Dundon, Alison. 2005. "The Sense of Sago: Motherhood and Migration in Papua New Guinea and Australia." *Journal of Intercultural Studies* 26, no. 1–2: 21–37. doi:10.1080/07256860500073997.

Durie, Mason. 2005. *Ngā tai Matatū = Tides of Māori endurance*. Melbourne: Oxford University Press.

Durutalo, Simone. 1992. "Anthropology and Authoritarianism in the Pacific Islands." In *Confronting the Margaret Mead Legacy: Scholarship, Empire, and the South Pacific*, edited by Lenora Foerstel and Angela M. Gilliam, 205–32. Philadelphia: Temple University Press.

Dwyer, Peter D. 1990. *The Pigs That Ate the Garden: A Human Ecology from Papua New Guinea*. Ann Arbor: University of Michigan Press.

Dwyer, Peter D., and Monica Minnegal. 2007. "Social Change and Agency among Kubo of Papua New Guinea." *Journal of the Royal Anthropological Institute* 13, no. 3: 545–62. doi:10.1111/j.1467-9655.2007.00442.x.

Dwyer, Peter D., and Monica Minnegal. 2014. "Where All the Rivers Flow West: Maps, Abstraction and Change in the Papua New Guinea Lowlands." *Australian Journal of Anthropology* 25, no. 1: 37–53. doi:10.1111/taja.12071.

Dyer, Michelle. 2017. "Eating Money: Narratives of Equality on Customary Land in the Context of Natural Resource Extraction in the Solomon Islands." *Australian Journal of Anthropology* 28, no. 1: 88–103. doi:10.1111/taja.12213.

Edelman, Lee. 2004. *No Future: Queer Theory and the Death Drive.* Durham, NC: Duke University Press.

Edelman, Marc, Carlos Oya, and Saturnino M. Borras Jr., eds. 2015. *Global Land Grabs: History, Theory and Method.* Abingdon, UK: Routledge.

Eggertsson, Sveinn. 2018. "Making Skins: Initiation, Sorcery, and Eastern Min Notions of Knowledge." *Oceania* 88, no. 2: 152–67. doi:10.1002/ocea.5191.

Ellen, Roy. 2006. "Local Knowledge and Management of Sago Palm (*Metroxylon sagu* Rottboell) Diversity in South Central Seram, Maluku, Eastern Indonesia." *Journal of Ethnobiology* 26, no. 2: 258–98. doi:10.2993/0278-0771(2006)26[258:LKAMOS]2.0.CO;2.

Ellen, Roy. 2011. "Sago as a Buffer against Subsistence Stress and as a Currency of Inter-Island Trade Networks in Eastern Indonesia." In *Why Cultivate? Anthropological and Archaeological Approaches to Foraging–Farming Transitions in Southeast Asia,* edited by Graeme Barker and Monica Janowski, 47–60. Cambridge, UK: McDonald Institute for Archaeological Research.

Elmslie, Jim. 2017. "The Great Divide: West Papuan Demographics Revisited; Settlers Dominate Coastal Regions but the Highlands Still Overwhelmingly Papuan." *Asia-Pacific Journal* 15, no. 2: 1–11. https://apjjf.org/-Jim-Elmslie/5005/article.pdf.

Elmslie, Jim, and Camellia Webb-Gannon. 2013. "A Slow-Motion Genocide: Indonesian Rule in West Papua." *Griffith Journal of Law and Human Dignity* 1, no. 2: 142–66. https://ro.uow.edu.au/cgi/viewcontent.cgi?article=5030&context=sspapers.

Erazo, Juliet, and Christopher Jarrett. 2017. "Managing Alterity from Within: The Ontological Turn in Anthropology and Indigenous Efforts to Shape Shamanism." *Journal of the Royal Anthropological Institute* 24, no. 1: 145–63. doi:10.1111/1467-9655 .12756.

Ernst, Thomas M. 1979. "Myth, Ritual, and Population among the Marind-Anim." *Social Analysis* 1: 34–53. www.jstor.org/stable/23159675.

Errington, Frederik, Tatsuro Fujikura, and Deborah Gewertz. 2013. *The Noodle Narratives: The Global Rise of an Industrial Food into the Twenty-First Century.* Berkeley: University of California Press.

Errington, Frederik, and Deborah Gewertz. 2004. *Yali's Question: Sugar, Culture, and History.* Chicago: University of Chicago Press.

Errington, Frederik, and Deborah Gewertz. 2008. "Pacific Island Gastrologies: Following the Flaps." *Journal of the Royal Anthropological Institute* 14, no. 3: 590–608. doi:10.1111/j.1467-9655.2008.00519.x.

Eves, Richard. 1998. *The Magical Body: Power, Fame and Meaning in a Melanesian Society.* Amsterdam: Harwood Academic Publishers.

Fabian, Johannes. 1983. *Time and the Other: How Anthropology Makes Its Object.* New York: Columbia University Press.

Fajans, Jane. 1983. "Shame, Social Action, and the Person among the Baining." *Ethos* 11, no. 33: 166–80. doi:10.1525/eth.1983.11.3.02a00050.

Fajans, Jane. 1985. "The Person in Social Context: The Social Character of Baining 'Psychology.'" In *Person, Self, And Experience: Exploring Pacific Ethnopsychologies,* edited by Geoffrey M. White and John Kirkpatrick, 367–400. Berkeley: University of California Press.

Fajans, Jane. 1988. "The Transformative Value of Food: A Review Essay." *Food and Food-ways* 3, no. 1–2: 143–66. doi:10.1080/07409710.1988.9961941.

Fajans, Jane. 1997. *They Make Themselves: Work and Play among the Baining of Papua New Guinea.* Chicago: University of Chicago Press.

Fanon, Franz. 2008. *Black Skin, White Masks.* Translated by Charles L. Markmann. London: Pluto Press.

Fearnside, Phillip M. 1997. "Transmigration in Indonesia: Lessons from Its Environmental and Social Impacts." *Environmental Management* 21, no. 4: 553–70. doi:10.1007 /s002679900049.

Feld, Steven. 1996. "Waterfalls of Song: An Acoustemology of Place Resounding in Bosavi, Papua New Guinea." In *Senses of Place*, edited by Steven Feld and Keith H. Basso, 91–135. Santa Fe: School of American Research Press.

Feld, Steven. 2003. "A Rainforest Acoustemology." In *The Auditory Culture Reader*, edited by Michael Bull and Les Back, 223–41. New York: Berg.

Feld, Steven, and Keith H. Basso, 1996. "Introduction." In *Senses of Place*, edited by Steven Feld and Keith H. Basso, 3–11. Santa Fe: School of American Research Press.

Ferguson, James, and Akhil Gupta. 2002. "Spatializing States: Toward an Ethnography of Neoliberal Governmentality." *American Ethnologist* 29, no. 4: 981–1002. doi:10.1525 /ae.2002.29.4.981.

Filer, Colin. 2007. "Local Custom and the Art of Land Group Boundary Maintenance in Papua New Guinea." In *Customary Land Tenure and Registration in Australia and Papua New Guinea: Anthropological Perspectives*, edited by James F. Weiner and Katie Glaskin, 135–73. Canberra: Australian National University Press.

Filer, Colin. 2011. "New Land Grab in Papua New Guinea." *Pacific Studies* 34, no. 2–3: 269–94.

Filer, Colin. 2013. "Asian Investment in the Rural Industries of Papua New Guinea: What's New and What's Not?" *Pacific Affairs* 86, no. 2: 305–25. doi:10.5509/2013862305.

Filer, Colin. 2017. "The Formation of a Land Grab Policy Network in Papua New Guinea." In *Kastom, Property and Ideology: Land Transformations in Melanesia*, edited by Siobhan McDonnell, Matthew G. Allen, and Colin Filer, 169–203. Canberra: Australian National University Press.

Fischler, Claude. 1979. "Gastro-nomie et gastro-anomie: Sagesse du corps et crise bioculturelle de l'alimentation moderne." *Communications* 31: 189–210. www.persee.fr /doc/comm_0588-8018_1979_num_31_1_1477.

Fitting, Elizabeth M. 2011. *The Struggle for Maize: Campesinos, Workers, and Transgenic Corn in the Mexican Countryside.* Durham, NC: Duke University Press.

Forest Peoples Programme. 2013. "Request for Further Consideration of the Situation of the Indigenous Peoples of Merauke, Papua Province, Indonesia, under the Committee on the Elimination of Racial Discrimination's Urgent Action and Early Warning Procedures." July 25. https://www.forestpeoples.org/sites/fpp/files/publication/2013/08 /cerduamifeejuly2013english.pdf.

Forest Peoples Programme, PUSAKA, and Sawit Watch. 2013. "*A Sweetness Like unto Death": Voices of the Indigenous Malind of Merauke, Papua.* Moreton-in-Marsh, UK: Forest Peoples Programme, Sawit Watch, PUSAKA, and Rights and Resources Institute.

Forest Peoples Programme, Sawit Watch, and Aliansi Masyarakat Adat Nusantara. 2012. "Request for Further Consideration of the Situation of the Indigenous Peoples of Merauke, Papua Province, Indonesia, and Indigenous Peoples in Indonesia in General, under the Committee on the Elimination of Racial Discrimination's Urgent Action and Early Warning Procedures." Forest People's Programme on Behalf of Additional Submitting Organizations. February 6. https://www.forestpeoples.org/sites/default/files/publication/2012/02/2012-cerd-80th-session-ua-update-final.pdf.

Foster, Laura A. 2017. *Reinventing Hoodia: Peoples, Plants, and Patents in South Africa.* Seattle: University of Washington Press.

Foster, Robert J. 1999. "Melanesianist Anthropology in the Era of Globalization." *Contemporary Pacific* 11, no. 1: 140–59. www.jstor.org/stable/23717415.

Foster, Robert J. 2008. *Coca-Globalization: Following Soft Drinks from New York to New Guinea.* New York: Palgrave Macmillan.

Foucault, Michel. 1978. *The History of Sexuality*, Vol. 1, *An Introduction.* New York: Pantheon Books.

Foucault, Michel. 1997. "The Masked Philosopher." In *Ethics, Subjectivity, and Truth*, edited by Paul Rabinow, translated by Robert Hurley, 1:321–28. New York: New Press.

Fox, James J. 1977. *Harvest of the Palm: Ecological Change in Eastern Indonesia.* Cambridge, MA: Harvard University Press.

Fox, Jefferson. 2002. "Siam Mapped and Mapping in Cambodia: Boundaries, Sovereignty, and Indigenous Conceptions of Space." *Society and Natural Resources* 15, no. 1: 65–78. doi:10.1080/089419202317174020.

Fox, Jefferson, Krisnawati Suryanata, Peter Hershock, and Albertus H. Pramono. 2008. "Mapping Boundaries, Shifting Power: The Socio-Ethical Dimensions of Participatory Mapping." In *Contentious Geographies: Environmental Knowledge, Meaning, Scale*, edited by Michael K. Goodman, Maxwell T. Boykof, and Kyle T. Evered, 203–18. New York: Routledge.

Frankel, Stephen. 1986. *The Huli Response to Illness.* Cambridge: Cambridge University Press.

Franky, Yafet L., and Selwyn Morgan, eds. 2015. *West Papua Oil Palm Atlas: The Companies Behind the Plantation Explosion.* Jakarta Selatan: PUSAKA. https://awasmifee.potager.org/uploads/2015/04/atlas-sawit-en.pdf.

Freeman, Elizabeth. 2010. *Time Binds: Queer Temporalities, Queer Histories.* Durham, NC: Duke University Press.

Fukuyama, Francis. 1989. "The End of History?" *National Interest* 16: 3–18. www.jstor.org/stable/24027184.

Gabriel, Jennifer, Paul N. Nelson, Colin Filer, and Michael Wood. 2017. "Oil Palm Development and Large-Scale Land Acquisitions in Papua New Guinea." In *Kastom, Property and Ideology: Land Transformations in Melanesia*, edited by Siobhan McDonnell, Matthew G. Allen, and Colin Filer, 205–50. Canberra: Australian National University Press.

Galvin, Shaila S. 2018. "Interspecies Relations and Agrarian Worlds." *Annual Review of Anthropology* 47: 233–49. doi:10.1146/annurev-anthro-102317-050232.

Garnaut, Ross, and Chris Manning. 1974. *Irian Jaya: The Transformation of a Melanesian Economy.* Canberra: Australian National University Press.

Gegeo, David W. 1998. "Indigenous Knowledge and Empowerment: Rural Development Examined from Within." *Contemporary Pacific* 10, no. 2: 289–315. www.jstor.org/stable /23706891.

Gegeo, David W. 2001. "Cultural Rupture and Indigeneity: The Challenge of (Re)Visioning 'Place' in the Pacific." *Contemporary Pacific* 13, no. 2: 491–507. doi:10.1353/cp.2001.0052.

Gegeo, David W., and Karen A. Watson-Gegeo. 2001. "'How We Know': Kwara'ae Rural Villagers Doing Indigenous Epistemology." *Contemporary Pacific* 13, no. 1: 55–88. doi:10.1353/cp.2001.0004.

Gegeo, David W., and Karen A. Watson-Gegeo. 2002. "Whose Knowledge? Epistemological Collisions in Solomon Islands Community Development." *Contemporary Pacific* 14, no. 2: 377–409. doi:10.1353/cp.2002.0046.

Gell, Alfred. 1975. *Metamorphosis of the Cassowaries: Umeda Society, Language and Ritual.* London: Athlone Press.

Gell, Alfred. 2006. "Vogel's Net: Traps as Artworks and Artworks as Traps." In *The Art of Anthropology: Essays and Diagrams*, edited by Eric Hirsch, 187–214. London: Berg.

George, Marianne. 1995. "Dreams, Reality, and the Desire and Intent of Dreamers as Experienced by a Fieldworker." *Anthropology of Consciousness* 6, no. 3: 17–33. doi:10.1525/ac.1995.6.3.17.

Gewertz, Deborah, and Frederik Errington. 1996. "On PepsiCo and Piety in a Papua New Guinea 'Modernity.'" *American Ethnologist* 23, no. 3: 476–93. doi:10.1525/ae .1996.23.3.02a00020.

Giay, Benny, and Chris Ballard. 2003. "Becoming Papuans: Notes towards a History of Racism in Tanah Papua." Paper presented at the annual meeting of the American Anthropological Association, Chicago, November 19–23, 2003.

Giay, Ligia J. 2016. "Papuans within Slavery, Slavery within Papuans: Traces of the Slavery Past in New Guinea, 1600s–1950s." PhD diss., Leiden University.

Gieryn, Thomas F. 2018. *Truth-Spots: How Places Make People Believe.* Chicago: University of Chicago Press.

Gietzelt, Dale. 1988. "The Indonesianization of West Papua." *Oceania* 59, no. 3: 201–21. doi:10.1002/j.1834-4461.1989.tb02322.x.

Gill, Simryn, and Michael Taussig. 2017. *Becoming Palm.* Singapore: NTU Centre for Contemporary Art Singapore and Sternberg Press.

Gilroy, Paul. 2017. "'Where Every Breeze Speaks of Courage and Liberty': Offshore Humanism and Marine Xenology, or, Racism and the Problem of Critique at Sea Level." *Antipode* 50, no. 1: 3–22. doi:10.1111/anti.12333.

Ginn, Franklin, Uli Beisel, and Maan Barua. 2014. "Flourishing with Awkward Creatures: Togetherness, Vulnerability, Killing." *Environmental Humanities* 4, no. 1: 113–23. doi:10.1215/22011919-3614953.

Ginting, Longgena, and Oliver Pye. 2013. "Resisting Agribusiness Development: The Merauke Integrated Food and Energy Estate in West Papua, Indonesia." *Austrian Journal of South-East Asian Studies* 6, no. 10: 160–82. doi:10.14764/10.ASEAS-6.1-9.

Giraud, Eva H. 2019. *What Comes after Entanglement? Activism, Anthropocentrism, and an Ethics of Exclusion.* Durham, NC: Duke University Press.

Glazebrook, Diana. 2008. *Permissive Residents: West Papuan Refugees Living in Papua New Guinea*. Canberra: Australian National University Press.

Global Forest Coalition. 2017. "The Big Four Drivers of Deforestation: Beef, Soy, Wood, and Palm Oil." *Forest Cover*, no. 55, April 2017. globalforestcoalition.org/forest-cover -55-big-four-drivers-of-deforestation/.

Glymph, Thavolia. 2012. *Out of the House of Bondage: The Transformation of the Plantation Household*. New York: Cambridge University Press.

Goldman, Laurence C., and Chris Ballard. 1998. *Fluid Ontologies: Myth, Ritual and Philosophy in the Highlands of Papua New Guinea*. Westport: Bergin and Garvey.

Goldstein, Ruth. 2019. "Ethnobotanies of Refusal: Methodologies in Respecting Plant(ed)-Human Resistance." *Anthropology Today* 3, no. 2: 18–22. doi:10.1111/1467-8322.12495.

Gómez-Barris, Macarena. 2017. *The Extractive Zone: Social Ecologies and Decolonial Perspectives*. Durham, NC: Duke University Press.

Goodale, Jane C. 1995. *To Sing with Pigs Is Human: The Concept of Person in Papua New Guinea*. Seattle: University of Washington Press.

Gordillo, Gastón R. 2002. "'The Breath of the Devils: Memories and Places of an Experience of Terror." *American Ethnologist* 29, no. 1: 33–57. doi:10.1525/ae.2002.29.1.33.

Gordillo, Gastón R. 2014. *Rubble: The Afterlife of Destruction*. Durham, NC: Duke University Press.

Gose, Peter. 2018. "The Semi-Social Mountain: Metapersonhood and Political Ontology in the Andes." *HAU: Journal of Ethnographic Theory* 8, no. 3: 488–505. doi:10.1086/701067.

Govindrajan, Radhika. 2018. *Animal Intimacies: Interspecies Relatedness in India's Central Himalayas*. Chicago: University of Chicago Press.

Greenough, Paul, and Anna L. Tsing. 2003. "Introduction." In *Nature in the Global South: Environmental Projects in South and Southeast Asia*, edited by Paul Greenough and Anna L. Tsing, 1–23. Durham, NC: Duke University Press.

Groark, Kevin. 2017. "Specters of Social Antagonism: The Cultural Psychodynamics of Dream Aggression among the Tzotzil Maya of San Juan Chamula (Chiapas, Mexico)." *Ethos* 45, no. 3: 1–28. doi:10.1111/etho.12174.

Grosz, Elizabeth A. 2008. *Chaos, Territory, Art: Deleuze and the Framing of the Earth*. New York: Columbia University Press.

Groube, Les. 1989. "The Taming of the Rain Forests: A Model for Late Pleistocene Forest Exploitation in New Guinea." In *Foraging and Farming: The Evolution of Plant Exploitation*, edited by David R. Harris and Gordon C. Hillman, 292–304. London: Routledge.

Guinness, Patrick. 2018. "The Unbounded Space and Moral Transgression: Capitalist Expansion in West New Britain." *Anthropological Forum* 28, no. 1: 32–44. doi:10.1080 /00664677.2018.1410779.

Haddon, Alfred C. 1891. *The Tugeri Head-Hunters of New Guinea*. Leiden: Brill.

Hadiz, Vedi R. 2010. *Localising Power in Post-Authoritarian Indonesia: A Southeast Asia Perspective*. Stanford: Stanford University Press.

Hage, Ghassan. 2016. "Questions Concerning a Future-Politics." *History and Anthropology* 27, no. 4: 465–67. doi:10.1080/02757206.2016.1206896.

Hagwood, Austin G. 2017. "Botanist, Hit-Man, Assassin, Leech: The Political Ecology of Plantation Sorcery in Papua New Guinea." MPhil diss., University of Cambridge.

Hale, Charles R. 2006. "Activist Research v. Cultural Critique: Indigenous Land Rights and the Contradictions of Politically Engaged Anthropology." *Cultural Anthropology* 21, no. 1: 96–120. doi:10.1525/can.2006.21.1.96.

Hall, Derek. 2011. "Land Grabs, Land Control, and Southeast Asian Crop Booms." *Journal of Peasant Studies* 38, no. 4: 837–57. doi:10.1080/03066150.2011.607706.

Hall, Matthew. 2011. *Plants as Persons: A Philosophical Botany*. Albany: State University of New York Press.

Halvaksz, Jamon A. 2013. "The Taste of Public Places: *Terroir* in Papua New Guinea's Emerging Nation." *Anthropological Forum* 23, no. 2: 142–57. doi:10.1080/00664677.2012.753868.

Halvaksz, Jamon A. 2015. "Forests of Gold: From Mining to Logging (and Back Again)." In *Tropical Forests of Oceania: Anthropological Perspectives*, edited by Joshua A. Bell, Paige West, and Colin Filer, 75–94. Acton: Australian National University Press.

Haraway, Donna J. 1988. "Situated Knowledges: The Science Question in Feminism and the Privilege of Partial Perspective." *Feminist Studies* 14, no. 3: 575–99. doi:10.2307/3178066.

Haraway, Donna J. 1991. *Simians, Cyborgs, and Women: The Reinvention of Nature*. London: Free Association.

Haraway, Donna J. 1992. "The Promises of Monsters: A Regenerative Politics for Inappropriate/d Others." In *Cultural Studies*, edited by Lawrence Grossberg, Carla Nelson, and Paula A. Treichler, 295–337. New York: Routledge.

Haraway, Donna J. 1997. *Modest_Witness@Second_Millennium. FemaleMan©_Meets_OncoMouseTM: Feminism and Technoscience*. New York: Routledge.

Haraway, Donna J. 2003. *The Companion Species Manifesto: Dogs, People, and Significant Otherness*. Chicago: Prickly Paradigm Press.

Haraway, Donna J. 2008. *When Species Meet*. Minneapolis: University of Minnesota Press.

Haraway, Donna J. 2011. "Speculative Fabulations for Technoculture's Generations: Taking Care of Unexpected Country." *Australian Humanities Review* 50. http://australianhumanitiesreview.org/2011/05/01/speculative-fabulations-for-technocultures-generations-taking-care-of-unexpected-country/.

Haraway, Donna J. 2013. "A Manifesto for Cyborgs: Science, Technology, and Socialist Feminism in the 1980s." In *Coming to Terms: Feminist, Theory, Politics*, edited by Elizabeth Weed, 173–205. Abingdon: Routledge.

Haraway, Donna J. 2015. "Anthropocene, Capitalocene, Plantationocene, Chthulucene: Making Kin." *Environmental Humanities* 6, no. 1: 159–65. doi:10.1215/22011919-3615934.

Haraway, Donna J. 2016. *Staying with the Trouble: Making Kin in the Chthulucene*. Durham, NC: Duke University Press.

Haritaworn, Jinthana. 2015. "Decolonizing the Non/Human." GLQ: A Journal of Lesbian and Gay Studies 21, no. 2–3: 210–13. muse.jhu.edu/article/582030.

Harley, John B. 1988. "Maps, Knowledge and Power." In *The Iconography of Landscape: Essays on the Symbolic Representation, Design and Use of Past Environments*, edited by Denis Cosgrove and Stephen Daniels, 277–312. Cambridge: Cambridge University Press.

Hartigan, John. 2017. *Care of the Species: Races of Corn and the Science of Plant Biodiversity*. Minneapolis: University of Minnesota Press.

Hartman, Saidiya. 1997. *Scenes of Subjection: Terror, Slavery, and Self-Making in Nineteenth Century America*. Oxford: Oxford University Press.

Hartog, François. 2015. *Regimes of Historicity: Presentism and Experiences of Time*. Translated by Saskia Brown. New York: Columbia University Press.

Harvey, David. 2000. *Spaces of Hope*. Berkeley: University of California Press.

Hatley, James. 2000. *Suffering Witness: The Quandary of Responsibility after the Irreparable*. Albany: State University of New York Press.

Hau'ofa, Epeli. 1981. *Mekeo: Inequality and Ambivalence in a Village Society*. Canberra: Australian National University Press.

Hau'ofa, Epeli. 2008. *We Are the Ocean: Selected Works*. Honolulu: University of Hawai'i Press.

Hayano, David M. 1989. "Like Eating Money: Card Gambling in a Papua New Guinea Highlands Village." *Journal of Gambling Behavior* 5: 231–45. doi:10.1007/BF01024389.

Hayano, David M. 1990. *Road through the Rain Forest: Living Anthropology in Highland Papua New Guinea*. Prospect Heights: Waveland Press.

Hayward, Eva. 2010. "Fingeryeyes: Impressions of Cup Corals." *Cultural Anthropology* 25, no. 4: 577–99. doi:10.1111/j.1548-1360.2010.01070.x.

Head, Lesley, Jennifer Atchison, and Alison Gates. 2012. *Ingrained: A Human Bio-Geography of Wheat*. London: Ashgate.

Heckler, Serena L. 2004. "Tedium and Creativity: The Valorization of Manioc Cultivation and Piaroa Women." *Journal of the Royal Anthropological Institute* 10, no. 2: 241–59. www.jstor.org/stable/3804150.

Henderson, Janice, and Daphne J. Osborne. 2000. "The Oil Palm in All Our Lives: How This Came About." *Endeavour* 24, no. 2: 63–68. doi:10.1016/s0160-9327(00)01293-x.

Herdt, Gilbert, and Michele Stephen, eds. 1989. *The Religious Imagination in New Guinea*. New Brunswick: Rutgers University Press.

Hernawan, Budi J. 2015. "Torture as a Mode of Governance: Reflections on the Phenomenon of Torture in Papua, Indonesia." In *From "Stone Age" to "Real-Time": Exploring Papuan Temporalities, Mobilities and Religiosities*, edited by Martin Slama and Jenny Munro, 195–220. Canberra: Australian National University Press.

Hernawan, Budi J., and Anthony van den Broek. 1999. "Dialog nasional Papua, sebuah kisah 'memoria passionis.'" *Tifa Irian*, March 8. P. 8.

Hess, Sabine. 2006. "Strathern's Melanesian 'Dividual' and the Christian 'Individual': A Perspective from Vanua Lava, Vanuatu." *Oceania* 76, no. 3: 285–96. 10.1002/j.1834-4461.2006.tb03058.x.

Hetherington, Kregg. 2013. "Beans before the Law: Knowledge Practices, Responsibility, and the Paraguayan Soy Boom." *Cultural Anthropology* 28, no.1: 65–85. doi:10.1111/j.1548-1360.2012.01173.x.

High, Holly. 2014. *Fields of Desire: Poverty and Policy in Laos*. Singapore: National University of Singapore Press.

Hitchcock, Garrick. 2009. "Manuscript XX: William Dammköhler's Third Encounter with the Tugeri (Marind-Anim)." *Journal of Pacific History* 44, no. 1: 89–97. doi:10.1080/00223340902900894.

Hobart, Hi'ilei J. K., and Tamara Kneese. 2020. "Radical Care: Survival Strategies for Uncertain Times." *Social Text* 38, no. 1(142): 1–16. doi:10.1215/01642472-7971067.

Hodges, Matt. 2008. "Rethinking Time's Arrow: Bergson, Deleuze and the Anthropology of Time." *Anthropological Theory* 8, no. 4: 399–429. doi:10.1177/1463499608096646.

Hollan, Douglas. 2004. "The Anthropology of Dreaming: Selfscape Dreams." *Dreaming* 14, no. 2–3: 170–82. doi:10.1037/1053-0797.14.2-3.170.

Howes, David. 2005. "Skinscapes: Embodiment, Culture, and Environment." In *The Book of Touch*, edited by Constance Classen, 27–39. New York: Berg.

Hugh-Jones, Christine. 1979. *From the Milk River: Spatial and Temporal Processes in Northwest Amazonia.* Cambridge: Cambridge University Press.

Hugh-Jones, Stephen. 1979. *The Palm and the Pleiades: Initiation and Cosmology in Northwest Amazonia.* Cambridge: Cambridge University Press.

Hugh-Jones, Stephen. 2019. "Rhetorical Antinomies and Radical Othering: Recent Reflections on Responses to an Old Paper Concerning Human-Animal Relations in Amazonia." *HAU: Journal of Ethnographic Theory* 9, no.1: 162–71. doi:10.1086/703873.

Hunt, Sarah. 2013. "Ontologies of Indigeneity: The Politics of Embodying a Concept." *cultural geographies* 21, no. 1: 27–32. doi:10.1177/1474474013500226.

Hustak, Carla, and Natasha Myers. 2012. "Involutionary Momentum: Affective Ecologies and the Sciences of Plant/Insect Encounters." *differences* 23, no. 3: 74–118. doi:10.1215/10407391-1892907.

Hviding, Edvard. 2003. "Contested rainforests, NGOs, and projects of desire in Solomon Islands." *International Social Science Journal* 55, no. 178: 539–54. doi:10.1111/j.0020-8701.2003.05504003.x.

Hviding, Edvard. 2015a. "Big Money in the Rural: Wealth and Dispossession in Western Solomons Political Economy." *Journal of Pacific History* 50, no. 4: 473–85. doi:10.1080/00223344.2015.1101818.

Hviding, Edvard. 2015b. "Non-Pristine Forests: A Long-Term History of Land Transformation in the Western Solomons." In *Tropical Forests of Oceania: Anthropological Perspectives*, edited by Joshua A. Bell, Paige West, and Colin Filer, 51–74. Canberra: Australian National University Press.

Hviding, Edvard, and Tim Bayliss-Smith. 2018. *Islands of Rainforest: Agroforestry, Logging and Eco-Tourism in Solomon Islands.* London: Routledge.

Hviding, Edvard, and Geoffrey White, eds. 2015. *Pacific Alternatives: Cultural Politics in Contemporary Oceania.* London: Sean Kingston.

Indonesia Investments. 2017. "Palm Oil." June 26, 2017. https://www.indonesia-investments.com/business/commodities/palm-oil/item166.

Ingold, Tim. 2007. *Lines: A Brief History.* Abingdon, UK: Routledge.

Ingold, Tim. 2011. *Being Alive: Essays on Movement, Knowledge and Description.* Abingdon, UK: Routledge.

Ingold, Tim. 2017. *Correspondences*. Aberdeen: University of Aberdeen Press.

Isenring, Richard. 2017. *Adverse Health Effects Caused by Paraquat: A Bibliography of Documented Evidence*. Zurich: Public Eye, Pesticide Action Network, and PAN Asia Pacific. February.

Ito, Takeshi, Noer F. Rachman, and Laksmi Savitri. 2014. "Power to Make Land Dispossession Acceptable: A Policy Discourse Analysis of the Merauke Integrated Food and Energy Estate (MIFEE), Papua, Indonesia." *Journal of Peasant Studies* 41, no. 1: 29–50. doi:10.1080/03066150.2013.873029.

Ives, Sarah. 2014. "Farming the South African 'Bush': Ecologies of Belonging and Exclusion in Rooibos Tea." *American Ethnologist* 41, no. 4: 698–713. doi:10.1111/amet.12106.

Ives, Sarah. 2017. *Steeped in Heritage: The Racial Politics of South African Rooibos Tea*. Durham, NC: Duke University Press.

Jacka, Jerry K. 2007. "'Our Skins Are Weak': Ipili Modernity and the Demise of Discipline." In *Embodying Modernity and Postmodernity: Ritual, Praxis, and Social Change in Melanesia*, edited by Sandra C. Bamford, 39–68. Durham, NC: Carolina Academic Press.

Jacka, Jerry K. 2015. *Alchemy in the Rainforest: Politics, Ecology, and Resilience in a New Guinea Mining Area*. Durham, NC: Duke University Press.

Jacka, Jerry K. 2018. "Riverine Disposal of Mining Wastes in Porgera: Capitalist Resource Development and Metabolic Rifts in Papua New Guinea." In *Island Rivers: Fresh Water and Place in Oceania*, edited by John R. Wagner and Jerry K. Jacka, 109–36. Canberra: Australian National University Press.

Jacka, Jerry K. 2019. "Resource Conflicts and the Anthropology of the Dark and the Good in Highlands Papua New Guinea." *Australian Journal of Anthropology* 30, no. 1: 35–52. doi:10.1111/taja.12302.

Jackson, Shona. 2012. *Creole Indigeneity: Between Myth and Nation in the Caribbean*. Minneapolis: University of Minnesota Press.

Jackson, Zakiyyah I. 2013. "Animal: New Directions in the Theorization of Race and Posthumanism." *Feminist Studies* 39, no. 3: 669–85. www.jstor.org/stable/23719431.

Jackson, Zakiyyah I. 2015. "Outer Worlds: The Persistence of Race in Movement 'Beyond the Human.'" *GLQ: A Journal of Lesbian and Gay Studies* 21, no. 2–3: 215–18. www .muse.jhu.edu/article/582032.

Janowski, Monica. 2003. *The Forest, Source of Life: The Kelabit of Sarawak*. London: British Museum.

Janur, Katharina. 2015. "Letakkan Kapsul Waktu di Merauke, Jokowi Tuliskan 7 Harapan." LIPUTAN6, December 30, 2015. www.liputan6.com/news/read/2400786 /letakkan-kapsul-waktu-di-merauke-jokowi-tuliskan-7-harapan.

Jegathesan, Mythri. 2019. *Tea and Solidarity: Tamil Women and Work in Postwar Sri Lanka*. Washington: University of Washington Press.

Jegathesan, Mythri. 2021. "Black Feminist Plots before the Plantationocene and Anthropology's 'Regional Closets.'" *Feminist Anthropology*. doi:10.1002/fea2.12037.

joannemariebarker, and Teresia K. Teaiwa. 1994. "Native InFormation." *Inscriptions* 7: 16–41. https://culturalstudies.ucsc.edu/inscriptions/volume-7/joannemariebarker -teresia-teaiwa/.

Jolly, Margaret. 1991. "Gifts, Commodities and Corporeality: Food and Gender in South Pentecost, Vanuatu." *Canberra Anthropology* 14, no. 1: 45–66. doi:10.1080/03149099109508475.

Jolly, Margaret. 1999. "Another Time, Another Place." *Oceania* 69, no. 4: 282–99. doi:10.1002/j.1834-4461.1999.tb00374.x.

Jolly, Margaret. 2018. "Horizons and Rifts in Conversations about Climate Change in Oceania." In *Pacific Futures: Past and Present*, edited by Warwick Anderson, Miranda Johnson, and Barbara Brookes, 17–48. Honolulu: University of Hawai'i Press.

Jones, Gwilym I. 1989. *From Slaves to Palm Oil: Slave Trade and Palm Oil Trade in the Bight of Biafra*. Cambridge, UK: African Studies Centre.

Juillerat, Bernard. 1996. *Children of the Blood: Society, Reproduction and Cosmology in New Guinea*. Translated by Nora Scott. Oxford, UK: Berg.

Kabutaulaka, Tarcisius. 2015. "Re-Presenting Melanesia: Ignoble Savages and Melanesian Alter-Natives." *Contemporary Pacific* 27, no. 1: 110–45. doi:10.1353/cp.2015.0027.

Kahn, Miriam. 1988. "'Men Are Taro' (They Cannot Be Rice): Political Aspects of Food Choices in Wamira, Papua New Guinea." *Food and Foodways* 3, no. 1–2: 41–57. doi:10.1080/07409710.1988.9961936.

Kahn, Miriam. 1994. *Always Hungry, Never Greedy: Food and the Expression of Gender in a Melanesian Society*. Long Grove, IL: Waveland Press.

Kahn, Miriam. 1996. "Your Place and Mine: Sharing Emotional Landscapes in Wamira, Papua New Guinea." In *Senses of Place*, edited by Steven Feld and Keith H. Basso, 167–96. Santa Fe: School of American Research Press.

Kahn, Miriam, and Lorraine Sexton. 1988. "The Fresh and the Canned: Food Choices in the Pacific." *Food and Foodways* 3, no. 1–2: 1–18. doi:10.1080/07409710.1988.9961934.

Karma, Filep. 2014. *Seakan Kitorang Setengah Binatang: Rasialisme Indonesia Di Tanah Papua*. Jayapura: Cetakan Pertama.

Keane, Webb. 2008. "Others, Other Minds, and Others' Theories of Other Minds: An Afterword on the Psychology and Politics of Opacity Claims." *Anthropological Quarterly* 81, no. 2: 473–82. doi:10.1353/anq.0.0000.

Kelley, Robin D. G. 2002. *Freedom Dreams: The Black Radical Imagination*. Boston: Beacon Press.

Kemal, Anwar. 2011. "UN CERD Formal Communication to the Permanent Mission of Indonesia Regarding Allegations of Threatening and Imminent Irreparable Harm for Indigenous Peoples in Merauke District Related to the MIFEE Project." September 2. http://www.forestpeoples.org/sites/fpp/files/publication/2011/09/cerduaindonesia02092011fm.pdf.

Kempf, Wolfgang, and Elfriede Hermann. 2003. "Dreamscapes: Transcending the Local in Initiation Rites among the Ngaing of Papua New Guinea." In *Dream Travelers: Sleep Experiences and Culture in the Western Pacific*, edited by Richard I. Lohmann, 61–86. New York: Palgrave Macmillan.

Kim, Claire J. 2015. *Dangerous Crossings: Race, Species, and Nature in a Multicultural Age*. Cambridge: Cambridge University Press.

Kimmerer, Robin W. 2014. *Braiding Sweetgrass: Indigenous Wisdom, Scientific Knowledge, and the Teachings of Plants*. Minneapolis: Milkweed Press.

King, Tiffany L. 2016. "The Labor of (Re)reading Plantation Landscapes Fungible(ly)." *Antipode* 48, no. 4: 1022–39. doi:10.1111/anti.12227.

Kirksey, Eben S. 2002. "Anthropology and Colonial Violence in West Papua." *Cultural Survival Quarterly* 26, no. 3. www.culturalsurvival.org/publications/cultural-survival -quarterly/anthropology-and-colonial-violence-west-papua.

Kirksey, Eben S. 2012. *Freedom in Entangled Worlds: West Papua and the Architecture of Global Power*. Durham, NC: Duke University Press.

Kirksey, Eben S. 2015. *Emergent Ecologies*. Durham, NC: Duke University Press.

Kirksey, Eben S. 2017. "Lively Multispecies Communities, Deadly Racial Assemblages, and the Promise of Justice." *South Atlantic Quarterly* 116, no. 1: 195–206. doi:10.1215/00382876-3749614.

Kirksey, Eben S., and Kiki van Bilsen. 2002. "A Road to Freedom: Mee Articulations and the Trans-Papua Highway." *Bijdragen Tot de Taal-, Land- En Volkenkunde* 158, no. 4: 837–54. www.jstor.org/stable/27867996.

Kirksey, Eben S., and Stefan Helmreich. 2010. "The Emergence of Multispecies Ethnography." *Cultural Anthropology* 25, no. 4: 545–76. doi:10.1111/j.1548-1360.2010.01069.x.

Kirsch, Stuart. 2001. "Property Effects: Social Networks and Compensation Claims in Melanesia." *Social Anthropology* 9, no. 2: 147–63. doi:10.1111/j.1469-8676.2001.tb00143.x.

Kirsch, Stuart. 2002. "Rumour and Other Narratives of Political Violence in West Papua." *Critique of Anthropology* 22, no. 1: 53–79. doi:10.1177/0308275X020220010301.

Kirsch, Stuart. 2006. *Reverse Anthropology: Indigenous Analysis of Social and Environmental Relations in New Guinea*. Stanford: Stanford University Press.

Kirsch, Stuart. 2010. "Ethnographic Representation and the Politics of Violence in West Papua." *Critique of Anthropology* 30, no. 1: 3–22. doi:10.1177/0308275X09363213.

Kirsch, Stuart. 2018. *Engaged Anthropology: Politics beyond the Text*. Berkeley: University of California Press.

Knauft, Bruce M. 1989. "Bodily Images in Melanesia: Cultural Substances and Natural Metaphors." In *Fragments for a History of the Human Body*, edited by Michel Feher, Ramona Naddaff, and Nadia Tazi, 198–279. New York: Zone.

Knauft, Bruce M. 1993. *South Coast New Guinea Cultures: History, Comparison, Dialectic*. Cambridge: Cambridge University Press.

Knauft, Bruce M. 1999. *From Primitive to Postcolonial in Melanesia and Anthropology*. Ann Arbor: University of Michigan Press.

Knauft, Bruce M., ed. 2002a. *Critically Modern: Alternatives, Alterities, Anthropologies*. Bloomington: Indiana University Press.

Knauft, Bruce M. 2002b. *Exchanging the Past: A Rainforest World of Before and After*. Chicago: University of Chicago Press.

Knauft, Bruce M. 2019. "Good Anthropology in Dark Times: Critical Appraisal and Ethnographic Application." *Australian Journal of Anthropology* 30, no. 1: 3–17. doi:10.1111 /taja.12300.

Koczberski, Gina, and George N. Curry. 2004. "Divided Communities and Contested Landscapes: Mobility, Development and Shifting Identities in Migrant

Destination Sites in Papua New Guinea." *Asia Pacific Viewpoint* 45, no. 4: 357–71. doi:10.1111/j.1467-8373.2004.00252.x.

Koczberski, Gina, and George N. Curry. 2005. "Making a Living: Land Pressures and Changing Livelihood Strategies among Oil Palm Settlers in Papua New Guinea." *Agricultural Systems* 85, no. 3: 324–39. doi:10.1016/j.agsy.2005.06.014.

Kohn, Eduardo. 2013. *How Forests Think: Toward an Anthropology beyond the Human.* Berkeley: University of California Press.

Kosek, Jake. 2010. "Ecologies of Empire: On the New Uses of the Honeybee." *Cultural Anthropology* 25, no. 4: 650–78. doi:10.1111/j.1548-1360.2010.01073.x.

Koselleck, Reinhart. 1985. *Futures Past: On the Semantics of Historical Time.* Translated by Keith Tribe. Cambridge, MA: MIT Press.

Koselleck, Reinhart. 2002. *The Practice of Conceptual History: Timing History, Spacing Concepts.* Translated by Todd S. Presner. Stanford: Stanford University Press.

Kovach, Margaret. 2009. *Indigenous Methodologies: Characteristics, Conversations, and Contexts.* Toronto: University of Toronto Press.

Langwick, Stacey Ann. 2018. "A Politics of Habitability: Plants, Healing, and Sovereignty in a Toxic World." *Cultural Anthropology* 33, no. 3: 415–43. doi:10.14506/ca33.3.06.

Lattas, Andrew. 1991. "Sexuality and Cargo Cults: The Politics of Gender and Procreation in West New Britain." *Cultural Anthropology* 6, no. 2: 230–56. doi:10.1525/can.1991.6.2.02a00070.

Lattas, Andrew. 1993. "Sorcery and Colonialism: Illness, Dreams and Death as Political Languages in West New Britain." *Journal of the Royal Anthropological Institute* 28, no. 1: 51–77. doi:10.2307/2804436.

Lawrence, Peter. 1964. *Road Belong Cargo: A Study of the Cargo Movement in the Southern Madang District, New Guinea.* Manchester: Manchester University Press.

Leach, James. 2003. *Creative Land: Place and Procreation on the Rai Coast of Papua New Guinea.* New York: Berghahn.

Lear, Jonathan. 2006. *Radical Hope: Ethics in the Face of Cultural Devastation.* Cambridge, MA: Harvard University Press.

Leenhardt, Maurice. 1979. *Do Kamo: Person and Myth in the Melanesian World.* Translated by Basia M. Gulati. Chicago: University of Chicago Press.

Lefebvre, Henri. 1991. *The Production of Space.* Translated by Donald Nicholson-Smith. Cambridge, UK: Basil Blackwell.

Lempert, William. 2018. "Generative Hope in the Postapocalyptic Present." *Cultural Anthropology* 33, no. 2: 202–12. doi:10.14506/ca33.2.04.

Lewis-Jones, Kay E. 2016. "Introduction: People and Plants." *Environment and Society* 7, no. 1: 1–7. doi:10.3167/ares.2016.070101.

Li, Tanya M. 2014. *Land's End: Capitalist Relations on an Indigenous Frontier.* Durham, NC: Duke University Press.

Li, Tanya M. 2017a. "After the Land Grab: Infrastructural Violence and the 'Mafia System' in Indonesia's Oil Palm Plantation Zones." *Geoforum* 96: 328–37. doi:10.1016/j.geoforum.2017.10.012.

Li, Tanya M. 2017b. "The Price of Un/Freedom: Indonesia's Colonial and Contemporary Plantation Labor Regimes." *Comparative Studies in Society and History* 59, no.2: 245–76. doi:10.1017/S0010417517000044.

Liboiron, Max, Manuel Tironi, and Nerea Calvillo. 2018. "Toxic Politics: Acting in a Permanently Polluted World." *Social Studies of Science* 48, no. 3: 331–49. doi:10.1177/0306312718783087.

Lindstrom, Lamont. 1993. *Cargo Cult: Strange Stories of Desire from Melanesia and Beyond*. Honolulu: University of Hawai'i Press.

Lipset, David. 1997. *Mangrove Man: Dialogics of Culture in the Sepik Estuary*. Cambridge: Cambridge University Press.

Lipset, David. 2014. "Living Canoes: Vehicles of Moral Imagination among the Murik of Papua New Guinea." In *Vehicles: Cars, Canoes, and Other Metaphors of Moral Imagination*, edited by David Lipset and Richard Handler, 21–47. New York: Berghahn.

Lipset, David. 2016. "The Knotted Person: Death, the Bad Breast, and Melanesian Modernity among the Murik, Papua New Guinea." In *Mortuary Dialogues: Death Ritual and the Reproduction of Moral Community in Pacific Modernities*, edited by David Lipset and Eric K. Silverman, 81–109. New York: Berghahn.

Lipset, David. 2017. *Yabar: The Alienations of Murik Men in a Papua New Guinea Modernity*. New York: Palgrave Macmillan.

LiPuma, Edward. 2000. *Encompassing Others: The Magic of Modernity in Melanesia*. Ann Arbor: University of Michigan Press.

Little, Christopher A. J. L. 2011. "How Asabano Children Learn; or, Formal Schooling amongst Informal Learners." *Oceania* 81, no. 2: 148–66. doi:10.1002/j.1834-4461.2011 .tb00100.x.

Locke, John. [1689] 2016. *Two Treatises of Government. Second Treatise*. Edited by Lee Ward. Bloomington, IN: Hackett.

Lohmann, Roger I., ed. 2003. *Dream Travelers: Sleep Experiences and Culture in the Western Pacific*. New York: Palgrave Macmillan.

Love, Heather. 2009. *Feeling Backward: Loss and the Politics of Queer History*. Cambridge, MA: Harvard University Press.

Lowe, Celia. 2006. *Wild Profusion: Biodiversity Conservation in an Indonesian Archipelago*. Princeton: Princeton University Press.

Luhrmann, Tanya M. 2012. "Toward an Anthropological Theory of Mind." *Suomen Antropologi: Journal of the Finnish Anthropological Society* 36, no. 4: 1–69.

Lyons, Kristina. 2016. "Decomposition as Life Politics: Soils, Selva, and Small Farmers under the Gun of the U.S.-Colombia War on Drugs." *Cultural Anthropology* 31, no. 1: 58–81. doi:10.14506/ca31.1.04.

MacGregor, William M. 1893a. "Despatch Reporting Expedition Undertaken with Object of Meeting Tugeri Invaders." In *Annual Report on British New Guinea from 1st July 1891 to 30th June 1892*, 49–50. Brisbane: Government Printer.

MacGregor, William M. 1893b. "Despatch Respecting Expedition Undertaken to Repel Tugeri Invasion." In *Annual Report on British New Guinea from 1st July 1891 to 30th June*, 50–53. Brisbane: Government Printer.

MacKay, Fergus. 2011a. "Request for Consideration of the Situation of Indigenous Peoples in Merauke, Papua Province, Indonesia, under the United Nations Committee on the Elimination of Racial Discrimination's Urgent Action and Early Warning Procedures." UN Committee on the Elimination of Racial Discrimination, Seventy-Ninth Session, August 8–September 2. https://www.forestpeoples.org/en/topics/un -human-rights-system/news/2011/08/request-consideration-situation-indigenous -peoples-merauk.

MacKay, Fergus. 2011b. "Request for Urgent Assistance to Address the Imminent Threat to the Right to Food of the Indigenous Peoples in Merauke, Papua Province, Indonesia." Forest Peoples Program on Behalf of Additional Submitting Organizations. August 9. https://www.forestpeoples.org/en/topics/un-human-rights-system/publication /2011/letter-un-special-rapporteur-right-food-re-request-ur.

MacLean, Neil. 1994. "Freedom or Autonomy: A Modern Melanesian Dilemma." *Journal of the Royal Anthropological Institute* 29, no. 3: 66–88. doi:10.2307/2804348.

MacLeod, Jason, Rosa Moiwend, and Jasmine Pilbrow. 2016. "A Historic Choice: West Papua, Human Rights and Pacific Diplomacy at the Pacific Island Forum and Melanesian Spearhead Group." Pasifika. https://www.ulmwp.org/wp-content/uploads/2016 /09/WP_PIF_MSG_Report_Online_RLR-1.pdf.

Mageo, Jeannette M. 2003. "Race, Postcoloniality, and Identity in Samoan Dreams." In *Dreaming and the Self: New Perspectives on Subjectivity, Identity, and Emotion*, edited by Jeannette M. Mageo, 75–96. Albany: State University of New York Press.

Mageo, Jeannette M. 2004. "Toward a Holographic Theory of Dreaming." *Dreaming* 14, no. 2–3: 15–69. doi:10.1037/1053-0797.14.2-3.151.

Maggio, Jay. 2007. "'Can the Subaltern Be Heard?': Political Theory, Translation, Representation, and Gayatri Chakravorty Spivak." *Alternatives: Global, Local, Political* 32, no. 4: 419–43. doi:10.1177/030437540703200403.

Majnep, Ian S., and Ralph Bulmer. 1977. *Birds of My Kalam Country*. Auckland: Auckland University Press.

Majnep, Ian S., and Ralph Bulmer. 2007. *Animals the Ancestors Hunted: An Account of the Wild Mammals of the Kalam Area, Papua New Guinea*, edited by Robin L. Hide and Andrea Pawley. Adelaide: Crawford House Publishing.

Malabou, Catherine. 2012. *The Ontology of the Accident: An Essay on Destructive Plasticity*. Translated by Carolyn Shread. Malden, MA: Polity Press.

Malinowski, Bronislaw. 1935. *Coral Gardens and Their Magic: A Study of the Methods of Tilling the Soil and of Agricultural Rites in the Trobriand Islands*. London: Allen and Unwin.

Manderson, Lenore, ed. 1986. *Shared Wealth and Symbol: Food, Culture, and Society in Oceania and Southeast Asia*. London: Cambridge University Press.

Manikmas, Made O. A. 2010. "Merauke Integrated Rice Estate: Kebangkitan Ketahanan Dan Kemandirian Pangan Dari Ufuk Timur Indonesia." *Analisis Kebijakan Pertanian* 8, no. 4: 323–38. doi:10.21082/akp.v8n4.2010.323-338.

Manning, Chris, and Michael Rumbiak. 1989. *Economic Development, Migrant Labor and Indigenous Welfare in Irian Jaya 1970–84*. Canberra: Australian National University Press.

Marder, Michael. 2011. "Vegetal Anti-Metaphysics: Learning from Plants." *Continental Philosophy Review* 44, no. 4: 469–89. doi:10.1007/s11007-011-9201-x.

Marder, Michael. 2016. *Grafts: Writings on Plants*. Minneapolis: University of Minnesota Press.

Massumi, Brian. 2015. *Politics of Affect*. Cambridge, UK: Polity Press.

Mawyer, Alexander. 2018. "Unflowing Pasts, Lost Springs and Watery Mysteries in Eastern Polynesia." In *Island Rivers: Fresh Water and Place in Oceania*, edited by John R. Wagner and Jerry K. Jacka, 83–108. Canberra: Australian National University Press.

Mbembe, Achille. 2003. "Necropolitics." Translated by Steven Corcoran. *Public Culture* 15, no. 1: 11–40. doi:10.1215/08992363-15-1-11.

McCarthy, John. 2010. "Processes of Inclusion and Adverse Incorporation: Oil Palm and Agrarian Change in Sumatra, Indonesia." *Journal of Peasant Studies* 37, no. 4: 821–50. doi:10.1080/03066150.2010.512460.

McCarthy, John, and Moira Moeliono. 2012. "The Post-Authoritarian Politics of Agrarian and Forest Reform in Indonesia." In *Routledge Handbook of Southeast Asian Politics*, edited by Richard Robison, 242–59. New York: Taylor and Francis.

McCarthy, John F., Jacqueline A. C. Vel, and Suraya Afiff. 2012. "Trajectories of Land Acquisition and Enclosure: Development Schemes, Virtual Land Grabs, and Green Acquisitions in Indonesia's Outer Islands." *Journal of Peasant Studies* 39, no. 2: 521–49. doi:10.1080/03066150.2012.671768.

McDonnell, John E. 2020. "The Merauke Integrated Food and Energy Estate (MIFEE): An Ecologically Induced Genocide of the Malind Anim." *Journal of Genocide Research*. doi:10.1080/14623528.2020.1799593.

McDonnell, Siobhan, Matthew G. Allen, and Colin Filer, eds. 2017. *Kastom, Property and Ideology: Land Transformations in Melanesia*. Canberra: Australian National University Press.

McGibbon, Rodd. 2004. *Plural Society in Peril: Migration, Economic Change, and the Papua Conflict*. Washington, DC: East-West Center.

McGranahan, Carole. 2016. "Theorising Refusal: An Introduction." *Cultural Anthropology* 31, no. 3: 319–25. doi:10.14506/ca31.3.01.

McKittrick, Katherine. 2006. *Demonic Grounds: Black Women and the Cartographies of Struggle*. Minneapolis: University of Minnesota Press.

McKittrick, Katherine. 2013. "Plantation Futures." *Small Axe* 17, no. 3(42): 1–15. doi:10.1215/07990537-2378892.

McNeill, John R. 2012. *Mosquito Empires: Ecology and War in the Greater Caribbean, 1620–1914*. Cambridge: Cambridge University Press.

Meigs, Anna S. 1984. *Food, Sex, and Pollution: A New Guinea Religion*. New Brunswick: Rutgers University Press.

Merleau-Ponty, Maurice. 1968. *The Visible and the Invisible: Followed by Working Notes*. Translated by Alphonso Lingis. Evanston, IL: Northwestern University Press.

Meyer, Manulani Aluli. 2001. "Our Own Liberation: Reflections on Hawaiian Epistemology." *Contemporary Pacific* 13, no. 1: 124–48. doi:10.1353/cp.2001.0024.

Millar, Kathleen M. 2014. "The Precarious Present: Wageless Labor and Disrupted Life in Rio de Janeiro, Brazil." *Cultural Anthropology* 29, no. 1: 32–53. doi:10.14506/ca29.1.04.

Miller, Theresa L. 2019. *Plant Kin: A Multispecies Ethnography in Indigenous Brazil*. Austin: University of Texas Press.

Mintz, Sidney W. 1985. *Sweetness and Power: The Place of Sugar in Modern History*. New York: Penguin Books.

Mitchell, Timothy. 2002. "Can the Mosquito Speak?" In *Rule of Experts: Egypt, Techno-Politics, Modernity*, 19–53. Berkeley: University of California Press.

Mitman, Gregg. 2019. "Reflections on the Plantationocene: A Conversation with Donna Haraway and Anna Tsing." Transcript. *Edge Effects*, last modified October 12, 2019. https://edgeeffects.net/haraway-tsing-plantationocene/.

Mittermaier, Amira. 2011. *Dreams That Matter: Egyptian Landscapes of the Imagination*. Berkeley: University of California Press.

Monbiot, George. 1989. *Poisoned Arrows: An Investigative Journey through Indonesia*. London: Michael Joseph.

Mondragón, Carlos. 2004. "Of Winds, Worms and *Mana*: The Traditional Calendar of the Torres Islands, Vanuatu." *Oceania* 74, no. 4: 289–308. doi:10.1002/j.1834-4461.2004.tb02856.x.

Mondragón, Carlos. 2009. "A Weft of Nexus: Changing Notions of Space and Geographical Identity in Vanuatu, Oceania." In *Boundless Worlds: An Anthropological Approach to Movement*, edited by Peter W. Kirby, 115–33. Oxford, UK: Berghahn.

Mondragón, Carlos. 2018. "A Source of Power, Disquiet and Biblical Purport: The Jordan River in Santo, Vanuatu." In *Island Rivers: Fresh Water and Place in Oceania*, edited by John R. Wagner and Jerry K. Jacka, 59–82. Canberra: Australian National University Press.

Moore, Donald S., Anand Pandian, and Jake Kosek. 2003. "The Cultural Politics of Race and Nature: Terrains of Power and Practice." In *Race, Nature, and the Politics of Difference*, edited by Donald Moore, Jake Kosek, and Anand Pandian, 1–70. Durham, NC: Duke University Press.

Moore, Jason W. 2015. *Capitalism in the Web of Life: Ecology and the Accumulation of Capital*. New York: Verso.

Morauta, Louise, Ann Chowning, Current Issues Collective (B. Kaspou and Others), Angela M. Gilliam, Fritz Hafer, Diane Kayongo-Male, Hal B. Levine, Robert F. Maher, Khalil Nakleh, Jacob L. Simet, A. J. Strathern, C. A. Valentine, Bettylou Valentine, and John Waiko. 1979. "Indigenous Anthropology in Papua New Guinea [and Comments and Reply]." *Current Anthropology* 20, no. 3: 561–76. doi:10.1086/202325.

Morgan, Jennifer. 2004. *Laboring Women: Reproduction and Gender in New World Slavery*. Philadelphia: University of Pennsylvania Press.

Morton, Timothy. 2013. *Hyperobjects: Philosophy and Ecology after the End of the World*. Minneapolis: University of Minnesota Press.

Mosko, Mark S. 2009. "The Fractal Yam: Botanical Imagery and Human Agency in the Trobriands." *Journal of the Royal Anthropological Institute* 15, no. 4: 679–700. doi:10.1111/j.1467-9655.2009.01579.x.

Mosko, Mark S. 2013. "Dividuals, Individuals, or Possessive Individuals? Recent Transformations of North Mekeo Commoditization, Personhood, and Sociality." In *Engaging with Capitalism: Cases from Oceania*, edited by Fiona McCormack and Kate Barclay, 167–98. Bingley, UK: Emerald Books.

Mote, Okto, and Danilyn Rutherford. 2001. "From Irian Jaya to Papua: The Limits of Primordialism in Indonesia's Troubled East." *Indonesia* 72: 115–40. doi:10.2307/3351483.

Munn, Nancy D. 1992. *The Fame of Gawa: A Symbolic Study of Value Transformation in a Massim (Papua New Guinea) Society*. Durham, NC: Duke University Press.

Muñoz, José E. 2015. "Theorizing Queer Inhumanisms: The Sense of Brownness." GLQ: *A Journal of Lesbian and Gay Studies* 21, no. 2–3: 209–10. muse.jhu.edu/article/581600.

Munro, Jenny. 2013. "The Violence of Inflated Possibilities: Education, Transformation, and Diminishment in Wamena, Papua." *Indonesia* 95: 25–46. doi:10.5728 /indonesia.95.0025.

Munro, Jenny. 2015a. "'Now We Know Shame': Malu and Stigma among Highlanders in the Papuan Diaspora." In *From 'Stone Age' to "Real-Time": Exploring Papuan Temporalities, Mobilities and Religiosities*, edited by Martin Slama and Jenny Munro, 169–94. Canberra: Australian National University Press.

Munro, Jenny. 2015b. "The President and the Papua Powder Keg." *New Mandala*, June 24, 2015. www.newmandala.org/the-president-and-the-papua-powder-keg/.

Munro, Jenny. 2018. *Dreams Made Small: The Education of Papuan Highlanders in Indonesia*. New York: Berghahn.

Munro, Jenny. 2020. "Global HIV Interventions and Technocratic Racism in a West Papuan NGO." *Medical Anthropology* 39, no. 8: 704–19. doi:10.1080/01459740.2020.1739036.

Murphy, Michelle. 2015. "Chemical Infrastructures of the St Clair River." In *Toxicants, Health and Regulation since 1945*, edited by Nathalie Jas and Soraya Boudia, 103–15. London: Routledge.

Murphy, Michelle. 2017. "Alterlife and Decolonial Chemical Relations." *Cultural Anthropology* 32, no. 4: 494–503. doi:10.14506/ca32.4.02.

Musharbash, Yasmine. 2013. "Night, Sight, and Feeling Safe: An Exploration of Aspects of Warlpiri and Western Sleep." *Australian Journal of Anthropology* 24, no. 1: 48–63. doi:10.1111/taja.12021.

Myers, Natasha. 2015. "Conversations on Plant Sensing: Notes from the Field." *Nature-Culture* 3: 35–66. www.natcult.net/wp-content/uploads/2018/12/PDF-natureculture-03 -03-conversations-on-plant-sensing.pdf.

Myers, Natasha. 2017a. "From the Anthropocene to the Planthroposcene: Designing Gardens for Plant/People Involution." *History and Anthropology* 28, no. 3: 297–301. doi:10.1080/02757206.2017.1289934.

Myers, Natasha. 2017b. "Ungrid-able Ecologies: Decolonizing the Ecological Sensorium in a 10,000 Year-Old NaturalCultural Happening." *Catalyst: Feminism, Theory, Technoscience* 3, no. 2: 1–24. doi:10.28968/cftt.v3i2.28848.

Narokobi, Bernard. 1976. "Art and Nationalism." *Gigibori* 3, no. 1: 12–15.

Narokobi, Bernard. 1980. *The Melanesian Way: Total Cosmic Vision of Life (and His Critics and Supporters)*. Boroko: Institute of Papua New Guinea Studies.

Narokobi, Bernard. 1983. *Life and Leadership in Melanesia*. Suva, Fiji: University of the South Pacific.

Nash, June. 1993. *We Eat the Mines and the Mines Eat Us: Dependency and Exploitation in Bolivian Tin Mines*. New York: Columbia University Press.

Neimanis, Astrida. 2009. "Bodies of Water, Human Rights and the Hydrocommons." *Topia: Canadian Journal of Cultural Studies* 21: 161–82. doi:10.3138/topia.21.161.

Nelson, Paul N., Jennifer Gabriel, Colin Filer, Murom Banabas, Jeffrey A. Sayer, George N. Curry, Gina Koczberski, and Oscar Venter. 2013. "Oil Palm and Deforestation in New Guinea." *Conservation Letters* 7, no. 3: 188–95. doi:10.1111/conl.12058.

Nerenberg, Jacob. 2018. "Terminal Economy: Politics of Distribution in Highland Papua, Indonesia." PhD diss., University of Toronto.

Nihill, Michael. 2001. "Pain and 'Progress': Revisiting Botol Sorcery in the Southern Highlands of Papua New Guinea." *Social Analysis* 45, no. 1: 103–21. www.jstor.org/stable/23169993.

Nimuendajú, Curt. 1939. *The Apinayé*. Washington, DC: Catholic University of America Press.

Nimuendajú, Curt. 1946. *The Eastern Timbira*. Berkeley: University of California Press.

Nixon, Rob. 2011. *Slow Violence and the Environmentalism of the Poor*. Cambridge, MA: Harvard University Press.

Obeyesekere, Gananath. 1992. *The Apotheosis of Captain Cook: European Mythmaking in the Pacific*. Princeton: Princeton University Press.

Ogden, Laura A., Billy Hall, and Kimiko Tanita. 2013. "Animals, Plants, People, and Things: A Review of Multispecies Ethnography." *Environment and Society* 4, no. 1: 5–24. doi:10.3167/ares.2013.040102.

O'Hanlon, Michael. 1989. *Reading the Skin: Adornment, Display and Society among the Wahgi*. London: British Museum Publications.

O'Hanlon, Michael, and Linda Frankland. 2003. "Co-Present Landscapes: Routes and Rootedness as Sources of Identity in Highlands New Guinea." In *Landscape, Memory and History: Anthropological Perspectives*, edited by Pamela J. Stewart and Andrew J. Strathern, 166–88. Sterling, VA: Pluto Press.

Ohnuki-Tierney, Emiko. 1993. *Rice as Self: Japanese Identities through Time*. Princeton: Princeton University Press.

Ohtsuka, Ryutaro. 1983. *Oriomo Papuans: Ecology of Sago-Eaters in Lowland Papua*. Tokyo: University of Tokyo Press.

Ondawame, Otto J. 2006. "West Papua: The Discourse of Cultural Genocide and Conflict Resolution." In *Cultural Genocide and Asian State Peripheries*, edited by Barry Sautman, 103–38. New York: Palgrave Macmillan.

Oosterwal, Gottfried. 1963. "A Cargo Cult in the Mamberamo Area." *Ethnology* 2, no.1: 1–14. doi:10.2307/3772964.

Ortiz Fernández, Fernando. 1947. *Cuban Counterpoint: Tobacco and Sugar*. Translated by Harriet de Ortís. Durham, NC: Duke University Press.

Ortner, Sherry B. 2016. "Dark Anthropology and Its Others: Theory since the Eighties." *HAU: Journal of Ethnographic Theory* 6, no. 1: 47–73. doi:10.14318/hau6.1.004.

Overweel, Jeroen A. 1992. *The Marind in a Changing Environment: A Study on Social-Economic Change in Marind Society to Assist in the Formulation of a Long-Term Strategy for the Foundation for Social, Economic, and Environmental Development (YAPSEL)*. Merauke: YAPSEL.

Pálsson, Gísli. 2016. "Unstable Bodies: Biosocial Perspectives on Human Variation." *Sociological Review Monographs* 64, no. 1: 100–16. doi:10.1002/2059-7932.12015.

Pandya, Vishvajit. 1990. "Movement and Space: Andamanese Cartography." *American Ethnologist* 17, no. 4: 775–97. doi:10.1525/ae.1990.17.4.02a00100.

Panoff, Françoise, and Françoise Barbira-Freedman, eds. 2018. *Maenge Gardens: A Study of Maenge Relationships to Domesticates.* Marseille: Pacific-Credo Publications.

Paredes, Alyssa. 2022. "We Are Not Pests." In *The Promise of Multispecies Justice*, edited by Sophie Chao, Karin Bolender, and Eben S. Kirksey. Durham, NC: Duke University Press.

Parreñas, Juno S. 2018. *Decolonizing Extinction: The Work of Care in Orangutan Rehabilitation.* Durham, NC: Duke University Press.

Pascoe, Bruce. 2018. *Dark Emu: Aboriginal Australia and the Birth of Agriculture.* Broome, Australia: Magabala Books.

Pedersen, Morten A. 2011. *Not Quite Shamans: Spirit Worlds and Political Lives in Northern Mongolia.* Ithaca: Cornell University Press.

Peluso, Nancy. 1995. "Whose Woods Are These? Counter-Mapping Forest Territories in Kalimantan, Indonesia." *Antipode* 27, no. 4: 383–406. doi:10.1111/j.1467-8330.1995.tb00286.x.

Peluso, Nancy. 1996. "Fruit Trees and Family Trees in an Anthropogenic Forest: Ethics of Access, Property Zones, and Environmental Change in Indonesia." *Comparative Studies in Society and History* 38, no. 3: 510–48. doi:10.1017/S0010417500020041.

Peña, Devon G. 1997. *The Terror in the Machine: Technology, Work, Gender, and Ecology on the US-Mexico Border.* Austin: University of Texas Press.

Perez, Craig Santos. 2013. "Facing Hawai'i's Future (Book Review)." *Kenyon Review*, July 10, 2013. www.kenyonreview.org/2013/07/facing-hawai'i's-future-book-review/.

Persoon, Gerard. 1992. "From Sago to Rice: Changes in Cultivation in Siberut, Indonesia." In *Bush Base, Forest Farm: Culture, Environment, and Development*, edited by Elisabeth Croll and David Parkin, 187–99. London: Routledge.

Plumwood, Val. 2002. *Environmental Culture: The Ecological Crisis of Reason.* London: Routledge.

Pollock, Nancy J. 1992. *These Roots Remain: Food Habits in Islands of the Central and Eastern Pacific since Western Contact.* Laie, HI: Institute of Polynesian Studies.

Pouwer, Jan. 1999. "The colonisation, decolonisation and recolonisation of West New Guinea." *Journal of Pacific History* 34, no. 2: 157–79. doi:10.1080/00223349908572900.

Pouwer, Jan. 2010. *Gender, Ritual and Social Formation in West Papua: A Configurational Analysis Comparing Kamoro and Asmat.* Leiden: KITLV Press.

Povinelli, Elizabeth A. 1993. *Labor's Lot: The Power, History, and Culture of Aboriginal Action.* Chicago: University of Chicago Press.

Povinelli, Elizabeth A. 2002. *The Cunning of Recognition: Indigenous Alterities and the Making of Australian Multiculturalism.* Durham, NC: Duke University Press.

Povinelli, Elizabeth A. 2009. "Beyond Good and Evil, Whither Liberal Sacrificial Love?" *Public Culture* 21, no. 1: 77–100. doi:10.1215/08992363-2008-022.

Povinelli, Elizabeth A. 2011. *Economies of Abandonment: Social Belonging and Endurance in Late Liberalism.* Durham, NC: Duke University Press.

Povinelli, Elizabeth A. 2016. *Geontologies: A Requiem to Late Liberalism*. Durham, NC: Duke University Press.

Pratt, Mary L. 2007. *Imperial Eyes: Travel Writing and Transculturation*. New York: Routledge.

Prom, Meta, and Jeremy Ironside. 2005. "Effective Maps for Planning Sustainable Land Use and Livelihoods." In *Mapping Communities: Ethics, Values, Practice*, edited by Jefferson Fox, Krisnawati Suryanata, and Peter Hershock, 29–42. Honolulu: East-West Center.

Puar, Jasbir K. 2017. *The Right to Maim: Debility, Capacity, Disability*. Durham, NC: Duke University Press.

Puig de la Bellacasa, María. 2009. "Touching Technologies, Touching Visions. The Reclaiming of Sensorial Experience and the Politics of Speculative Thinking." *Subjectivity* 28, no. 1: 297–315. doi:10.1057/sub.2009.17.

Puig de la Bellacasa, María. 2012. "'Nothing Comes without Its World': Thinking with Care." *Sociological Review* 60, no. 2: 197–216. doi:10.1111/j.1467-954X.2012.02070.x.

Puig de la Bellacasa, María. 2016. "Ecological Thinking, Materialist Spirituality, and the Poetics of Infrastructure." In *Boundary Objects and Beyond: Working with Leigh Star*, edited by Geoffrey C. Bowker, Stefan Timmermans, Adele E. Clarke, and Ellen Balka, 47–68. Cambridge, MA: MIT Press.

Pye, Oliver, and Jayati Bhattacharya, eds. 2013. *The Palm Oil Controversy in Southeast Asia: A Transnational Perspective*. Singapore: Institute of Southeast Asian Studies–Yusof Ishak Institute.

Rajan, Kaushik S., ed. 2012. *Lively Capital: Biotechnologies, Ethics, and Governance in Global Markets*. Durham, NC: Duke University Press.

Ramos, Alcida R. 2012. "The Politics of Perspectivism." *Annual Review of Anthropology* 41: 481–94. doi:10.1146/annurev-anthro-092611-145950.

Rancière, Jacques. 1999. *Disagreement: Politics and Philosophy*. Translated by Julie Rose. Minneapolis: University of Minnesota Press.

Rappaport, Roy A. 1984. *Pigs for the Ancestors: Ritual in the Ecology of a New Guinea People*. New Haven: Yale University Press.

Read, Kenneth E. 1995. "Morality and the Concept of the Person among the Gahuku-Gama." *Oceania* 25, no. 4: 233–82. doi:10.1002/j.1834-4461.1955.tb00651.x.

Reed, Christina. 2015. "Plastic Age: How It's Reshaping Rocks, Oceans, and Life." *New Scientist*, January 28, 2015. www.newscientist.com/article/mg22530060-200-plastic-age-how-its-reshaping-rocks-oceans-and-life/.

Reza, Aditya. 2015. "Begini Mimpi Jokowi Tentang Indonesia 2085." TEMPO, December 30. https://www.tempo.co/search?q=begini+mimpi#gsc.tab=0&gsc.q=begini%20mimpi&gsc.page=1.

Rifkin, Mark. 2017. *Beyond Settler Time: Temporal Sovereignty and Indigenous Self-Determination*. Durham, NC: Duke University Press.

Rijksen, Herman D., and Gerard Persoon. 1991. "Food from Indonesia's swamp forest: ideology or rationality?" *Landscape and Urban Planning* 20, no. 1–3: 95–102. doi:10.1016/0169-2046(91)90097-6.

Rival, Alain, and Patrice Levang. 2014. *Palms of Controversies: Oil Palm and Development Challenges*. Bogor, Indonesia: Center for International Forestry Research.

Rival, Laura M. 1998. "Androgynous Parents and Guest Children: The Huaorani Couvade." *Journal of the Royal Anthropological Institute* 4, no. 4: 619–42.
doi:10.2307/3034825.

Rival, Laura M. 2002. *Trekking through History: The Huaorani of Amazonian Ecuador.*
New York: Columbia University Press.

Robbins, Joel. 1998. "Becoming Sinners: Christianity and Desire among the Urapmin of
Papua New Guinea." *Ethnology* 37, no. 4: 299–316. doi:10.2307/3773784.

Robbins, Joel. 2001. "Secrecy and the Sense of an Ending: Narrative, Time, and Everyday
Millenarianism in Papua New Guinea and in Christian Fundamentalism." *Comparative Studies in History and Society* 43, no. 3: 525–51. doi:10.1017/S0010417501004212.

Robbins, Joel. 2004. *Becoming Sinners: Christianity and Moral Torment in a Papuan New
Guinea Society.* Berkeley: University of California Press.

Robbins, Joel. 2013. "Beyond the Suffering Subject: Toward an Anthropology
of the Good." *Journal of the Royal Anthropological Institute* 19, no. 3: 447–62.
doi:10.1111/1467-9655.12044.

Robbins, Joel, and Alan Rumsey. 2008. "Introduction: Cultural and Linguistic Anthropology and the Opacity of Other Minds." *Anthropological Quarterly* 81, no. 2: 407–20.
www.jstor.org/stable/30052755.

Roberts, Elizabeth F. S. 2017. "What Gets Inside: Violent Entanglements and Toxic
Boundaries in Mexico City." *Cultural Anthropology* 32, no. 4: 592–619. doi:10.14506
/ca32.4.07.

Roberts, Jody A. 2010. "Reflections of an Unrepentant Plastiphobe: Plasticity and the
STS Life." *Science as Culture* 19, no. 1: 101–20. doi:10.1080/09505430903557916.

Robison, Richard, and Vedi R. Hadiz. 2004. *Reorganising Power in Indonesia: The Politics
of Oligarchy in an Age of Markets.* New York: Routledge.

Rose, Deborah B. 2004. *Reports from a Wild Country: Ethics of Decolonisation.* Sydney:
University of New South Wales Press.

Rose, Deborah B. 2011. *Wild Dog Dreaming: Love and Extinction.* Charlottesville: University of Virginia Press.

Rose, Deborah B. 2012. "Multispecies Knots of Ethical Time." *Environmental Philosophy*
9, no. 1: 127–40. doi:10.5840/envirophil2012918.

Rose, Deborah B. 2017. "Shimmer: When All You Love Is Being Trashed." In *Arts of
Living on a Damaged Planet: Ghosts and Monsters of the Anthropocene*, edited by
Anna L. Tsing, Nils Bubandt, Elaine Gan, and Heather A. Swanson, G51–64. Min-
neapolis: University of Minnesota Press.

Rose, Deborah B., and Thom van Dooren. 2017. "Keeping Faith with the Dead: Mourning
and De-extinction." *Australian Zoologist* 38, no. 3: 375–78. doi:10.7882/AZ.2014.048.

Rubenstein, Mary-Jane. 2008. *Strange Wonder: The Closure of Metaphysics and the
Opening of Awe.* New York: Columbia University Press.

Rumsey, Alan, and James F. Weiner, eds. 2001. *Emplaced Myth: Space, Narrative, and
Knowledge in Aboriginal Australia and Papua New Guinea.* Honolulu: University of
Hawai'i Press.

Rusert, Britt. 2019. "Naturalizing Coercion: The Tuskegee Experiments and the Labo-
ratory of Life of the Plantation." In *Captivating Technology: Race, Carceral Tech-*

noscience, and *Liberatory Imagination in Everyday Life*, edited by Ruha Benjamin, 25–49. Durham, NC: Duke University Press.

Russell, Andrew, and Elizabeth Rahman, eds. 2015. *The Master Plant: Tobacco in Lowland South America*. New York: Bloomsbury.

Rutherford, Danilyn. 1996. "Of Birds and Gifts: Reviving Tradition on an Indonesian Frontier." *Cultural Anthropology* 11, no. 4: 577–616. doi:10.1525/can.1996.11.4.02a00060.

Rutherford, Danilyn. 2000. "The White Edge of the Margin: Textuality and Authority in Biak, Irian Jaya, Indonesia." *American Ethnologist* 27, no. 2: 312–39. doi:10.1525 /ae.2000.27.2.312.

Rutherford, Danilyn. 2003. *Raiding the Land of the Foreigners: The Limits of the Nation on an Indonesian Frontier*. Princeton: Princeton University Press.

Rutherford, Danilyn. 2012. *Laughing at Leviathan: Sovereignty and Audience in West Papua*. Chicago: University of Chicago Press.

Rutherford, Danilyn. 2015. "Demonstrating the Stone-Age in Dutch New Guinea." In *From "Stone Age" to "Real-Time": Exploring Papuan Temporalities, Mobilities and Religiosities*, edited by Martin Slama and Jenny Munro, 39–58. Canberra: Australian National University Press.

Sahlins, Marshall. 1983. "Other Times, Other Customs: The Anthropology of History." *American Anthropologist* 85, no. 3: 517–44. doi:10.1525/aa.1983.85.3.02a00020.

Sahlins, Marshall. 1985. *Islands of History*. Chicago: University of Chicago Press.

Said, Edward W. 2012. *Reflections on Exile: And Other Literary and Cultural Essays*. London: Granta Books.

Saleh, Shofia, Bukti Bagja, Thontowi A. Suhada, and Hermawati Widyapratami. 2018. "Intensification by Smallholder Farmers Is Key to Achieving Indonesia's Palm Oil Targets." *World Resources Institute*, April 17. www.wri.org/blog/2018/04/intensification -smallholder-farmers-key-achieving-indonesia-s-palm-oil-targets.

Sartre, Jean-Paul. 1987. *The Family Idiot: Gustave Flaubert 1821–1857*. Vol. 2. Translated by Carol Cosman. Chicago: University of Chicago Press.

Schieffelin, Edward L. 1976. *The Sorrow of the Lonely and the Burning of the Dancers*. New York: St Martin's Press.

Scott, James C. 1985. *Weapons of the Weak: Everyday Forms of Peasant Resistance*. New Haven: Yale University Press.

Scott, James C. 1998. *Seeing Like a State: How Certain Schemes to Improve the Human Condition Have Failed*. New Haven: Yale University Press.

Scott, Michael W. 2007. *The Severed Snake: Matrilineages, Making Place, and a Melanesian Christianity in Southeast Solomon Islands*. Durham, NC: Carolina University Press.

Scott, Michael W. 2016. "To be Makiran is to see like Mr Parrot: the anthropology of wonder in Solomon Islands." *Journal of the Royal Anthropological Institute* 22, no. 3: 474–95. doi:10.1111/1467-9655.12442.

Sen, Debarati. 2017. *Everyday Sustainability: Gender Justice and Fair Trade Tea in Darjeeling*. Stanford: Stanford University Press.

Serpenti, Laurentius M. 1965. *Cultivators in the Swamps: Social Structure and Horticulture in a New Guinea Society (Frederik-Hendrik Island, West New Guinea)*. Assen, The Netherlands: Van Gorcum.

Serres, Michel. 1980. *Le Parasite*. Paris: Bernard Grasset.

Sexton, Lorraine. 1988. "'Eating' Money in Highland Papua New Guinea." *Food and Foodways* 3, no. 1–2: 119–42. doi:10.1080/07409710.1988.9961940.

Shapiro, Nicholas. 2015. "Attuning to the Chemosphere: Domestic Formaldehyde, Bodily Reasoning, and the Chemical Sublime" *Cultural Anthropology* 30, no. 3: 368–93. doi:10.14506/ca30.3.02.

Shapiro, Nicholas, and Eben S. Kirksey. 2017. "Chemo-Ethnography: An Introduction." *Cultural Anthropology* 32, no. 4: 481–93. doi:10.14506/ca32.4.01.

Shiva, Vandana. 1993. *Monocultures of the Mind: Perspectives on Biodiversity and Biotechnology*. London: Zed Books.

Shotwell, Alexis. 2016. *Against Purity: Living Ethically in Compromised Times*. Minneapolis: University of Minnesota Press.

Sillitoe, Paul. 1983. *Roots of the Earth: Crops in the Highlands of Papua New Guinea*. Sydney: New South Wales University Press.

Sillitoe, Paul. 2003. *Managing Animals in New Guinea: Preying the Game in the Highlands*. London: Routledge.

Silva, Denise F. 2017. "1 (Life) ÷ 0 (Blackness) = ∞ − ∞ or ∞ / ∞: On Matter Beyond the Equation of Value." *E-Flux* 79. www.e-flux.com/journal/79/94686/1-life-0-blackness-or-on-matter-beyond-the-equation-of-value.

Silverman, Eric K. 2013. "After *Cannibal Tours*: Cargoism and Marginality in a Post-Touristic Sepik River Society." *Contemporary Pacific* 25, no. 2: 221–57. doi:10.1353/cp.2013.0031.

Silverman, Eric K. 2018. "The Sepik River, Papua New Guinea: Nourishing Tradition and Modern Catastrophe." In *Island Rivers: Fresh Water and Place in Oceania*, edited by John R. Wagner and Jerry K. Jacka, 187–222. Canberra: Australian National University Press.

Simpson, Audra. 2014. *Mohawk Interruptus: Political Life across the Borders of Settler States*. Durham, NC: Duke University Press.

Simpson, Audra. 2016. "Consent's Revenge." *Cultural Anthropology* 31, no. 3: 326–33. doi:10.14506/ca31.3.02.

Simpson, Audra. 2017. "The ruse of consent and the anatomy of 'refusal': cases from indigenous North America and Australia." *Postcolonial Studies* 20, no. 1: 18–33. doi:10.1080/13688790.2017.1334283.

Sirait, Martua, Sukirno Prasodjo, Nancy Podger, Alex Flavelle, and Jefferson Fox. 1994. "Mapping Customary Land in East Kalimantan, Indonesia: A Tool for Forest Management." *Ambio* 23, no. 7: 411–17. www.jstor.org/stable/4314246.

Slama, Martin, and Jenny Munro. 2015. "From 'Stone-Age' to 'Real-Time': Exploring Papuan Temporalities, Mobilities and Religiosities—An Introduction." In *From "Stone Age" to "Real-Time": Exploring Papuan Temporalities, Mobilities and Religiosities*, edited by Martin Slama and Jenny Munro, 1–38. Canberra: Australian National University Press.

Sletto, Bjørn I. 2009. "We Drew What We Imagined": Participatory Mapping, Performance, and the Arts of Landscape Making." *Current Anthropology* 50, no. 4: 443–76. doi:10.1086/593704.

Sloterdijk, Peter. 2011. *Bubbles. Spheres Vol. 1, Microspherology*. Translated by Wieland Hoban. Los Angeles: Semiotext(e).

Soemarwoto, Otto. 1985. "Constancy and Change in Agroecosystems." In *Cultural Values and Human Ecology in Southeast Asia*, edited by Karl L. Hutterer, Terry A. Rambo, and George Lovelace, 205–48. Ann Arbor: University of Michigan Press.

Sousanis, Nick. 2015. *Unflattening*. Cambridge, MA: Harvard University Press.

Spillers, Hortense. 1987. "Mama's Baby, Papa's Maybe: An American Grammar Book." *Diacritics* 17, no. 2: 64–81. doi:10.2307/464747.

Spivak, Gayatri C. 1999. *A Critique of Postcolonial Reason: Toward a History of the Vanishing Present*. Cambridge, MA: Harvard University Press.

Star, Susan L. 1991. "Power, Technologies and the Phenomenology of Conventions: On Being Allergic to Onions." In *A Sociology of Monsters: Essays on Power, Technology and Domination*, edited by John Law, 26–56. London: Routledge.

Star, Susan L. 1999. "The Ethnography of Infrastructure." *American Behavioral Scientist* 43, no. 3: 377–91. doi:10.1177/00027649921955326.

Star, Susan L., and James R. Griesemer. 1989. "Institutional Ecology, 'Translations,' and Boundary Objects: Amateurs and Professionals in Berkeley's Museum of Vertebrate Zoology, 1907–39." *Social Studies of Science* 19, no. 3: 387–420. doi:10.1177/030631289019003001.

Stasch, Rupert. 2001. "Giving up Homicide: Korowai Experience of Witches and Police (West Papua)." *Oceania* 72 no. 1: 33–52. doi:10.1002/j.1834-4461.2001.tb02763.x.

Stasch, Rupert. 2009. *Society of Others: Kinship and Mourning in a West Papuan Place*. Berkeley: University of California Press.

Stasch, Rupert. 2013. "The poetics of village space when villages are new: Settlement form as history making in Papua, Indonesia." *American Ethnologist* 40, no. 3: 555–70. doi:10.1111/amet.12039.

Stasch, Rupert. 2016. "Singapore, Big Village of the Dead: Cities as Figures of Desire, Domination, and Rupture among Korowai of Indonesian Papua." *American Anthropologist* 118, no. 2: 258–69. doi:10.1111/aman.12525.

Stengers, Isabelle. 2010. *Cosmopolitics I*. Translated by Robert Bononno. Minneapolis: University of Minnesota Press.

Stengers, Isabelle. 2015. *In Catastrophic Times: Resisting the Coming Barbarism*. Translated by Andrew Goffey. London: Open Humanities Press.

Stephen, Michele. 1982. "'Dreaming Is Another Power!': The Social Significance of Dreams among the Mekeo of Papua New Guinea." *Oceania* 53, no. 3: 106–22. doi:10.1002/j.1834-4461.1982.tb01532.x.

Stephen, Michele. 1995. *A'aisa's Gifts: A Study of Magic and the Self*. Berkeley: University of California Press.

Stewart, Pamela J., and Andrew J. Strathern. 2000. "Accident, Agency, and Liability in New Guinea Highlands Compensation Practices." *Bijdragen Tot de Taal-, Land- En Volkenkunde* 165, no. 2: 275–95. www.jstor.org/stable/27865620.

Stewart, Pamela J., and Andrew J. Strathern. 2001. *Humors and Substances: Ideas of the Body in New Guinea*. Westport: Bergin and Garvey.

Stewart, Pamela J., and Andrew J. Strathern. 2003a. "Dreaming and Ghosts among the Hagen and Duna of the Southern Highlands, Papua New Guinea." In *Dream*

Travelers: Sleep Experiences and Culture in the Western Pacific, edited by Richard I. Lohmann, 43–60. New York: Palgrave Macmillan.

Stewart, Pamela J., and Andrew J. Strathern, eds. 2003b. *Landscape, Memory and History: Anthropological Perspectives*. Sterling: Pluto Press.

Stewart-Harawira, Makere. 2005. *The New Imperial Order: Indigenous Responses to Globalization*. New York: Zed Books.

Stewart-Harawira, Makere. 2012. "Returning the Sacred: Indigenous Ontologies in Perilous Times." In *Radical Human Ecology: Intercultural and Indigenous Approaches*, edited by Rose Roberts and Lewis Williams, 94–109. New York: Taylor and Francis.

Stewart-Harawira, Makere. 2013. "Challenging Knowledge Capitalism: Indigenous Research in the 21st Century." *Socialist Studies* 9, no. 1: 39–51. doi:10.18740/S43S3V.

Stewart-Harawira, Makere. 2018. "Indigenous Resilience and Pedagogies of Resistance: Responding to the Crisis of Our Age." In *Resilient Systems, Resilient Communities*, edited by Jordan B. Kinder and Makere Stewart-Harawira, 158–79. Edmonton: University of Alberta.

Stoler, Ann L. 1985. *Capitalism and Confrontation in Sumatra's Plantation Belt, 1879–1979*. New Haven: Yale University Press.

Stoler, Ann L. 2001. "Tense and Tender Ties: The Politics of Comparison in North American History and (Post) Colonial Studies." *Journal of American History* 88, no. 3: 829–65. doi:10.2307/2700385.

Stoler, Ann L. 2013. *Imperial Debris: On Ruins and Ruination*. Durham, NC: Duke University Press.

Stoler, Ann L. 2016. *Duress: Imperial Durabilities in Our Times*. Durham, NC: Duke University Press.

Stoler, Ann L. 2018. "The Politics of "Gut Feelings": On Sentiment in Governance and the Law." *KNOW: A Journal on the Formation of Knowledge* 2, no. 2: 207–28. doi:10.1086/699009.

Strathern, Andrew J. 1973. "Kinship, Descent and Locality: Some New Guinea Examples." In *The Character of Kinship*, edited by Jack Goody, 21–33. Cambridge: Cambridge University Press.

Strathern, Marilyn. 1988. *The Gender of the Gift: Problems with Women and Problems with Society in Melanesia*. Berkeley: University of California Press.

Strathern, Marilyn. 1996. "Cutting the Network." *Journal of the Royal Anthropological Institute* 2, no. 3: 517–35. doi:10.2307/3034901.

Strathern, Marilyn. 2004. "Introduction: Rationales of Ownership." In *Rationales of Ownership: Transactions and Claims to Ownership in Contemporary Papua New Guinea*, edited by Lawrence Kalinoe and James Leach, 1–12. Wantage, UK: Sean Kingston.

Strong, Thomas P. 2004. "Pikosa: Loss and life in the Papua New Guinea highlands." PhD diss., Princeton University.

Strong, Thomas P. 2007. "'Dying Culture' and Decaying Bodies." In *Embodying Modernity and Postmodernity: Ritual, Praxis, and Social Change in Melanesia*, edited by Sandra C. Bamford, 105–23. Durham, NC: Carolina Academic Press.

Sundberg, Juanita. 2013. "Decolonizing Posthumanist Geographies." *cultural geographies* 21, no. 1: 33–47. doi:10.1177/1474474013486067.

Swadling, Pamela. 1996. *Plumes from Paradise: Trade Cycles in Outer Southeast Asia and their Impact on New Guinea and Nearby Islands until 1920.* Boroko: Papua New Guinea National Museum and Robert Brown.

TallBear, Kim. 2011. "Why Interspecies Thinking Needs Indigenous Standpoints." *Fieldsights*, November 18. https://culanth.org/fieldsights/why-interspecies-thinking-needs -indigenous-standpoints.

TallBear, Kim. 2015. "An Indigenous Reflection on Working Beyond the Human/Not Human." GLQ: *A Journal of Lesbian and Gay Studies* 21, no. 2–3: 230–35. muse.jhu.edu /article/582037.

TallBear, Kim. 2016. "Failed Settler Kinship, Truth and Reconciliation, and Science." Accessed March 26 2021. https://indigenoussts.com/failed-settler-kinship-truth-and -reconciliation-and-science/.

Tammisto, Tuomas 2016. "Enacting the Absent State: State-Formation on the Oil-Palm Frontier of Pomio (Papua New Guinea)." *Paideuma: Mitteilungen zur Kulturkunde* 62: 51–68. www.jstor.org/stable/44243086.

Tammisto, Tuomas. 2018a. "Life in the Village is Free: Socially Reproductive Work and Alienated Labour on an Oil Palm Plantation in Pomio, Papua New Guinea." *Suomen Anthropologi: Journal of the Finnish Anthropological Society* 43, no. 4: 19–35. doi:10.30676/jfas.v43i4.79476.

Tammisto, Tuomas. 2018b. "New Actors, Historic Landscapes: The making of a frontier place in Papua New Guinea." PhD Diss., University of Helsinki.

Tammisto, Tuomas. 2019. "Making Temporal Environments: Work, Places and History in the Mengen Landscape." In *Dwelling in Political Landscapes: Contemporary Anthropological Perspectives*, edited by Anu Lounela, Eeva Berglund, and Timo Kallinen, 247–63. Helsinki: Suomalaisen Kirjallisuuden Seura.

Tan, Koolin. 1979. "Logging the Swamp for Food." In *The Equatorial Swamp as Natural Resource. Proceedings of the Second International Sago Symposium. Kuala Lumpur, 15–17 September*, edited by Robert W. Stanton and Michiel Flach, 13–34. The Hague: Nijhoff.

Taussig, Michael. 1984. "Culture of Terror–Space of Death. Roger Casement's Putumayo Report and the Explanation of Torture." *Comparative Studies in Society and History* 26, no. 3: 467–97. doi:10.1017/S0010417500011105.

Taussig, Michael. 1987. *Shamanism, Colonialism, and the Wild Man: A Study in Terror and Healing.* Chicago: University of Chicago Press.

Taussig, Michael. 1992. *The Nervous System.* New York: Routledge.

Taussig, Michael. 2015. *The Corn Wolf.* Chicago: University of Chicago Press.

Taussig, Michael. 2018. *Palma Africana.* Chicago: University of Chicago Press.

Teaiwa, Katerina M. 2014. *Consuming Ocean Island: Stories of People and Phosphate from Banaba.* Bloomington: Indiana University Press.

Teaiwa, Teresia K. 2006. "On Analogies: Rethinking the Pacific in a Global Context." *Contemporary Pacific Studies* 18, no. 1: 71–87. doi:10.1353/cp.2005.0105.

Teaiwa, Teresia K. 2014. "The Ancestors We Get to Choose: White Influences I Won't Deny." In *Theorizing Native Studies*, edited by Audra Simpson and Andrea Smith, 43–55. Durham, NC: Duke University Press.

Tebay, Neles. 2005. *West Papua: The Struggle for Peace with Justice.* London: Catholic Institute for International Relations.

Tedlock, Barbara. 1981. "Quiché Maya Dream Interpretation." *Ethos* 9, no. 4: 313–30. doi:10.1525/eth.1981.9.4.02a00050.

Te Punga Somerville, Alice. 2018. "Inside Us the Unborn: Genealogies, Futures, Metaphors, and the Opposite of Zombies." In *Pacific Futures: Past and Present,* edited by Warwick Anderson, Miranda Johnson, and Barbara Brookes, 69–80. Honolulu: University of Hawai'i Press.

Thomas, Deborah A. 2016. "Time and the Otherwise: Plantations, Garrisons and Being Human in the Caribbean." *Anthropological Theory* 16, no. 2–3: 177–200. doi:10.1177/1463499616636269.

Thomas, Deborah A. 2019. *Political Life in the Wake of the Plantation: Entanglement, Witnessing, Repair.* Durham, NC: Duke University Press.

Throop, Jason. 2010. *Suffering and Sentiment: Exploring the Vicissitudes of Experience and Pain in Yap.* Berkeley: University of California Press.

Timmer, Jaap. 2011. "Cloths of Civilisation: Kain Timur in the Bird's Head of West Papua." *The Asia Pacific Journal of Anthropology* 12, no. 4: 383–401. doi:10.1080/14442 213.2011.587020.

Timmer, Jaap. 2015. "Papua Coming of Age: The Cycle of Man's Civilisation and Two Other Papuan Histories." In *From "Stone Age" to "Real-Time": Exploring Papuan Temporalities, Mobilities and Religiosities,* edited by Martin Slama and Jenny Munro, 95–124. Canberra: Australian National University Press.

Todd, Zoe. 2015. "Indigenizing the Anthropocene." In *Art in the Anthropocene: Encounters among Aesthetics, Politics, Environments and Epistemologies,* edited by Heather Davis and Etienne Turpin, 241–54. London: Open Humanities Press.

Todd, Zoe. 2017. "Fish, Kin and Hope: Tending to Water Violations in *amiskwaciwâskahikan* and Treaty Six Territory." *Afterall: A Journal of Art, Context, and Enquiry* 43: 102–7. doi:10.1086/692559.

Trask, Haunani-Kay. 1999. *From a Native Daughter: Colonialism and Sovereignty in Hawai'i.* Honolulu: University of Hawai'i Press.

Trouillot, Michel-Rolph. 1988. *Peasants and Capital: Dominica in the World Economy.* Baltimore: Johns Hopkins University Press.

Trouillot, Michel-Rolph. 1997. *Silencing the Past: Power and the Production of History.* New York: Beacon Press.

Trouillot, Michel-Rolph. 2002. "Culture on the Edges: Caribbean Creolization in Historical Context." In *From the Margins: Historical Anthropology and Its Futures,* edited by Brian K. Axel, 189–210. Durham, NC: Duke University Press.

Trouillot, Michel-Rolph. 2003. *Global Transformations: Anthropology and the Modern World.* New York: Palgrave Macmillan.

Tsing, Anna L. 1993. *In the Realm of the Diamond Queen.* Princeton: Princeton University Press.

Tsing, Anna L. 1994. "From the Margins." *Cultural Anthropology* 9, no. 3: 279–97. doi:10.1525/can.1994.9.3.02a00020.

Tsing, Anna L. 2000. "The Global Situation." *Cultural Anthropology* 15, no. 3: 327–60. doi:10.1525/can.2000.15.3.327.

Tsing, Anna L. 2003. "Natural Resources and Capitalist Frontiers." *Economic and Political Weekly* 38, no. 48: 5100–6. www.jstor.org/stable/4414348.

Tsing, Anna L. 2005. *Friction: An Ethnography of Global Connection.* Princeton: Princeton University Press.

Tsing, Anna L. 2009. "Supply Chains and the Human Condition." *Rethinking Marxism* 21, no. 2: 148–76. doi:10.1080/08935690902743088.

Tsing, Anna L. 2011. "Arts of Inclusion, or, How to Love a Mushroom." *Australian Humanities Review* 50: 5–22. http://australianhumanitiesreview.org/2011/05/01/arts-of -inclusion-or-how-to-love-a-mushroom/.

Tsing, Anna L. 2014a. "Blasted Landscapes (and the Gentle Arts of Mushroom Picking)." In *The Multispecies Salon*, edited by Eben Kirksey, 87–109. Durham, NC: Duke University Press.

Tsing, Anna L. 2014b. "More-Than-Human Sociality: A Call for Critical Description." In *Anthropology and Nature*, edited by Kirsten Hastrup, 27–42. Oxford, UK: Routledge.

Tsing, Anna L. 2015. *The Mushroom at the End of the World: On the Possibility of Life in Capitalist Ruins.* Princeton: Princeton University Press.

Tsing, Anna L. 2018. "Getting by in Terrifying Times." *Dialogues in Human Geography* 8, no. 1: 73–76. doi:10.1177/2043820617738836.

Tsing, Anna L., Andrew S. Mathews, and Nils Bubandt. 2019. "Patchy Anthropocene: Landscape Structure, Multispecies History, and the Retooling of Anthropology: An Introduction to Supplement 20." *Current Anthropology* 60, S20: S186–97. doi:10.1086/703391.

Tuck, Eve. 2009. "Suspending Damage: A Letter to Communities." *Harvard Educational Review* 79, no. 3: 409–27. doi:10.17763/haer.79.3.n0016675661t3n15.

Tuhiwai Smith, Linda. 2012. *Decolonizing Methodologies: Research and Indigenous Peoples.* London: Zed Books.

Turnbull, David. 2007. "Maps Narratives and Trails: Performativity, Hodology and Distributed Knowledges in Complex Adaptive Systems—An Approach to Emergent Mapping." *Geographical Research* 45, no. 2: 140–49. doi:10.1111/j.1745-5871.2007.00447.x.

Turner, Dale. 2006. *This Is Not a Peace Pipe: Towards a Critical Indigenous Philosophy.* Toronto: University of Toronto Press.

Tuzin, Donald. 1972. "Yam Symbolism in the Sepik: An Interpretative Account." *Southwestern Journal of Anthropology* 28, no. 3: 230–54. doi:10.1086/soutjanth.28.3.3629221.

Tuzin, Donald. 1992. "Sago Subsistence and Symbolism among the Ilahita Arapesh." *Ethnology* 31, no. 2: 103–14. doi:10.2307/3773615.

Tuzin, Donald. 1997. *The Cassowary's Revenge: The Life and Death of Masculinity in a New Guinea Society.* Chicago: University of Chicago Press.

Ulijaszek, Stanley J. 2002. "Sago, Economic Change, and Nutrition in Papua New Guinea." In *New Frontiers of Sago Palm Studies. Proceedings of the International Symposium on Sago*, edited by Keiji Kainuma, Masanori Okazaki, Yukio Toyoda, and John E. Cecil, 219–26. Tokyo: Universal Academy Press.

van Baal, Jan. 1966. *Dema: Description and Analysis of Marind-Anim Culture (South New Guinea)*. The Hague: Martinus Nijhoff.

van der Kroef, Justus M. 1952. "Some Head-Hunting Traditions of Southern New Guinea." *American Anthropologist* 54, no. 1: 221–35. www.jstor.org/stable/663912.

van der Veur, Paul W. 1972. "Dutch New Guinea." In *Encyclopaedia of Papua and New Guinea*, edited by Peter Ryan, 276–83. Melbourne: Melbourne University Press.

van Dooren, Thom. 2014. *Flight Ways: Life and Loss at the Edge of Extinction*. New York: Columbia University Press E-Book.

van Dooren, Thom, Eben S. Kirksey, and Ursula Münster. 2016. "Multispecies Studies: Cultivating Arts of Attentiveness." *Environmental Humanities* 8, no. 1: 1–23. doi:10.1215/22011919-3527695.

van Dooren, Thom, and Deborah B. Rose, eds. 2011. *Unloved Others: Death of the Disregarded in the Time of Extinctions*. Canberra: Australian National University Press.

van Dooren, Thom, and Deborah B. Rose. 2016. "Lively Ethography: Storying Animist Worlds." *Environmental Humanities* 8, no. 1: 77–94. doi:10.1215/22011919-3527731.

van Oosterhout, Dianne. 2000. "Tying the Time String Together: An End-of-Time Experience in Irian Jaya, Indonesia." *Ethnohistory* 47, no. 1: 67–99. doi:10.1215/00141801-47-1-67.

van Oosterhout, Dianne. 2001. "The Scent of Sweat: Notions of Witchcraft and Morality in Inanwatan." In *Humours and Substances: Ideas of the Body in New Guinea*, edited by Pamela J. Stewart and Andrew J. Strathern, 23–50. Westport: Bergin and Garvey.

Verdery, Katherine. 2002. "Seeing Like a Mayor: Or, How Local Officials Obstructed Romanian Land Restitution." *Ethnography* 3, no. 1: 5–33. doi:10.1177/14661380222231054.

Verschueren, Jan. 1970. "Marind-Anim Land Tenure." *New Guinea Research Bulletin* 38: 42–59.

Viveiros de Castro, Eduardo. 1998. "Cosmological Deixis and Amerindian Perspectivism." *Journal of the Royal Anthropological Institute* 4, no. 3: 469–88. doi:10.2307/3034157.

Viveiros de Castro, Eduardo. 2011. "Zeno and the Art of Anthropology: Of Lies, Beliefs, Paradoxes, and Other Truths." Translated by Antonia Walford. *Common Knowledge* 17, no. 1: 128–45. doi:10.1215/0961754X-2010-045.

Viveiros de Castro, Eduardo. 2012. *Cosmological Perspectivism in Amazonia and Elsewhere*. Chicago: HAU Books.

Viveiros de Castro, Eduardo. 2015. *The Relative Native: Essays on Indigenous Conceptual Worlds*. Translated by David Rodgers and Julia Sauma. Chicago: HAU Books.

Vizenor, Gerald R. 2000. *Fugitive Poses: Native American Indian Scenes of Absence and Presence*. Lincoln: University of Nebraska Press.

Vogel, L.C., and John Richens. 1989. "Donovanosis in Dutch South New Guinea: history, evolution of the epidemic and control." *Papua and New Guinea Medical Journal* 32, no. 3: 203–18.

von Poser, Anita. 2011. "Bosmun Foodways: Emotional Reasoning in a Papua New Guinea Lifeworld." In *The Anthropology of Empathy: Experiencing the Lives of Others in Pacific Societies*, edited by Douglas W. Hollan and Jason C. Throop, 169–94. New York: Berghahn.

von Poser, Anita. 2013. *Foodways and Empathy: Relatedness in a Ramu River Society, Papua New Guinea.* New York: Berghahn.

von Uexküll, Jacob. 1982. "The Theory of Meaning." *Semiotica* 42, no. 1: 25–82. doi:10.1515/semi.1982.42.1.25.

Wagner, John R. 2018. "Rivers of Memory and Forgetting." In *Island Rivers: Fresh Water and Place in Oceania,* edited by Roy Wagner and Jerry K. Jacka, 223–50. Canberra: Australian National University Press.

Wagner, John R., and Jerry K. Jacka, eds. 2018. *Island Rivers: Fresh Water and Place in Oceania.* Canberra: Australian National University Press.

Wagner, Roy. 1981. *The Invention of Culture.* Chicago: University of Chicago Press.

Wagner, Roy. 1991. "The Fractal Person." In *Big Men and Great Men: Personifications of Power in Melanesia,* edited by Maurice Godelier and Marilyn Strathern, 159–73. Cambridge: Cambridge University Press.

Wallis, George W. 1970. "Chronopolitics: The Impact of Time Perspectives on the Dynamics of Change." *Social Forces* 49, no.1: 102–8. doi:10.2307/2575743.

Wambacq, Judith, and Sjoerd van Tuinen. 2017. "Interiority in Sloterdijk and Deleuze." *Palgrave Communications* 3, no. 17072: 1–7. doi:10.1057/palcomms.2017.72.

Wangge, Hipolitus Y. R. 2014. "Jokowi: Hope for Papua?" *New Mandala,* November 24. https://www.newmandala.org/jokowi-hope-for-papua/.

Wardlow, Holly. 2006. *Wayward Women: Sexuality and Agency in a New Guinea Society.* Berkeley: University of California Press.

Watts, Vanessa. 2013. "Indigenous place-thought and agency amongst humans and non-humans (First Woman and Sky Woman go on a European world tour!)." *Decolonization: Indigeneity, Education, & Society* 2, no. 1: 20–34. https://jps.library.utoronto.ca/index.php/des/article/view/19145.

Weheliye, Alexander G. 2008. "After Man." *American Literary History* 20, no. 1–2: 321–36. doi:10.1093/alh/ajm057.

Weheliye, Alexander G. 2014. *Habeas Viscus: Racializing Assemblages, Biopolitics, and Black Feminist Theories of the Human.* Durham, NC: Duke University Press.

West, Paige. 2006a. *Conservation Is Our Government Now: The Politics of Ecology in Papua New Guinea.* Durham, NC: Duke University Press.

West, Paige. 2006b. "Environmental Conservation and Mining: Strange Bedfellows in the Eastern Highlands of Papua New Guinea." *Contemporary Pacific* 18, no. 2: 295–313. doi:10.1353/cp.2006.0031.

West, Paige. 2012. *From Modern Production to Imagined Primitive: The Social World of Coffee from Papua New Guinea.* Durham, NC: Duke University Press.

West, Paige. 2016. *Dispossession and the Environment: Rhetoric and Inequality in Papua New Guinea.* New York: New York University Press.

West, Paige. 2019. "Translation, Value, and Space: Theorizing an Ethnographic and Engaged Environmental Anthropology." *American Anthropologist* 107, no. 4: 632–42. doi:10.1525/aa.2005.107.4.632.

West, Paige. 2020. "Translations, Palimpsests, and Politics. Environmental Anthropology Now." *Ethnos,* 85, no. 1: 118–23. doi:10.1080/00141844.2017.1394347.

Whitehouse, Andrew. 2015. "Listening to Birds in the Anthropocene: The Anxious Semiotics of Sound in a Human-Dominated World." *Environmental Humanities* 6, no. 1: 53–72. doi:10.1215/22011919-3615898.

Whyte, Kyle P. 2017. "Is It Colonial Déjà Vu? Indigenous Peoples and Climate Injustice." In *Humanities for the Environment: Integrating Knowledge, Forging New Constellations of Practice*, edited by Joni Adamson and Michael Davis, 88–105. New York: Routledge.

Whyte, Kyle P. 2018a. "Indigenous Science (Fiction) for the Anthropocene: Ancestral Dystopias and Fantasies of Climate Change Crises." *Environment and Planning E: Nature and Space* 1, no. 1–2: 224–42. doi:10.1177/2514848618777621.

Whyte, Kyle P. 2018b. "What Do Indigenous Knowledges Do for Indigenous Peoples?" In *Traditional Ecological Knowledge: Learning from Indigenous Practices for Environmental Sustainability*, edited by Melissa K. Nelson and Daniel Shilling, 57–82. Cambridge: Cambridge University Press.

Willerslev, Rane. 2004. "Not Animal, Not Not-Animal: Hunting, Imitation and Empathetic Knowledge among the Siberian Yukaghirs." *Journal of the Royal Anthropological Institute* 10, no. 3: 629–52. doi:10.1111/j.1467-9655.2004.00205.x.

Wing, John R., with Peter King. 2005, August. "Genocide in West Papua? The Role of the Indonesian State Apparatus and a Current Needs Assessment of the Papuan People." University of Sydney, Centre for Peace and Conflict Studies, West Papua Project, and ELSHAM Jayapura, Papua.

Winichakul, Thongchai. 1994. *Siam Mapped: A History of the Geo-Body of a Nation*. Honolulu: University of Hawai'i Press.

Winter, Christine J. 2019a. "Decolonising Dignity for Inclusive Democracy." *Environmental Values* 28, no. 1: 9–30. doi:10.3197/096327119X15445433913550.

Winter, Christine J. 2019b. "Does Time Colonise Intergenerational Environmental Justice Theory?" *Environmental Politics* 29, no. 2: 278–96. doi:10.1080/09644016.2019.1569745.

Wolfe, Patrick. 1999. *Settler Colonialism and the Transformation of Anthropology: The Politics and Poetics of an Ethnographic Event*. London: Bloomsbury.

Wolfe, Patrick. 2006. "Settler Colonialism and the Elimination of the Native." *Journal of Genocide Research* 8, no. 4: 387–409. doi:10.1080/14623520601056240.

World Wildlife Fund. 2020. "8 Things to Know about Palm Oil." January 17, 2020. wwf .org.uk/updates/8-things-know-about-palm-oil.

Worsley, Peter. 1968. *The Trumpet Shall Sound: A Study of "Cargo" Cults in Melanesia*. New York: Schocken.

Wynter, Sylvia. 1971. "Novel and History, Plot and Plantation." *Savacou* 5: 95–102.

Wynter, Sylvia. 2003. "Unsettling the Coloniality of Being/Power/Truth/Freedom: Towards the Human, AfterMan, Its Overrepresentation—An Argument." *CR: The New Centennial Review* 3, no. 3: 257–337. doi:10.1353/ncr.2004.0015.

Wynter, Sylvia. 2015. "The Ceremony Found: Towards the Autopoetic Turn/Overturn, Its Autonomy of Human Agency and Extraterritoriality of (Self-)Cognition." In *Black Knowledges/Black Struggles: Essays in Critical Epistemology*, edited by Jason R. Ambroise and Sabine Broeck, 184–252. Liverpool: Liverpool University Press.

Young, Michael W. 1971. *Fighting with Food: Leadership, Value and Social Control in a Massim Society*. Cambridge: Cambridge University Press.

Yayasan PUSAKA and iLab. 2016. "Mata Papua." matapapua.org.

Yusoff, Kathryn. 2019. *A Billion Black Anthropocenes or None*. Minneapolis: University of Minnesota Press.

Zelenietz, Marty, and Shirley Lindenbaum. 1981. *Sorcery and Social Change in Melanesia*. Adelaide: University of Adelaide Press.

lethal capital: oil palm as, 161–62; oil palm plantations as, 211–12

Libra, 969, 168, 217; construction of, 175–76

Lipset, David, 259n9

LiPuma, Edward, 92

listening, to sago grove, 139

lively capital, 155; oil palm as, 161–62; oil palm plantations as, 211–12

logic of elimination, 233n27

love, 210

Love, Heather, 265n16

Lowe, Celia, 260n10

MacLeod, Jason, 19

Mageo, Jeanette, 193, 196–97

Mahuze, Darius, 28–29, 33–35, 50, 85–86; on resocialization, 39–40; on sago palm, 35–36

Mahuze, Paulus, 1–3

Mahuze clan, 16, 68

makhav, 255n5

Malabou, Catherine, 266n18

Malinowski, Bronislaw, 231n14

maps and mapping, 29, 51–54; colonialism and, 56; dead, 54–56; différance and, 68; with drones, 64–67; expeditions, 58; of Marind community, 58–64, 70–73; Marind community on government, 56–57; of MIFEE, 56; moving, 67–68; multispecies relationships and, 65–66; by NGOs, 52, 57–58, 61, 68, 72; ontology of, 72; participatory, 57–58, 71, 246n3, 246n4; pressure points embodied by, 71; risks of, 69–70; of sago grove, 136–40; sound in, 61–62, 73; temporality of, 70; as weapons, 69. *See also* counter-mapping

Marcia, 81, 132

Marcus, 54, 56, 193

Marder, Michael, 140, 229n6

Marina, 179–81

Marind Bian, 15

Marind community, 2; activism of, 109; children in, 77–78, 80–81, 106; clan structure of, 16, 151, 238n56; death in, 88–89; development impacting, 3–4; diet of, 16; as domesticated, 103; domesticates in, 101–2, 113; dream sharing sessions, 192–96, 199; under Dutch colonialism, 17–18; environmental care practices of, 132; ethnography of, 6, 238n55; freedoms denied to, 103; gastro-identity

of, 123, 256n8; gender in, 126–27, 259n9; the good life in, 12; on government maps, 56–57; historicity of, 27; hope in, 180–81, 217; human-vegetal relations of, 36–37, 212; language of, 227n2; maps of, 58–64, 70–73; marginalization of, 55; MIFEE and violation of rights of, 240n69; modernity and, 110, 233n26; multispecies relationships in, 82–83, 108–9; names in, 16; nature in, 211; on oil palm, 159–63, 204, 212, 215; oil palm arrival altering, 4–6, 11, 143–44, 170, 205, 263n6; oil palm plantations constraining, 42–43, 211; on opacity of mind, 86, 90–91, 93; on plastic goods, 106–8; posthumanism of, 9; racial stereotypes of, 180; religion in, 16–17, 26; roads for, 38–41; sago in, 119, 122–24, 130–31, 133–34, 139, 255n2; self-determination of, 178; self-representation of, 26; on settlers, 104–5; temporality of, 178, 179–80; villages in, 241n71

Marind-deg, 15

Maring, 92

Marius, 86–87, 89–90

Massumi, Brian, 86, 140

Mata Papua, 58, 61

material-semiosis, 261n20

Matthias, 123

Mbaraka, Romanus, 35, 167, 176

McCarthy, John, 237n48

McGranahan, Carole, 178

McKittrick, Katherine, 55

Melaleuca, 15, 122–23

Melanesia, 8, 40, 88, 229n4, 234n30; development in, 259n8

Melanesian Way, 236n45

memoria passionis, 104

Merauke, 1, 42, 227n1; multispecies relationships in, 209–10; plant life in, 15

Merauke Board for Planning and Regional Development, 52

Merauke City, 17, 25, 40, 44, 143

Merauke Forestry Agency, 67

Merauke Integrated Food and Energy Estate (MIFEE), 2–3, 14, 19–21, 148, 174; implementation of, 143–44; maps of, 56; Marind rights violated by, 240n69

Merauke Regency, 2

merdeka, 109

Merleau-Ponty, Maurice, 82, 86

messav, 188, 194

MIFEE. *See* Merauke Integrated Food and Energy Estate (MIFEE)

military garrisons, 38, 45–48

Millar, Kathleen, 179

Miller, Theresa, 8, 232n20

mimetic recovery, 193

Mina, 195, 203, 204

mind, opacity of, 86, 90–91, 93

minimal manipulation, strategies of, 101

Mira, 110

Miranda, 45, 54

Miriam, 153, 155

Mittermaier, Amira, 197, 268n13

modernity, 39; cruel optimism of, 181–82; Marind community relationship to, 110, 233n26; oil palm plantations and, 150; temporality of, 265n12

modern totems, 168, 175

Moiwend, Rosa, 19

Mondragón, Carlos, 243n2

monocrops, 43–44, 49, 105, 155; critiques of, 153; disease vulnerability of, 153–54; ecologies of, 147

morality, skin and wetness as idioms of, 86

Moran-Thomas, Amy, 267n12

Morton, Timothy, 107

moving maps, 67–68

Muli Strait, 15

multiplicity, 163

multispecies relationships, 147, 151; anim identity and, 91–94; under capitalism, 180–81; as community of faith, 98; desirability of, 214; dialogical, 208; maps and, 65–66; Marind community and, 82–83, 108–9; in Merauke, 209–10; in mythical time, 171; ontology of, 101; pedagogy, 81; perspective in, 65–66; as pharmakon, 208; reciprocity of, 206; relatedness in, 36, 84, 101, 112; restrained care in, 101; sago and, 120, 131–33; wetness in, 250n9. *See also* human-vegetal relations; skinship

multispecies studies, 231n19

Munro, Jenny, 245n15

Musharbash, Yasmine, 267n8

Muting, 15

Muyu, 16, 147

Myers, Natasha, 232n21

mythical time, multispecies relationships in, 171

Narokobi, Bernard Mullu, 236n45

Nataliya, 153

neoliberalism, 155

Nevermann, Hans, 238n60

Ngaing, 197

NGOs. *See* nongovernmental organizations (NGOs)

night watching, 192

Nikolaus, 109

Nixon, Rob, 50

nongovernmental organizations (NGOs), 3, 5, 21, 26, 39, 68, 180, 187; mapping by, 52, 57–58, 61, 68, 72

oil palm: as abu-abu, 157, 159–60, 163; as agent, 212–14; awkward flourishing and, 206–7; under capitalism, 155; care towards, 152–53, 159; children, 144, 186; collectives, 157; death and, 188; defining, 237n54; dispersed ontology of, 160–61; ecologies of, 43–44; growth and, 174; history of, 151–52; home of, 156; as immoral, 213; insects and, 154; as kin, 158–59; as lethal and lively capital, 161–62; limits of, 204; Marind community altered by arrival of, 4–6, 11, 143–44, 170, 205, 263n6; Marind community on, 159–63, 204, 212, 215; onto-epistemology of, 214–15; ontology of, 8, 11–12, 49, 94, 145, 151, 163, 208; origins of, 151; parasites, 154; as pressure point, 163; reverse domestication of, 162; sago palm as opposed to, 11, 13, 144–45, 158, 163, 210; as settler, 105; skin of, 159; spies and, 47–48; surveillance and, 156; temporality of, 152, 170–71, 174, 177–78; temporality of arrival of, 6, 12; as victim, 152; violence unleashed by, 162; wetness of, 152–53, 158; as zombie, 155. *See also* dreams, oil palm

oil palm plantations: as abu-abu force, 43, 49; as colonizing force, 150; under Dutch colonialism, 20; expansion of, 173–74; in Indonesia, 20; as lively and lethal capital, 211–12; Marind community constrained by, 42–43, 211; modernity and, 150; as panopticon, 46–47; in Papua New Guinea, 21; as pressure points, 41–45; spread of, 5–6; toxicity of, 84–85; workers at, 158–59

oil palm seeds: growth of, 153; as pitiable, 157–58; proliferation of, 157; sago suckers compared with, 153

Printed and bound by CPI Group (UK) Ltd, Croydon, CR0 4YY

16/04/2025

14658729-0001